TAKING DEVELOPMENT SERIOUSLY: A FESTSCHRIFT FOR ANNETTE KARMILOFF-SMITH

This influential festschrift honours the legacy of Annette Karmiloff-Smith, a seminal thinker in the field of child development and a pioneer in developmental cognitive neuroscience. The current volume brings together many of the researchers, collaborators and students who worked with Prof. Karmiloff-Smith to show how her ideas have influenced and continue to influence their own research.

Over four parts, each covering a different phase or domain of Karmiloff-Smith's research career, leading developmental psychologists in cognition, neuroscience and computer science reflect on her extensive contribution, from her early work with Piaget in Geneva to her innovative research project investigating children with Down syndrome to understand the mechanisms of Alzheimer's disease. The chapters provide a mix of cutting-edge science and reminiscence, providing a fascinating insight into the historical contexts in which many of Annette's theoretical insights arose, including such ideas as the microgenetic approach, representational redescription and neuroconstructivism. The chapters also provide updates about how earlier theoretical ideas have stood the test of time, and present unpublished data from the early years of Annette's career.

Taking Development Seriously is essential reading for students and scholars in child development and developmental cognitive neuroscience.

Michael S. C. Thomas is Professor of Cognitive Neuroscience at Birkbeck, University of London, UK.

Denis Mareschal is Professor of Psychology at Birkbeck, University of London, UK.

Victoria C. P. Knowland is Lecturer in Speech and Language Science at University of Newcastle, UK.

TAKING DEVELOPMENT SERIOUSLY: A FESTSCHRIFT FOR ANNETTE KARMILOFF-SMITH

Neuroconstructivism and the Multi-Disciplinary Approach to Understanding the Emergence of Mind

Edited by Michael S. C. Thomas, Denis Mareschal, and Victoria Knowland

Routledge
Taylor & Francis Group

LONDON AND NEW YORK

First published 2021
by Routledge
2 Park Square, Milton Park, Abingdon, Oxon OX14 4RN

and by Routledge
52 Vanderbilt Avenue, New York, NY 10017

Routledge is an imprint of the Taylor & Francis Group, an informa business

British Library Cataloguing-in-Publication Data
A catalogue record for this book is available from the British Library

Library of Congress Cataloging-in-Publication Data
A catalog record has been requested for this book

ISBN: 978-1-138-33404-5 (hbk)
ISBN: 978-1-138-33405-2 (pbk)
ISBN: 978-0-429-44559-0 (ebk)

Typeset in Bembo
by MPS Limited, Dehradun

This volume is dedicated to Annette Karmiloff-Smith, one of the most incredible women to grace Developmental Science.

Our irreplaceable friend and colleague Annette Karmiloff-Smith passed away on the 19th of December 2016. Annette was an inspirational scientist and inspirational teacher but above all an inspirational woman. A true pioneer in the study of child development, she touched the lives of many across the world with her writings and teachings.

In December 2016, the Centre for Brain and Cognitive Development, of which she was a faculty member, set up a webpage to allow her many international friends and colleagues to express what she meant to them and how much she had influenced their lives.

We invite you to read through these comments for a deeper sense of who she was as a person, as a researcher, as a friend, as a mother, and as woman.

https://cbcdsite.wordpress.com/2016/12/19/annette/

We all miss you.

CONTENTS

In their introductory chapter the editors highlight the numerous honours and awards given to Annette Karmiloff-Smith over the years for her many significant and substantial contributions to the science and theory of development. Here I briefly note the key features of her work. Historically, development has been largely ignored on the basis of its supposed focus on age differences and on development as an outcome on the influence of non-developmental features. Cognitive deficits in developmental disorders were assumed to be equivalent to cases of brain damage with both intact and impaired modules.

Annette's approach was quite different, although she accepted that development was bound to be influenced by non-developmental features and thus the importance of an interplay among genetic and environmental factors. As she put it, in order to study developmental disorders, it was essential to take development itself seriously; she proposed that modularity was the outcome of development rather than a precursor of it. What was needed was developmental studies following individuals over time, together with details of normal functioning and comparison of atypical trajectories, such as seen in Down's syndrome and Williams syndrome (both genetic disorders).

In order to carry out these investigations, Annette needed teams of colleagues spanning a range of skills, and so she set them up. Her success in that was facilitated by her good working relationships and friendships with colleagues (as exemplified by chapters in this Festschrift).

What is very clear is that Annette's creative theorising and ingenious experiments provide an excellent model for all of us to follow and will shape cognitive and developmental studies for many years to come.

Sir Michael Rutter

CONTRIBUTORS

Mike Anderson was a colleague of Annette at the Cognitive Development Unit in London from 1983 to 1988. He moved to Western Australia in 1990, being appointed Head of the School of Psychology at UWA 2003–2006, and Winthrop Professor in 2006. Having retired in 2017, Mike is currently an Emeritus Professor at Murdoch University, Australia.

Daniel Ansari is a Professor and Canada Research Chair in Developmental Cognitive Neuroscience and Learning in the Department of Psychology and the Faculty of Education at Western University in Ontario, Canada, where he heads the Numerical Cognition Laboratory. Ansari and his team explore the developmental trajectory underlying both the typical and atypical development of numerical and mathematical skills, using both behavioural and neuroimaging methods.

Frank D. Baughman runs the Neurocognitive Developmental Laboratory in the School of Psychology, at Curtin University in Perth, Australia. Frank's research interests lie in understanding the mechanisms that influence cognitive variability. This includes the study of variability in both typical and atypical development, using a mix of behavioural and computational methodologies.

April A. Benasich is the Elizabeth H. Solomon Professor of Developmental Cognitive Neuroscience and the Director of the Infancy Studies Laboratory at the Center for Molecular and Behavioural Neuroscience, Rutgers University, Newark, USA. Her research uses converging paradigms and prospective longitudinal studies to explore the human developing brain, the dynamics of early brain plasticity and the role of attention and sensory recruitment in the construction of brain networks crucial to normative language and cognitive development.

Daniel Brady completed his PhD at Goldsmiths in 2016. He worked with Annette on the infant stream of the London Down Syndrome (LonDownS) project during and after his PhD, and is currently working as a Research Technician at the University of Reading, UK.

Jean-Paul Bronckart is an Honorary Professor at the University of Geneva, Switzerland, where he fulfilled his academic career. From 1969 to 1976, he was an assistant to Jean Piaget at the International Center of Genetic Epistemology, and then he held, from 1976 to 2012, a full professor position in language didactics.

Tessa M. Dekker was inspired by Annette's multidisciplinary approach to development. Her lab in the Division of Psychology and Language Sciences, at University College London, studies how genes, brains, and experience interact in early life to shape how we come to see the world, using carefully designed model-based neuroimaging and behavioural studies to untangle mechanisms of visual development in children with normal sight and early eye disease.

Hana D'Souza is the Beatrice Mary Dale Research Fellow at Newnham College and an affiliated lecturer in the Department of Psychology, University of Cambridge, UK. Her research focuses on the development of infants and toddlers with neurodevelopmental disorders of known genetic origin, particularly Down syndrome and William syndrome. As part of the LonDownS Consortium, she has been investigating individual differences and interactions between various domains and levels of description across development in Down syndrome.

Jamie Edgin is an Associate Professor and Director of the Cognition and Neural Systems Program in the Department of Psychology at the University of Arizona, USA. Her area of expertise is sleep and memory development in typical and atypical development, including work with individuals with Down syndrome and autism. She currently directs the University of Arizona Public Policy Fellowship, focused on policy development and advocacy for persons with developmental disabilities. Annette and Jamie shared a passion for sleep science and communicating the importance of sleep and the study of disability to our communities.

Emily K. Farran leads the Cognition, Genes & Developmental Variability Lab at the University of Surrey, UK. The aim of her research is to characterise both typical and atypical development of cognitive functions within a neuroconstructavist framework, with a special focus on visuo-spatial processing and how it relates to mathematical ability in school-aged children.

Susan Goldin-Meadow is the Beardsley Ruml Distinguished Service Professor of Psychology and Comparative Human Development at the Department of Psychology, University of Chicago, USA. She is a developmental psychologist recognised for her work on the emergence of language, and on role that co-language gestures play in communicating, thinking, and learning.

Mark A. Good is a professor of neuroscience at the School of Psychology, University of Cardiff, UK. His research interests lie in using preclinical models of brain pathology to understand the functional role of the hippocampus in typically developing individuals and those with neurodegenerative disease such as Alzheimer's disease and frontotemporal dementia.

Katherine Hughes is a teaching fellow at the Department of Psychology, Goldsmith College, University of London, UK. She was one of Annette's final PhD students, completing the childhood-adolescent span of the LonDowns life span investigation of risk of Alzheimer's in Down syndrome.

Mark H. Johnson is a Professor of Experimental Psychology and Head of the Department of Psychology at the University of Cambridge, UK. His laboratory currently focuses on typical, at-risk, and atypical functional brain development in human infants and toddlers using a variety of different brain imaging, cognitive, behavioural, genetic and computational modelling techniques.

Victoria C. P. Knowland is a lecturer in speech and language sciences at the School of Education, Communication & Language Sciences, Newcastle University, UK. She studied for a PhD at Birkbeck College under Annette Karmiloff-Smith and Michael Thomas. Her research interests lie in explaining variability in language development. She is currently exploring the role that sleep plays in language acquisition.

Kang Lee is a Canada Research Chair in the Department of Applied Psychology and Human Development, Dr. Eric Jackman Institute of Child Study, at the University of Toronto, Canada. As a developmental psychologist, Lee studies moral development and the development of lying in children. He is a leader in his field and has made significant contributions not only to the literature about moral development, but also to real world practices.

Yonata Levy is a Professor of Developmental Neuropsychology in the Psychology Department and Hadassah-Hebrew University Medical School, Hebrew University of Jerusalem, Israel. Her research centers on first language acquisition in children with neurodevelopmental disorders. Recently, her work has focused on definitional issues of neurodevelopmental disorders.

Denis Mareschal is a Professor of Psychology in the Department of Psychological Sciences and Director of the Centre for Brain and Cognitive Development, Birkbeck College, University of London, UK. His research focuses on all aspects of perceptual and cognitive development in infancy and childhood.

Jay L. McClelland studied cognitive and conceptual development and learning using neural-network- based computer simulation models at the Center for Mind, Brain, Computation and Technology, Department of Psychology, Stanford University, USA. Together with David E. Rumelhart, he led the team producing the two-volume work *Parallel Distributed Processing: Explorations in the Micro-Structure of Cognition,* published in 1986. His current work focuses on the nature and acquisition of advanced cognitive abilities, including mathematical cognition.

Parag K. Mital is a computational artist and interdisciplinary researcher. He completed his PhD in Arts and Computational Technologies at Goldsmiths,

University of London, UK. His is currently CTO & Head of Research at Hypersurfaces. His research interests include mixed reality, pattern recognition, audiovisual signal processing, audiovisual cognition, and information retrieval.

Kim Plunkett is a member of faculty in the Department of Experimental Psychology, University of Oxford, UK and a Fellow of St. Hugh's College. In 1992 he founded the Oxford Babylab, a research facility for the experimental investigation of linguistic and cognitive development in babies and young children. His primary interest is in understanding the mechanisms of change that drive linguistic and cognitive development.

Luca Rinaldi is an Assistant Professor at the Department of Brain and Behavioural Sciences of the University of Pavia, Italy. By combining different methods and techniques, his main research investigates how our brain adjusts prior sensorimotor and linguistic experience to construct memory representations. In particular, his work tries to shed light on the specific mechanisms that help structure our semantic memory, with a focus on how this memory system develops over the lifespan.

Gaia Scerif is a Professor of Developmental Cognitive Neuroscience at the Department of Experimental Psychology, University of Oxford, UK, and a fellow of St Catherine's College. Her research focuses on the development of attentional control and underlying attentional difficulties, from their neural correlates to their outcomes on emerging cognitive abilities.

Aaron Sloman earned a degree in Maths and Physics in Cape Town in 1957. He became a multidisciplinary seeker after explanations of mathematical cognition, in the Universities of Oxford, Hull, Sussex and Birmingham (UK). He is currently an Honorary Professor at the School of Computer Science, University of Birmingham, UK.

Tim Smith is a cognitive psychologist at the Department of Psychological Sciences, Birkbeck College, University of London, UK. His interests are in naturalistic audiovisual (AV) cognition, how we perceive AV media (e.g. TV, cinema, videogames), and how media shapes the developing brain. He uses a range of techniques in his research including behavioural experiments, cognitive/ computational modelling, signal processing, electrophysiology, questionnaires, qualitative methods, and eye tracking.

Michael S. C. Thomas is a Professor of Cognitive Neuroscience in the Department of Psychological Sciences and Director of the Centre for Educational Neuroscience, Birkbeck College, University of London, UK. His research employs a multi-disciplinary approach to investigate cognitive variability, and his translational work seeks to harness new findings in the neuroscience of learning to improve educational outcomes.

Frances K. Wiseman is a programme leader in animal models for the UK Dementia Research Institute and Alzheimer's Research UK, and a senior research fellow at University College London, UK. Her work aims to understand how genes on chromosome 21 modify disease, with the goal of developing novel treatments for dementia, particularly via the use of mouse models.

Frank Wiesemann has a PhD in Bioinorganic Chemistry. He took up a position in product design and research with Procter & Gamble in 1995 and was promoted to Principal Scientist in 2006. Since then he has worked on projects as diverse as product sustainability and safety, consumer understanding and the impact of sleep on development.

Katherine Wolfert is a PhD candidate in the Behavioral and Neural Sciences Graduate Program at the Center for Molecular and Behavioral Neuroscience, Rutgers University, USA. Her research in the Infancy Studies Laboratory, under the supervision of April A. Benasich, focuses on the development of the auditory and visual systems, exploring how sensory processing skills support language learning.

1

ANNETTE KARMILOFF-SMITH: SCIENTIST, MOTHER AND FRIEND

Michael S. C. Thomas, Denis Mareschal, and Victoria C. P. Knowland

Professor Annette Karmiloff-Smith was a highly influential developmental and cognitive scientist who made significant contributions to our understanding of typical and atypical development. She was born 18th July, 1938, and passed away 19th December, 2016, aged 78. Annette was a seminal thinker in the field of child development and the recipient of many awards. These achievements include elected member to Academia Europaea (1991); Fellow of the British Academy (1993); British Psychological Society Book Award (1994) for *Beyond Modularity: A Developmental Perspective on Cognitive Science*; elected Fellow of the Academy of Medical Sciences (1999); European Latsis prize for Cognitive Science (2002; one of only two women to achieve this accolade); Commander of the British Empire (CBE; 2004) for services to cognitive development; elected Fellow of the Cognitive Science Society (2008); Lifetime Achievement award, British Psychological Society (2009); honorary doctorate from the University of Amsterdam (2010); Mattei International Prize for Psychological Sciences, Union Psychological Society (2012); William Thierry Preyer Award for Excellence in Research on Human Development, European Association of Developmental Science (2017; joint with Mark Johnson); British Psychological Society Distinguished Contributions Award (2017).

As the British philosopher Maggie Boden describes her, Annette was "theoretically deep and experimentally ingenious. Her work highlighted fundamental questions about the nature and origins of representation in various domains (language, vision, physical action). She combined psychological theorising with computational modelling, clinical investigations, and—partly inspired by her husband Mark Johnson—cognitive neuroscience, too".

One of the final book projects that Annette set in motion, shortly before she passed away in 2016, was the publication of her collected papers (Karmiloff-Smith, Thomas, & Johnson, 2018, *Thinking Developmentally from Constructivism to Neuroconstructivism: Selected Works of Annette Karmiloff-Smith*). The current volume should be seen as a companion to that book, in that it brings together many of the researchers, collaborators, and students who worked with Annette, to show how her ideas have influenced and continue to influence their own research. The chapters provide a mix of cutting-edge science and reminiscence, providing a fascinating insight into the historical contexts in which many of Annette's theoretical insights arose, including such ideas as *the microgenetic approach, representational redescription*, and *neuroconstructivism*. The chapters also provide updates about how earlier theoretical ideas have stood the test of time, and present unpublished data from the early years of Annette's career.

Annette began her scientific career in Geneva, Switzerland. She was working as a simultaneous interpreter for the United Nations, but was bored because she was always repeating other people's thoughts and was not allowed, as interpreter, to have any of her own. She decided to go back to university and was exploring a local bookshop for inspiration when one day, she caught sight of Jean Piaget, whose photo she recognised from a textbook. She followed him back to the university buildings, deciding then to audit his course. Her eyes were opened. To her, Piaget's approach to development offered so much more than observation. It included epistemology, logic, philosophy of mind, and the philosophy of science. It was there that she discovered her absolute passion for research.

Annette went on to complete her PhD at the University of Geneva in Piaget's *International Centre for Genetic Epistemology*, under the supervision of Bärbel Inhelder and Hermine Sinclair-De Zwart, successfully defending her thesis in 1977. Her first article *If you want to get ahead, get a theory!* was published in 1975 (Karmiloff-Smith & Inhelder, 1975). One of the innovations of her early work was the development of the *microgenetic method*. The key innovation was not just to focus on macro-development – change across age – but also micro-development, "children's spontaneous organizing activity in goal-oriented tasks with relatively little intervention from the experimenter [where] the focus is not on success or failure per se but on the interplay between action sequences and children's theories-in-action" as they explore and figure out how, in this case, physical systems like the balance beam work (Karmiloff-Smith & Inhelder, 1975, p. 196).

Her work on cognitive and language development led her towards key insights in how psychological nativism, and particularly Jerry Fodor's thesis that the mind is composed of innate modules (Fodor, 1983), could be combined with the Piagetian idea that cognition is constructed through an active, experience-driven process of development. Annette brought together these ideas in her seminal 1992 book, *Beyond Modularity*, where she sought to advance "cognitive development as a serious theoretical science contributing to the discussion of *how* the human mind is organised internally, and not as merely a cute empirical database about *when* external behaviour can be observed" (1992, p. xiii). Annette's solution to the opposing nativist and constructivist positions was to accept some degree of Fodor's modularity as an endstate in adults, but now to argue that this was always the outcome of a domain-general developmental process of *modularisation* from a less specialised initial state. "I hypothesize that if the human mind/brain ends up with any modular structure, then this is the result of a process of modularization as development proceeds" (Karmiloff-Smith, 1994, p. 695). Emergent specialisation would be complemented by *representational redescription*, with sharing of information occurring beyond mastery of skills to generate the (perhaps unique degree of) cognitive flexibility observed in humans. With some prescience, Annette argued that the degree of specialisation in the infant would only be established with the emergence of technologies for measuring on-line brain activation with neonates and young infants.

After a number of years working on typical development, with a two-year gap working in the Palestinian refugee camps in Beirut, in 1982 she moved to London to the Medical Research Council's Cognitive Development Unit. There, several colleagues were working on autism and Down syndrome, which stimulated her interest in atypical development.

In 1998, she set up her own MRC-funded Neurocognitive Development Unit at the Institute of Child Health in London (which subsequently became part of University College London). By then, she was convinced that the study of development had to embrace a multidisciplinary approach, combining research at several levels of description, including genes, brains, cognition, behaviour, and the environment. This was reflected in her co-authorship of the influential book, *Rethinking Innateness: A Connectionist Perspective on Development*, published in 1996. Her new unit brought together researchers using behavioural methods, brain imaging, genetics, and computational modelling (particularly Parallel Distributed Processing/connectionist approaches), to study infants, children, adolescents and adults. The primary focus was on developmental disorders, including seminal work on the rare genetic disorder Williams syndrome.

Annette developed an approach called *neuroconstructivism*, to integrate Piagetian theory with new findings on functional brain development. An understanding of developmental mechanism was again to the fore, as in her 1998 paper, *Development itself is the key to understanding developmental disorders*. She argued that what seems to be the 'natural' organisation of the brain in adults is actually a result of development itself – developmental disorders are not the 'knocking out' of specific abilities but affect the dynamics of neurodevelopment, as the child interacts with the world. Early low-level differences in infancy have potentially cascading effects across development. In terms of experimental design, she emphasised the need for longitudinal studies to chart the process of development, and advocated research that spanned different developmental disorders and different emerging domains. To study development disorders, one had to take the process of development seriously.

After joining Birkbeck, University of London, as a Professorial Research Fellow in 2006, her work focused on understanding the complex epigenetic interactions involved in brain organisation across early development. Her final Wellcome-Trust-funded research project was part of the interdisciplinary London Down Syndrome (LonDownS) Consortium, which combined researchers at University College London, Birkbeck, and King's College London. The project used Down syndrome (DS) as a model for Alzheimer's Disease. It brought together a multidisciplinary team of human geneticists, cellular biologists, psychiatrists, psychologists, neuroscientists, and mouse geneticists, with the goal of investigating potential early risk and protective factors for the emergence of dementia in DS. Annette's focus was to link variation in early development in DS with differences in the risk of dementia, a condition that is elevated in adults with the syndrome. "What I'm doing is kind of counterintuitive", she said of the project. "I'm arguing that we can study babies to understand Alzheimer's disease in adults" (Karmiloff-Smith, 2016). Indeed, she confessed her ambition was to construe dementia as a kind of developmental disorder, if one viewed develop as a lifespan concept (Thomas et al., 2020). As ever, the framing was intended to encourage researchers to increase their focus on mechanisms of change.

As someone committed to the communication of science to a broader audience, Annette also wrote several extremely successful books directed at a lay audience, including some co-authored with her daughter Kyra (*Baby It's You: A unique Insight into the First Three Years of the Developing Baby*, 1994; *Everything Your Baby Would Ask*, 1998; *Understanding Your Baby: A Parent's Guide to Early Child Development*, 2015). Her commitment to the wider dissemination of science also led her to work with industry to incorporate the findings of academic research on cognitive development into product design. This included her collaboration with Proctor and Gamble in the design of Pampers nappies; and at the end of her career, in response to the startling pace of change in people's engagement with technology, research considering the influence of screen time on infant development.

On a personal level, Annette was immensely supportive to students and junior colleagues, and was an inspiring mentor. She was also a keen advocate of women in science. Her advice to women starting out in a career in science captures Annette's spirit perfectly: "Don't forget that you are a woman! You can be a top scientist and a mother, if you want to (I have two daughters and seven grandchildren and love it), and you can be feminine and intelligent! Above all, only do research if you feel passionate about it – you have to work very hard (eight days a week), it is physically and psychologically demanding, and the data aren't always kind! But it is such fun" (Borovsky, 2005).

Overview of chapters

Part one: Geneva and Beyond Modularity

The chapters in this volume span four parts. Part One contains contributions from Annette's colleagues and collaborators from her time in Geneva in the 1970s and in London in the 1980s. These span Piaget, the microgenetic approach, typical development, implicit and explicit knowledge, and her influential book, *Beyond Modularity*.

In Chapter 2, **Susan Goldin-Meadow** takes us back to the time when both she and Annette were students in Geneva, working within the Piagetian school of research. SGM describes an innovative project she and Annette undertook together exploring the constraints that shape children's understanding of relative clauses – were these syntactic, with difficulty influenced by unusual word orders or the embedding of phrases? Or were they semantic, such as whether the same concept had consistent or inconsistent thematic roles across phrases? (e.g., in the sentence *the squirrel licks the cat that pushes the monkey*, the cat plays the role of patient in the main clause but the role of agent in the subordinate relative clause). It turned out that the semantic role-switch was particularly difficult for children to master, pointing towards the importance of semantic limiting factors in child language development (a constraint that fades away by adulthood as conceptual abilities increase). SGM shows how this project influenced her later thinking in work on *homesign*, the gestural language that deaf children invent to communicate with hearing parents in the absence of formal sign language.

In Chapter 3, **Jean-Paul Bronckart** gives an insight into the historical context in which Annette's early theoretical ideas emerged. This was the University of Geneva in the 1970s, the home of Jean Piaget's research project at the *Centre international d'épistémologie génétique* (International Centre of Genetic Epistemology). JPB gives a sketch of the leading figures of the time, such as Bärbel Inhelder and Hermina Sinclair-De Zwart (Annette's PhD supervisors) and the dynamics among them – not least those generated by the character of the great Jean Piaget himself. JPB examines the problems encountered by Piagetian constructivism as early as the 1950s and various attempts to solve them, which enables him to sketch out the relationship that Annette had with Piaget and his close collaborators.

Mike Anderson was a colleague of Annette's back from her days at the MRC Cognitive Development Unit in London, where his own interest was in intelligence. In Chapter 4, **Frank Baughman** (who trained with Annette 2005–2009) and **Mike Anderson** pursue what it means to *take development seriously* on the question of intelligence (see also Chapter 19). For Annette, taking development seriously meant rejecting static snapshots of abilities at different ages and providing an explanation of how change takes place; and it meant requiring methods that focus on multiple, nested influences across time. FB and MA comment that devising a research project that truly takes development seriously is a complex endeavour, and they explain how they set about this challenge in their work on intelligence. They contrast two different models explaining how intelligence alters across development, the *Minimal Cognitive Architecture*, and a more recent dynamical systems approach that formally models causal relationships between developing cognitive components over time. They then use the second of these approaches as a framework to consider, via a set of computational simulations, how variations in abilities might interact with developmental change. They show how a key effect of this interaction is to increase the extent to which variations in ability are correlated across cognitive domains (the so-called general factor of intelligence). Crucially, they show that this pattern is also influenced by different possible cognitive architectures.

Kang Lee travelled from China to work with Annette in London in 1989, the year the Tiananmen Square massacre took place. In Chapter 5, KL reflects on

Annette's committed role as a mentor to her students and researchers. He shares several stories about how he was mentored by Annette and the lessons he learned from his experience as her student. Among them, he learned that being a mentor is a solemn responsibility, because the seemingly routine act of accepting students into our labs may completely transform their lives – as it did for him.

In Chapter 6, the philosopher **Aaron Sloman** charts the trajectory of his thinking on a kaleidoscope of issues, and how those have been influenced by Annette's work in *Beyond Modularity*. Through points of agreement and disagreement with Annette, AS discusses his work on Kantian aspects of the development of mathematical competences, his Meta-Configured Genome project exploring the possible evolutionary origins of knowledge of necessary truths and impossibilities, the influence of developmental robotics and artificial intelligence on developmental theory, and even the mathematical knowledge required by trapeze artists and spider monkeys! Throughout, his touchstone is Annette's notion of *representational redescription* articulated in *Beyond Modularity* (most notably, in the chapter, 'The Child as Mathematician'), a book which AS concludes should be "compulsory reading for all researchers working in the areas of intelligent robotics, cognitive robotics, AI learning systems, and more generally computational cognitive science".

Part two: Rethinking Innateness

Part Two moves on to consider Annette's collaboration in a co-authored book published in 1996, *Rethinking Innateness*, which recast theories of cognitive development through the lens of developmental cognitive neuroscience and the computations that can be carried out by neural networks (the Parallel Distributed Processing [PDP] or connectionist approach). In Chapter 7, **Mark Johnson** reflects on the circumstances that led to the writing of the book, which brought together Jeff Elman, Elizabeth Bates, MJ, Annette, Dominic Parisi, and Kim Plunkett. Although Annette was originally suspicious of connectionist models of cognitive development in *Beyond Modularity* – because she could not see how they could accommodate representational redescription – she later embraced their utility for focusing on mechanisms of developmental change. In this chapter, MJ briefly assesses the current evidence for, and status of, some of the primary claims and ideas from *Rethinking Innateness*, with the benefit of 25 years' hindsight. He covers recent changes in connectionism (including the emergence of deep neural networks), an increasing focus on the importance of the infants' proactive novelty-seeking behaviour for the self-organisation of his or her nervous system, and the gaps that still exist in understanding the key role of social interaction in the earliest stages of postnatal brain development.

The next two chapters trace new lines of thinking with respect to Annette's idea of *representational description*, from two researchers whose research has been shaped by PDP modelling. In Chapter 8, **Kim Plunkett**, one of the co-authors of *Rethinking Innateness*, presents some of his recent work on vocabulary development in young children. Specifically, he considers Annette's contention, from her 1986 paper, *From meta-process to conscious access* (Karmiloff-Smith, 1986), that progressive explicitisation and restructuring of mental representations are driven not by failure (negative feedback mechanisms) but by success (positive feedback mechanisms). KP considers how success in word learning in toddlers leads to a reorganisation of representations of both word meaning (semantics) and word sounds (phonology). He reveals, in a series of careful lexical-priming experiments, how vocabulary acquisition first involves the establishment of excitatory pathways between lexical items that are related to each other (such as *dog* and *cat*), and then a process of reorganisation, where inhibitory processes emerge to control unconstrained spread of excitation in the system. Notably, he argues that this reorganisation is sometimes accompanied by a temporary worsening in language performance.

In Chapter 9, **Jay McClelland** concentrates on Annette's reservations about the ability of PDP/connectionist networks to capture *representational redescription*. Specifically, Annette argued (as in her co-authored paper with Andy Clark in 1993) that such models only capture implicit learning of regularities in the physical transitions of objects in the world or in the structures of language: the models do not capture the phase when children move beyond mastery to knowledge that is both more flexible and available to conscious access. JM shows how this insight initially led neural network modellers to think about ways to capture the relationship between explicit and implicit learning through separate processing pathways, such as the work of Axel Cleeremans in the 1990s and early 2000s. Some two decades later, JM shows how his student, Andrew Lampinen, has returned to the idea of *representational redescription* from a PDP perspective. In new work, Lampinen has explored how the knowledge encoded in the connection weights of a neural network can be made accessible by representing this connectivity as a pattern of activation in another separate network, and that such knowledge may then be manipulable. The second network might be able to map between activation states that represent different patterns of connectivity in the first network, reasoning, as it were, about strategies. This new avenue provides a way to start thinking about the explicitisation and manipulation of previously implicit knowledge, perhaps enabled and triggered by language. It presents a vocabulary to progress Annette's ideas within a formal computational approach.

Part three: Neuroconstructivism and developmental disorders

Part Three moves on to consider researchers who have been influenced by Annette's work from 1998 onwards on neuroconstructivism and developmental disorders. This stage of Annette's career was marked by the publication of her seminal paper, *The Key to Understanding Developmental Disorders is Development Itself*, and the extension of her research to incorporate multidisciplinary cognitive neuroscience methods such as brain imaging and genetics.

In Chapter 10, **April Benasich** and **Katherine Wolfert** show how Annette's 1998 paper points to a multi-level and dynamic explanatory framework that encourages prospective and longitudinal studies. Such studies can reveal how alternative developmental pathways might lead to different phenotypic outcomes. B&W comment that Annette's 1998 paper seems as current and important today as it was when first published, perhaps, they say, because its theme of "taking development seriously" mirrors a trend in contemporary neuroscience toward taking complex systems seriously. To do so is to embrace the subtleties of interactions between the brain's structure and function, genetics and changing environments. B&W then describe a range of studies from the Benasich lab that follow this approach with infants and toddlers. The studies address the early neural processes necessary for normal cognitive and language development, and then consider the impact of disordered processing on later outcomes, either for children at risk of language-based learning impairments or autism (by virtue of their family history), or in very low birth-weight pre-term babies. They detail prospective studies that follow infants from 2 through 60 months and employ the latest computational neuroscience techniques, including electrophysiology, magnetic resonance imaging, diffusion tensor imaging, and eye-gaze tracking, to uncover the developmental roots of disorders and the dynamics of their unfolding.

In Chapter 11, **Gaia Scerif** takes us back to a summer research assistant position she undertook with Annette in the Neurocognitive Development Unit at the Institute of Child Health in London in 2000. She recalls an early project that cuts to the heart of whether developmental disorders can be characterised by domain-specific or domain-general deficits, and shows how a study utilising cross-domain comparisons helped answer this question. The project investigated numerical skills and language skills in Williams syndrome, the disorder that was then the focus of Annette's work. From this early study, GS draws out the three key lessons that she then applied in her subsequent PhD on the typical and atypical development of the attention system (supervised by Annette and Jon Driver). These lessons were a focus on atypical development, a call for clarity about theory and hypothesis making, and an interest in mechanistic understanding. GS then charts how these lessons shaped her subsequent research, including a longitudinal investigation of the atypical development of attention and executive control in fragile X syndrome, and a deepening understanding that attentional processes must be characterised in the context of the cognitive domains in which they are manifested, rather than viewed as a type of independent 'homunculus'.

In Chapter 12, **Daniel Ansari** also recalls the time when he was studying for his PhD under Annette's supervision at the London Institute of Child Health. He describes his experience of co-authoring of a paper with Annette that was a turning point in his research career. This paper explored atypical trajectories of number development from a neuroconstructivist perspective and was published in 2002. It focused on the insights offered by developmental disorders into the cognitive building blocks of numerical cognition, including the relationship between analogue representations of quantity and symbolic representations of number. The subsequent influential paper (Ansari & Karmiloff-Smith, 2002, *Atypical trajectories of number development: a neuroconstructivist perspective*) finished with a set of outstanding questions for the research field at that point in time. These included the relationship between dyscalculia and dyslexia, and the apparent mismatch of young children's poor understanding of the meaning of counting with the (then) new evidence suggesting preverbal infants already possess numerical abilities. If you already understand number, why is it so hard to learn to count? Some 18 years later, DA has the opportunity to revisit these outstanding questions and provides some fascinating answers.

In Chapter 13, **Michael Thomas** and **Daniel Brady** focus on another of the key ideas at the heart of Annette's thinking, *modularity*. MT's first co-authored publication with Annette at the Institute of Child Health was a paper entitled *Quo Vadis Modularity in the 1990s?*. The paper, published in 1999, assessed the status of the concept of modularity in developmental theory, and its application to explaining developmental deficits. Again, the chapter provides an opportunity to revisit an old debate. Modularity is the idea that cognition is largely composed of independent, domain-specific cognitive components, as revealed by the possibility of selective damage in adulthood. Annette felt strongly that modularity had been inappropriately applied to the study of developmental disorders through the 1980s and 1990s. Modularity promulgated static explanations of cognitive deficits in developmental disorders, as if they were analogous to cases of adult brain damage with intact and impaired modules – but this gave no role for development (Thomas & Karmiloff-Smith, 2002). As we have seen, Annette proposed that modularity was the outcome of development rather than a precursor to it. MT and DB chart changes in the theoretical concept of modularity over 30 years. They argue that advances in neuroscience have caused modularity to be recast from an *a priori* design principle, which it would make good sense for potential cognitive systems to employ, to a *data-driven* concept based on how the brain is actually working. They illustrate the latter point by presenting data from Annette's final project on infant development in DS, which uses graph theoretical analyses of electrophysiological brain imaging data to explore functional network structure in the brain activity of young children. The take-home message is that since modularity is now a data-driven concept, it becomes an empirical issue whether it is a property that alters across development or indeed varies with the level of intellectual ability.

In Chapter 14, **Emily Farran** describes her programme of research applying neuroconstructivist principles to the study of spatial cognition in developmental disorders. These principles include the use of cross-domain associations within disorders (such as between spatial cognition and motor skills, verbal skills, numerical skills, face recognition, and reasoning); cross-syndrome comparisons of abilities (such as between Williams syndrome (WS), DS, and autism); the use of trajectory analysis to study developmental change, either in cross-sectional or longitudinal studies; and the origin of individual variation within disorders. EF also looks to the future, with recommendations to progress research in the field, such as addressing the surprising lack of longitudinal data for rare genetic syndromes by combining data across laboratories that test the same population. She gives the example of the new Williams Syndrome Development (WISDom) group, led by Jo van Herwegen, which is compiling longitudinal data across labs, including many of Annette's collaborators. Lastly, EF illustrates Annette's sense of humour: when their edited volume *Neurodevelopmental Disorders Across the Lifespan: A Neuroconstructivist Approach* was published in 2012, it coincided with the month that EF's first daughter was born. Annette asked EF to weigh the book, and then put together a slide for her own talks that year announcing the birth of both book and baby, along with their birthweights and names!

In Chapter 15, **Yonata Levy** describes her theoretical journey from a generative, Chomskian view of language acquisition in children with neurodevelopmental disorders to embracing a neuroconstructivist perspective, prompted by her interactions with Annette across a series of sabbatical visits to London. YL confesses that she is not sure if Annette realised that she had brought YL around to neuroconstructivism, that she was no longer a generativist. YL presents her chapter as a belated debt to Annette. YL addresses a tantalising puzzle: neuroconstructivism, stressing the interactivity of brain regions across development, would seem to predict that in different genetic disorders, language development should look different. Yet reports of qualitatively different patterns of language acquisition across different disorders are few. YL then presents the results of her work following the early language development of young children with WS and DS in a longitudinal study. She confirms little evidence of overt atypicality but reports differences in age of onset and rate of transition through phases of language acquisition. In the second half of the chapter, she then reviews evidence from neuroscience to argue that age and rate of change are actually key aspects of neurocognitive development, and disruptions to timings themselves constitute patterns of atypical development. She concludes from her work on language acquisition that chronological age should be a key factor in our attempts to understand neurodevelopmental disorders.

Part four: New avenues

Part Four finishes with Annette's most recent work, including her innovative research on the LonDownS project, research on sleep and cognitive development, the design of videos for babies, the effect of screen time on children's development, and her influential role in collaborating with industry to develop products shaped by research on cognitive development and to communicate this research to parents.

In Chapter 16, **Hana D'Souza and colleagues** give an insight into Annette's thinking on her last project with the LonDownS Consortium. Recall, the consortium is a large interdisciplinary collaboration between human geneticists, cellular biologists, psychiatrists, psychologists, neuroscientists, and mouse geneticists, whose aim is to understand the link between DS and Alzheimer's disease. HD and colleagues focus on Annette's interest in mouse modelling, as a further tool to help characterise atypical developmental mechanisms at multiple levels of description (see also D'Souza, Karmiloff-Smith, Mareschal, & Thomas, 2020; D'Souza, Lathan, Karmiloff-Smith, & Mareschal, 2020; D'Souza, Mason et al., 2020; Glennon et al., 2020; Karmiloff-Smith, 2016; Startin et al., 2020; Thomas et al., 2020, for related work from the project). One of Annette's goals was to nurture crosstalk between mouse modelling and human phenotyping because, although mouse models of DS hold great promise for advancing our understanding of the aetiology of cognitive phenotypes in individuals with DS – and may indeed ultimately lead to targeted interventions – the strength of each mouse model critically depends on how well the tasks used with the mice map onto neural and cognitive processes of interest in humans. The chapter draws out the challenges in aligning mouse models with human behaviour. It describes how in the LonDownS project, memory tasks were designed for infants and toddlers with DS to echo those already used with mice. The authors present data from their children with DS and compare them with patterns observed in memory tasks used with a trisomy 21 mouse model – the *Tc1* mouse (Hall et al., 2016) – specifically, the observation that these mice exhibit typical-looking immediate memory for object recognition, impaired short-term memory, but then typical-looking long-term memory.

In Chapter 17, **Katherine Hughes** and **Jamie Edgin** cover Annette's involvement in sleep research in the latter part of her career. The authors present work that they view as an unfinished scientific conversation with Annette. KH was one of Annette's final graduate students, while JE began a personal friendship and science collaboration with Annette when Annette initiated her final studies of DS in 2012. The focus of this collaboration was to understand the role of sleep in development, and in particular, in the ongoing construction of representational change and memory formation. Indeed, returning to her ideas of representational redescription, Annette speculated in 2015 that this process is likely to occur via brain activity during sleep (Karmiloff-Smith, 2015). As applied to developmental disorders, disruptions to sleep may feed into altered trajectories of cognitive development. Annette explored this hypothesis through identifying correlations between sleep and memory development in infancy (Pisch, Wiesemann, & Karmiloff-Smith, 2019) and through cross-syndrome comparisons of sleep patterns across different disorders, such as WS and DS, and their links to memory consolidation (Ashworth, Hill, Karmiloff-Smith, & Dimitriou, 2015). KH and JE review this emerging literature, and show how Annette's work motivated a sequence of studies in the Edgin lab. The authors identify directions for future research and theory, such as the need to incorporate a more complete understanding of the role of the developing hippocampus into our explanations of cognitive development.

In Chapter 18, **Tim Smith**, **Parag Mital** and **Tessa Dekker** present data on how infants interact with educational videos, collected using sensitive eye-tracking measures of attention. Annette cared deeply about engaging with the debate around screen time and viewed it as essential that researchers should influence this debate in a positive way. For her, this meant empirically investigating the bidirectional relationship between how the developing visual system responds to screen

media and how designers intuitively tailor their media to developmental constraints to create infant-directed TV (Karmiloff-Smith, 2012). She proposed a set
of design principles that would optimise the infant's interactions with the content,
such as attracting attention through movement, and not using cluttered backgrounds. The study the authors present contrasts infant-looking behaviour for a
video designed according to Annette's principles against that elicited by another
commercially available educational video targeted at infants. The study was designed and carried out by Annette and the authors (TD and TS) from 2011 to
2015. In the second half of the chapter, the authors then describe the ongoing
Toddler Attentional Behaviours and Learning with Touchscreens (TABLET)
Project, developed by Annette with TS and Rachael Bedford at King's College
London. The project aims to test whether infant exposure to touchscreen devices
at 12 months-of-age is associated with long-term developmental differences at 18
and 42 months-of-age in domains including attention, temperament, developmental milestones, language, executive function and sleep. As the authors quote
Annette, "we live in a media-saturated world. It is far better that parents know
how to choose the right television or DVD programmes for their children than to
make them ashamed at even thinking of ever using screen exposure".

In Chapter 19, **Luca Rinaldi** discusses his work in collaboration with Annette
to extend the neuroconstructivist theoretical framework to intelligence, that is,
individual differences in cognitive ability within the normal range (Rinaldi &
Karmiloff-Smith, 2017). LR lauds Annette's unique ability to step back from the
instantaneous interpretation of modularised processes to look at the whole picture
of cognitive development. The chapter reviews evidence showing how the
neuroconstructivist approach can explain the rise and fall of intelligence over
development, in terms of a dynamic interaction between the developing system
itself and the environmental factors involved at different times across ontogenesis.
It covers phenomena such as the stability of IQ measures over development, how
IQ differences are paralleled by neural changes, such as differential rates of
thickening and thinning of the cortex, and the environmental measures that associate with IQ differences, such as socioeconomic status. The new approach to
intelligence is linked to neuroconstructivist principles such as interactive specialisation of cortical structures, proactivity, embodiment, ensocialment, and constraints operating at multiple levels of description (Mareschal et al., 2007). LR
concludes that Annette's pioneering spirit is still shaping mainstream research on
intellectual development.

In Chapter 20, **Frank Wiesemann** describes the collaboration between
Annette and Procter and Gamble's Baby Care division, which worked towards
two objectives: educating parents through Pampers' advertising and brand communication, and influencing the design of nappies and wipes to better answer to
babies' needs. This collaboration began around the year 2000 when Annette
became a member of the "Pampers Institute", a group of experts from different
scientific areas related to babies' health and development. FW recalls how passionate Annette was about communicating new scientific insights into infant

development to parents and baby-care professionals, enabling them to better care for their babies. Indeed, she taught Procter and Gamble's scientists how to learn directly from babies, leveraging methods that were only evolving at the beginning of the millennium, such as eye-tracking for monitoring babies' attention or actigraphy for monitoring sleep. This collaboration led, for example, to the design and launch of an advertising campaign called *See the World through Babies' Eyes*, which celebrated babies' development; and later, to the joint supervision of PhD students on unique projects which spanned academic and commercial interests. In FW's view, Annette opened the eyes of millions of parents to the amazing development of their babies, and inspired better nappies that support healthy development.

References

Ansari, D., & Karmiloff-Smith, A. (2002). Atypical trajectories of number development: A neuroconstructivist perspective. *Trends in Cognitive Sciences*, *6*(12), 511–516. DOI: 10. 1016/s1364-6613(02)02040-5

Ashworth, A., Hill, C. M., Karmiloff-Smith, A., & Dimitriou, D. (2015). The importance of sleep: Attentional problems in school-aged children with Down syndrome and Williams syndrome. *Behavioral Sleep Medicine*, *13*(6), 455–471.

Borovsky, A. (2005). Column: Interview with Annette Karmiloff-Smith. *Cognitive Science Online, Department of Cognitive Science, UCSD*, *3*(1), 2005. http://cogsci-online.ucsd. edu/column_archive/CSO3-1-interview.pdf. Retrieved April 27, 2020.

Clark, A., & Karmiloff-Smith, A. (1993). The cognizer's innards: A psychological and philosophical perspective on the development of thought. *Mind & Language*, *8*(4), 487–519. https://doi.org/10.1111/j.1468-0017.1993.tb00299.x

D'Souza, H., Karmiloff-Smith, A., Mareschal, D., & Thomas, M. S. C. (2020). *The Down syndrome profile emerges gradually over the first years of life*. Manuscript submitted for publication.

D'Souza, H., Lathan, A., Karmiloff-Smith, A., & Mareschal, D. (2020). Down syndrome and parental depression: A double hit on early expressive language development. *Research in Developmental Disabilities*, *100*, 103613. https://www.sciencedirect.com/ science/article/pii/S0891422220300433

D'Souza, H. D., Mason, L., Mok, K. Y., Startin, C. M., Hamburg, S., Hithersay, R., Asaad Baksh, R., Hardy, J., Strydom, A., & Thomas, M. S. C. (2020). Differential Associations of Apolipoprotein E ε4 Genotype With Attentional Abilities Across the Life Span of Individuals With Down Syndrome. *JAMA Netw Open*, *3*(9), e2018221. DOI: 10.1001/ jamanetworkopen.2020.18221

Elman, J., Karmiloff-Smith, A., Bates, E., Johnson, M., Parisi, D., & Plunkett, K. (1996). *Rethinking Innateness: A Connectionist Perspective on Development*. Cambridge, MA: MIT Press.

Farran, E.K., & Karmiloff-Smith, A. (Eds.), (2012). *Neurodevelopmental Disorders Across the Lifespan: A Neuroconstructivist Approach*. Oxford: Oxford University Press.

Fodor, J. A. (1983). *Modularity of Mind: An Essay on Faculty Psychology*. Cambridge, MA: MIT Press.

Glennon, J. M., D'Souza, H., Mason, L., Karmiloff-Smith, A., & Thomas, M. S. C. (2020). Visuo-attentional correlates of Autism Spectrum Disorder (ASD) in children with Down syndrome: A comparative study with children with idiopathic ASD. *Research in Developmental Disorders*, *104*, 103678. DOI: 10.1016/j.ridd.2020.103678

Hall, J. H., Wiseman, F. K., Fisher, E. M. C., Tybulewicz, V. L. J., Harwood, J. L., & Good, M. A. (2016). Tc1 mouse model of trisomy-21 dissociates properties of short- and long-term recognition memory. *Neurobiology of Learning and Memory, 130,* 118–128. https://doi.org/10.1016/j.nlm.2016.02.002

Karmiloff-Smith, A. (1986). From meta-processes to conscious access: Evidence from children's metalinguistic and repair data. *Cognition, 23*(2), 95–147.

Karmiloff-Smith, A. (1992). *Beyond Modularity: A. Developmental Perspective on Cognitive Science.* Cambridge, MA: MIT Press.

Karmiloff-Smith, A. (1994). *Baby It's You: A Unique Insight Into the First Three Years of the Developing Baby.* London: Ebury Press, Random House.

Karmiloff-Smith, A. (1994). Précis of beyond modularity: A developmental perspective on cognitive science. *Behavioral and Brain Sciences, 17*(4), 693–707.

Karmiloff-Smith, A. (1998). Development itself is the key to understanding developmental disorders. *Trends in Cognitive Sciences, 2*(10), 389–398. DOI: 10.1016/s1364-6613(98) 01230-3

Karmiloff-Smith, A. (2012). 'TV is bad for children' – less emotion, more science please! In P. Adey & J. Dillon (Eds.), *Bad Education* (p. 231–244). Open University Press.

Karmiloff-Smith, A. (2015). An alternative to domain-general or domain-specific frameworks for theorizing about human evolution and ontogenesis. *AIMS neuroscience, 2*(2), 91.

Karmiloff-Smith, A. (2016). *What can studying babies with Down syndrome possibly tell us about Alzheimer's disease in adults?* UCSD Dart Neuroscience Seminar, University of California at San Diego, April 2016. Retrieved from https://tdlc.ucsd.edu/research/DNS/videos/ Karmiloff-Smith.mp4, August 28, 2019.

Karmiloff-Smith, A., & Inhelder, B. (1975). If you want to get ahead, get a theory. *Cognition, 3*(3), 195–212. https://doi.org/10.1016/0010-0277(74)90008-0

Karmiloff-Smith, A., Thomas, M. S. C., & Johnson, M. H. (2018). *Thinking Developmentally from Constructivism to Neuroconstructivism: Selected Works of Annette Karmiloff-Smith.* London: Routledge.

Karmiloff, K., & Karmiloff-Smith, A. (1998). *Everything Your Baby Would Ask If Only He/ She Could Talk.* London: Cassell/Ward Lock

Karmiloff, K., & Karmiloff-Smith, A. (2015). *Understanding Your Baby: A Parent's Guide to Early Child Development.* London: Hamlyn

Mareschal, D., Johnson, M., Sirios, S., Spratling, M., Thomas, M. S. C., & Westermann, G. (2007). *Neuroconstructivism: How the Brain Constructs Cognition.* Oxford: Oxford University Press.

Pisch, M., Wiesemann, F., & Karmiloff-Smith, A. (2019). Infant wake after sleep onset serves as a marker for different trajectories in cognitive development. *Journal of Child Psychology and Psychiatry, 60*(2), 189–198.

Rinaldi, L., & Karmiloff-Smith, A. (2017). Intelligence as a developing function: A neuroconstructivist approach. *Journal of Intelligence, 5,* 18. https://doi.org/10.3390/ jintelligence5020018

Startin, C. M., D'Souza, H., Ball, G., Hamburg, S., Hithersay, R., Hughes, K. M., … & LonDownS Consortium. (2020). Health comorbidities and cognitive abilities across the lifespan in Down syndrome. *Journal of Neurodevelopmental Disorders, 12*(1), 4. https:// link.springer.com/article/10.1186/s11689-019-9306-9

Thomas, M. S. C., & Karmiloff-Smith, A. (1999). Quo vadis modularity in the 1990s? *Learning and Individual Differences, 10*(3), 245–250.

Thomas, M. S. C., & Karmiloff-Smith, A. (2002). Are developmental disorders like cases of adult brain damage? Implications from connectionist modelling. *Behavioural and Brain Sciences*, *25*(6), 727–750.

Thomas, M. S. C., Ojinaga Alfageme, O., D'Souza, H., Patkee, P. A., Rutherford, M. A., Mok, K. Y., Hardy, J. the LonDownS Consortium, & Karmiloff-Smith, A. (2020). A multi-level developmental approach to exploring individual differences in Down syndrome: genes, brain, behaviour, and environment. *Research in Developmental Disorders*, *104*, 103638. DOI: 10.1016/j.ridd.2020.103638

2

THE COGNITIVE UNDERPINNINGS OF RELATIVE CLAUSE COMPREHENSION IN CHILDREN

Susan Goldin-Meadow and Annette Karmiloff-Smith

I first met Annette Karmiloff-Smith in 1969 when we were both students at the University of Geneva. It was a life-changing time for us. Annette hadn't yet decided to commit to studying psychology – she had been a simultaneous interpreter at the UN in Geneva but found that the job was not intellectually stimulating, and a chance encounter with Piaget at a bookstore had led her to dabble in psychology. She was working on her licence, equivalent to a master's degree, at the University of Geneva, when we met. I was doing my junior year abroad from Smith College and hadn't committed to anything yet (see Figure 2.1).

Annette and I met in Prof. Mimi Sinclair's lab, the Piagetian expert on language. I didn't know what I was doing, but I knew I was interested in language. Annette and I were part of a team exploring English- and French-speaking children's understanding of the relative clause. The team was supposed to write up a joint report on the project. But the study had too many loose ends to satisfy the two of us. So we went rogue.

We focused on English-speakers and narrowed the window of the study. We redesigned the stimuli, collected new data, and wrote up our own report in June of 1970 (which is now in the Archives de Psychologie in Geneva) – The Relative Pronoun, co-authored by Goldin and Karmiloff (neither of us had hyphens in our names at that point in our lives). I submitted the paper to Smith College to fulfil my junior year abroad requirement; Annette submitted it as a fourth piece of research on the way to obtaining her licence in genetic psychology. We ended our report with a suggestion for future research and I went back to Smith for my senior year and did the project for my honors thesis – our hypotheses were supported by the data I collected in the US.

And there the matter sat. We didn't publish the work, nor did anyone else on the relative pronoun team. When I visited Annette a week before she passed away, in addition to a bright and bold red necklace that to me embodied Annette,

(a)

(b)

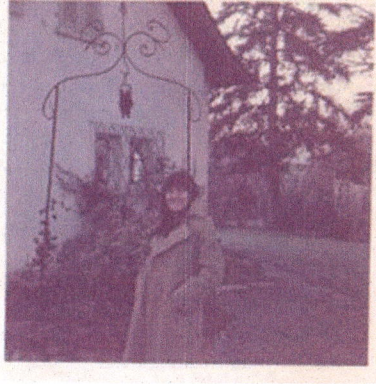

FIGURE 2.1 AK-S with her two-year-old daughter (left) and SG-M (right) in 1969 outside of Annette's house in Geneva where we often worked on our project.

I brought her a copy of the paper that we submitted in 1970. I remarked that we had done good work together when we were kids. Annette's comment was – "We should have published it". And so, at long last, we are.

This paper is a tribute to Annette, who more than anyone else inspired me to go to graduate school and become a developmental psychologist. Although we never worked together after this project, Annette and I were colleagues for each other and, most importantly, close friends for the next 47 years – a lifetime of love and respect.

The linguistic and cognitive underpinnings of relative clause comprehension

The relative clause (RC) is a subordinate clause (SC), which may be embedded within, or placed at the end of, the main clause (MC). The relative pronoun that introduces the SC refers to an antecedent in the MC. The relationship between the antecedent and the relative pronoun forms the link between the two clauses. The RC is defined in grammatical terms as a descriptive clause modifying the antecedent. The pronoun itself may act as either the subject or object of the SC; its antecedent may also assume the role of either subject or object, but of the MC. Our study focused on four types of sentences, each of which contained a transitive MC and a transitive SC (see examples in Table 2.1).

1. The Relative Subject Clause, RSC, where the referent of the relative pronoun *that* (i.e., the CAT in the first example in Table 2.1) is the subject in the SC (the cat is doing the pushing), and the object in the MC (the cat is being licked).
2. The Embedded Relative Subject Clause, [RSC], where the referent of the relative pronoun *that* (i.e., the MONKEY in the second example) is the

TABLE 2.1 Examples of the four types of relative clause tested in the study. The sentences are classified according to whether the relative pronoun *that* plays a subject or object role in the subordinate clause, whether the subordinate clause is embedded or not, and whether the referent of the relative pronoun *that* (the shared semantic element) plays the same role or different roles in the main and subordinate clauses

Type of relative clause (RC) construction	Examples of relative clause constructions*	Subject (S) or object (O) in relative clause**	Relative clause is embedded (+Emb) or not (−Emb)	Shared semantic element plays the same roles or different roles in main clause (MC) and subordinate clause (SC)
1. RSC	The squirrel licks the CAT [that pushes the monkey]	S	−Emb	Different Roles (Patient in MC switches to Agent in SC)
2. [RSC]	The MONKEY [that pushes the cat] licks the squirrel	S	+Emb	Same Roles (Agent in both MC and SC)
3. ROC	The cat licks the BEAR [that the pig pushes]	O	−Emb	Same Roles (Patient in both MC and SC)
4. [ROC]	The SQUIRREL [that the pig pushes] licks the bear	O	+Emb	Different Roles (Agent in MC switches to Patient in SC)

Notes
* The referent in CAPS plays a role in both the Main Clause and the Subordinate Clause; the Subordinate Clause is in brackets; the relative pronoun *that* in the Subordinate Clause refers back to the word in CAPS in the Main Clause.
** S indicates that the referent of the relative pronoun *that* (i.e., the term in CAPS in the Main Clause) plays the Subject role in the Subordinate Relative Clause; O indicates that the referent plays the Object role in the Subordinate Relative Clause.

subject of the SC (the monkey is doing the pushing) and the subject of the MC (the monkey is doing the licking); the square brackets around RSC mark the embedding.

3. The Relative Object Clause, ROC, where the referent of the relative pronoun *that* (i.e., the BEAR in the third example) is the object in the SC (the bear is being pushed) and the object in the MC (the bear is being licked).

4. The Embedded Relative Object Clause, [ROC], where the referent of the relative pronoun *that* (i.e., the SQUIRREL in the fourth example) is the

object in the SC (the squirrel is being pushed) and the subject in the MC (the squirrel is doing the licking); the square brackets around ROC again mark the embedding.

The purpose of our study was to determine whether these four types of RC sentences are acquired (in terms of comprehension) at the same moment in a child's development and, if not, which sentences were more difficult and why. The deeper question was to determine the extent to which the acquisition of these syntactic structures was dependent upon the development of cognitive structures.

In her book *Sentences Children Use*, Menyuk (1969) analysed RC sentences in terms of production. She noted that many children produce RC sentences of the following type (I):

a. I've got the book you want (an ROC sentence)
b. I saw the lady who was here yesterday (an RSC sentence)

However, even at age 7, children rarely produce sentences of this type (II):

c. The book that you want is on the table (an [ROC] sentence)
d. The lady who was here yesterday came back (an [RSC] sentence)

The factor that divides sentences (a) and (b) from (c) and (d) is embedding. The SC is not embedded within the MC in type I sentences, but is in type II sentences. As Menyuk points out, all four sentences contain two underlying sentences, one subordinate to the other, and, in this sense, are recursive. However, the type II sentences involve embedding the relative subordinate clause within the MC, which creates a distance between the subject of the MC (the word in CAPS in examples 2 and 4 in Table 2.1) and its predicate. This distance seems to be a stumbling block for young children, at least in production. Although 87% of Menyuk's participants (ages 3 to 7) used the RC construction, only 46% used the type II construction, and 66% of these children were in the 1st grade, the highest grade level included in the study (see also Limber, 1973). Extrapolating from production data, we might hypothesize that children ought to have particular difficulty understanding sentences in which the RC is embedded, that is, sentences like examples 2 and 4 in Table 2.1, where there is distance between the subject of the MC and its predicate. Indeed, Slobin (1971) hypothesized that this type of interruption in a sentence should make that sentence structure difficult for young children to acquire.

There is, however, another linguistic property that distinguishes these four sentences – whether the relative pronoun *that* is the subject of the subordinate clause (RSC sentences, examples 1 and 2 in the table) or the object of the subordinate clause (ROC sentences, examples 3 and 4). Note that whether the relative pronoun is a subject or object of the subordinate clause has implications for word order within the clause – the RC follows SVO word order in RSC sentences, but OSV word order in ROC sentences. If English-learning children use a linear comprehension strategy (i.e., if they assume that the words in the string reflect an SVO order, the canonical order in English) to figure out who is doing what in the SC, they will arrive at the correct interpretation in RSC sentences – in sentence 1, the cat is followed by the verb "push" and is, in fact, doing the pushing; similarly in sentence 2, the monkey is followed by the verb "push", and is doing the pushing. In contrast, applying a linear comprehension strategy to ROC sentences does not give a coherent interpretation – the bear is not followed by any verb at all in sentence 3, nor is the squirrel in sentence 4; in both examples, another semantic element (the pig) is doing the action, in this case, pushing. Assuming that children are likely to apply a linear comprehension strategy to the sentences they hear, we might hypothesize that children ought to have particular difficulty understanding sentences in which this strategy results in an incorrect interpretation, that is, in ROC sentences like examples 3 and 4 in Table 2.1. In fact, sentences with object relatives have been found to be more difficult for adults to process than matched sentences with subject relatives (see O'Grady, 2011, for review; and Dick *et al.,* 2001, and Dick, Wulfeck, Krupa-Kwiatkowski, & Bates, 2004, for descriptions of related phenomena in adults with aphasia and children with developmental language impairment).

Finally, we considered another factor, one that reflects the Piagetian tradition in Geneva. Piaget (1967; Piaget & Inhelder, 1966) describes a number of phenomena in which children have difficulty thinking about an entity playing two roles at the same time. For example, children have trouble understanding that object B, which is to the left of A, is, at the same time, to the right of C. Children at this broad developmental stage also have difficulty ordering a set of sticks of different lengths, and treating a particular flower not only as a tulip but also as a flower. If a child has difficulty understanding that an entity can play two different roles at the same time, that child is likely to have trouble interpreting sentences like examples 1 and 4 in Table 2.1. The semantic element that is shared across the MC and SC in example 1 (the cat) plays the subject/agent role in the SC (he's the pusher), but the object/patient role in the MC (he is the lickee). The roles that the shared element plays in the MC and SC are also different in example 4, but they are reversed: the squirrel plays the object/patient role in the SC (he is the pushee), but the agent/subject role in the MC (he is the licker). Sentences containing this type of role-switch are likely to be difficult for a young child to interpret correctly since the situations to which these sentences refer are difficult for the child to conceptualize. In contrast, in sentences 2 and 3, the semantic element shared across the MC and SC plays the same role in both clauses: the subject/agent role in example 2 (the monkey is the pusher and the licker), and the object/patient role in example 3 (the bear is the pushee and the lickee). If role-switch is a factor in children's comprehension of the RC, children should find it easier to understand sentences 2 and 3 than sentences 1 and 4.

These factors lead us to make different predictions about comprehension. If containing an ROC and/or an embedded RC both contribute to comprehension difficulty, example 1 should be the easiest sentence to understand (having neither property) and example 4 the hardest (having both), with example 2 (having Emb but not ROC) and 3 (having ROC but not Emb) somewhere in between. In contrast, if switching roles in the MC and SC contributes to comprehension difficulty, examples 1 and 4 should both be difficult to understand, and examples 2 and 3 should be relatively easy. Annette and I collected data in Geneva to explore these predictions.

The design of our study in Geneva

We tested 60 children, 10 in each age group: 4–5, 5–6, 6–7, 7–8, 8–9, and 9–10 years. The majority of children were pupils at the International School of Geneva, but 10 4-year-olds and 1 5-year-old attended the United Nations Nursery School, also in Geneva. Each child was presented with six sentences in five random orders. The six sentences included the four sentences with relative clauses displayed in Table 2.1, and two sentences with embedded adjectival clauses ("The monkey next to the cat licks the pig" and "The bear next to the pig pushes the monkey").

In all of the sentences, three animals took part in two actions in which the agent and patient could sensibly be reversed. As a result, the child had to be able to interpret the syntactic structure of the sentence in order to correctly act it out. Children were asked to act out each sentence with a set of stuffed animals; all of the animals could stand alone to avoid manipulation problems that might be caused by having three animals and two actions. Two durative verbs ("to lick" and "to push") were used for all sentences; "lick" was always the verb in the MC, "push" was the verb in the SC. One relative pronoun ("that") was used for all sentences.

The experimenter briefly talked about the toys and asked the child to name them. She then presented two demonstration sentences using "and" and acted out the sentences with the toys, emphasizing two successive actions. All of the verbs in the sentences were in the present tense, but if the child began to act out both parts of the sentence simultaneously, the experimenter repeated the sentence with the verb of the first part in the past tense in order to stress the two distinct and consecutive actions; otherwise, the sentence was presented only once with verbs in the present tense. The experimenter read each sentence according to the random order assigned to the child. Children selected from the five animals (pig, cat, bear, monkey, squirrel) those with which they wished to act out the sentence. After each acting out, the experimenter asked the child to describe what the animals just did.

All of the child's actions, hesitations, requests for repetitions, as well as sentence produced after the action were noted by a second experimenter. The experiment lasted approximately 5 minutes for older children, 15 minutes for the younger ones for whom lengthier presentations were essential as they were slow to respond.

The findings and next steps

Across all ages, children were correct on 71% of the embedded adjectival sentence, and even the 4-year-olds were correct on 65% of these sentences. The children could thus follow the instructions and perform two acts with three animals.

In contrast, across all ages, children were correct on only 42% of the RC sentences displayed in Table 2.1, and there was improvement with age (18%

4-year-olds, 38% 5-year-olds, 40% 6-year-olds, 33% 7-year-olds, 68% 8-year-olds, 55% 9-year-olds). For the most part, children produced verbal responses that mirrored their acting-out responses – they gave correct verbal responses on 35% of the relative clauses sentences (we considered a verbal response to be correct only when the child's action response was also correct; in other words, we did not consider the verbal response to be correct if the child repeated the experimenter's sentence after having acted out the sentence incorrectly).

As expected, performance was not uniform across the four types of RC sentences. Figure 2.2[1] presents the percentage of correct acting-out responses for the four relative clause sentences, classified according to the type of RC. Using a within-groups ANOVA to determine whether correct acting-out responses differed among the four types of sentences, we found a significant effect of sentence type, $F(3,36) = 5.23$, $p < .002$. Tukey HSD pairwise tests reveal significant differences at the $p = .05$ level between sentence types 1 and 2, between sentence types 2 and 4, and between sentence types 3 and 4; no other differences between pairs were reliable. We also used a non-parametric paired-samples sign test to explore differences between pairs of the four types of sentences. Eliminating ties, we found no differences between sentences 2 and 3 ($p = .41$, $N = 18$), nor between sentences 1 and 4 ($p = .50$, $N = 15$). However, sentence 1 differed from sentences 2 ($p = .012$, $N = 16$) and 3 ($p = .026$, $N = 17$), as did sentence 4 ($p = .005$, $N = 18$ and $p = .012$, $N = 16$, respectively).

All three of the factors outlined earlier lead to the prediction that sentence 4 should be difficult to comprehend, and it was. The interesting result is sentence 1, which (under one set of hypotheses) should be the easiest sentence to comprehend since it is an RSC that is not embedded. But, in fact, sentence 1 is difficult for the group as a whole and for each age group (the 4-year-olds produced no correct

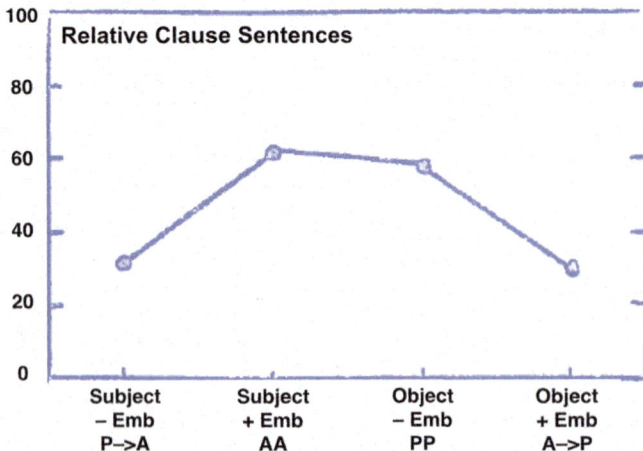

FIGURE 2.2   The percentage of correct acting-out responses produced by the 37 children in the Geneva study, classified according to the four types of relative clauses (see Table 2.1 for description of sentences).

responses on either sentence 1 or 4, but were correct on 30% and 40%, respectively, of sentences 2 and 3). This pattern of results suggests that role-switching is indeed an important factor in a child's comprehension of the relative clause – sentences that referred to a scene in which the shared element played a different role in each clause were more difficult to understand than sentences referring to a scene in which the shared element played the same role in the two clauses.

The role-switch hypothesis assumes that the semantic structure of the to-be-understood relative clause sentence presents a stumbling block for young children, not necessarily the syntactic structure of the sentence. At the end of our 1970 paper, Annette and I suggested that one way to test this hypothesis would be to change the syntactic structure of the four sentences while maintaining their semantic structure. My honors thesis at Smith College was designed to address this question.

Evidence for a cognitive bottleneck from coordinate sentences with passive clauses

We designed a study parallel to the one we conducted in Geneva, and included three different syntactic constructions, each reflecting the four types of semantic structures displayed in Table 2.1: (1) Relative clause sentences identical to the sentences presented to children in Geneva. (2) Coordinate sentences containing active and passive clauses; by using both types of clauses in a coordinate structure, we were able to create the same role structures found in the original relative clauses (see Table 2.2 for examples). Note that the semantic element shared across the two clauses (in CAPS) appears at the beginning of the sentence in each example, which could potentially ease the child's comprehension burden. (3) Coordinate sentences containing only active sentences; these sentences differed from the preceding category in that the semantic element that was shared across clauses (in CAPS) was presented twice, once in each clause, a manipulation that could also ease the child's comprehension burden (see Table 2.2).

Sentences were created by a random selection of animals and verbs for each position in the sentence, with the stipulation that three different animals and two different verbs appeared in each sentence; two sets of sentences were used. The three categories of sentences were not presented in blocks, but were instead presented together in a randomized order; 5 random orders of 12 sentences were used. The procedure was identical to the procedure we used in Geneva, with the exception that the child was not asked to give a verbal account after acting out the sentence with the animals. Because the pattern in Figure 2.2 was strongest in the younger children in our Genevan sample, we focused on children between 4 and 6 years in the US study – 20 children, ages 3.9 to 6.2 (median age = 4.9), were tested. Only two of the older children were attending kindergarten; the rest attended either a bi-weekly daycare center or a private nursery school.

As expected, the 20 children performed best on the coordinate sentences containing only active clauses (60% correct), next best on the coordinate sentences containing both active and passive clauses (44% correct), and least well on the relative

TABLE 2.2 Examples of the four types of coordinate sentences containing active and passive clauses, and the four types of coordinate sentences containing only active sentences. The sentences are classified according to whether the semantic element that appears in both clauses (in CAPS) plays the same roles or different roles in the two clauses

Type of coordinate construction	Examples of coordinate construction	Shared semantic element plays the same or different roles in the two coordinate clauses
Active and Passive Clauses		
1	The MONKEY licks the pig and is pushed by the mouse	Different Roles (Agent in 1st clause switches to patient in 2nd)
2	The MOUSE pushes the bear and licks the pig	Same Roles (Agent in both clauses)
3	The PIG is licked by the mouse and is pushed by the monkey	Same Roles (Patient in both clauses)
4	The BEAR is pushed by the monkey and licks the pig	Different Roles (Patient in 1st clause switches to Agent in 2nd)
Active Clauses		
1	The MONKEY licks the pig and the bear pushes the MONKEY	Different Roles (Agent in 1st clause switches to patient in 2nd)
2	The PIG pushes the mouse and the PIG licks the bear	Same Roles (Agent in both clauses)
3	The mouse licks the PIG and the monkey pushes the PIG	Same Roles (Patient in both clauses)
4	The pig licks the MOUSE and the MOUSE pushes the monkey	Different Roles (Patient in 1st clause switches to Agent in 2nd)

clause sentences (20%). Using a within-groups ANOVA, we found that children's performance on the three categories of sentences was significantly different, $F(2,19) = 15.02$, $p < .0001$. Tukey HSD pairwise comparisons revealed that performance on the relative clause sentences was significantly different at the $p < .01$ level from performance on the other two categories of sentences; the difference between the two types of coordinate sentences was not significant.

Despite the fact that the US sample performed worse on the relative clause sentences overall than the Geneva sample, they nevertheless displayed the same pattern with respect to the four types of RC sentences: They did better on sentences 2 and 3, the two sentences without role-switch (i.e., AA and PP), than on sentences 1 and 4, the two sentences with role-switch (i.e., P– > A and A– > P) (see Figure 2.3).

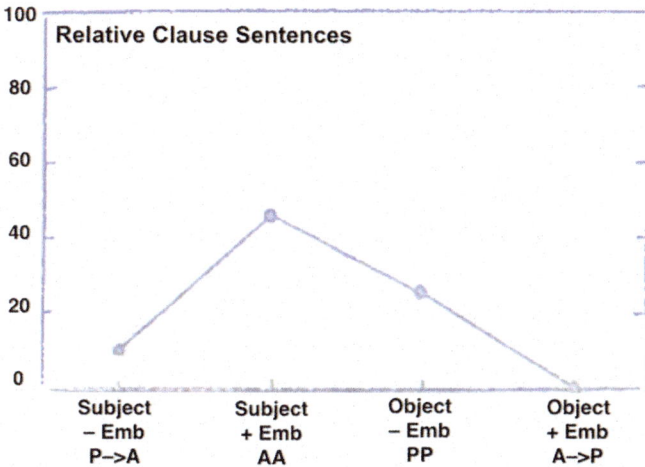

FIGURE 2.3   The percentage of correct acting-out responses produced by the 20 children in the US study, classified according to the four types of relative clauses (see Table 2.1 for description of sentences).

But did the children display this same pattern on sentences with a different syntactic structure? The answer is "yes" for coordinate sentences containing active and passive sentences (see Figure 2.4). Again, we see better performance on the two sentences without role-switch (AA and PP) than on the two sentences with role-switch (A→P and P→A).

However, the children displayed no differences across the four types of co-ordinate sentences containing only active sentences (see Figure 2.5), a surprising result given that the children's responses were not close to ceiling on these sentences.

Using a within-groups ANOVA, we found that the differences among the four sentence types were significant for the relative clause sentences, $F(3,19) = 6.07$, $p = .001$ and the coordinate sentences with active and passive clauses, $F(3,19) = 4.9$, $p = .004$, but not for the coordinate sentences with only active clauses, $F(3,19) = .31$, $p = .81$. Tukey HSD pairwise comparisons revealed differences between sentence types 1 and 2 (at $p < .05$) and between sentence types 2 and 4 (at $p < .01$) for relative clauses, and between sentence types 2 and 4 (at $p < .01$) for the coordinate sentences with active and passive clauses; none of the other pairwise differences was significant. We again used a non-parametric paired-samples sign test to explore differences between pairs of the four types of sentences. In the relative clauses, eliminating ties, we found no differences between the two RC sentences without role-switch ($p = .145$, $N = 8$); we could not carry out the test on the two sentences with role-switch because of the large number of ties (18 of the 20 children responded in the same way to both items, suggesting that responses to the two items did not differ). We did, however, find significant differences between the sentences with and without role-switch (2 vs. 1, $p = .02$, $N = 9$; 2 vs. 4, $p = .005$, $N = 9$; 3 vs. 4, $p = .031$, $N = 5$; the exception was 3 vs. 1, $p = .187$, $N = 5$).

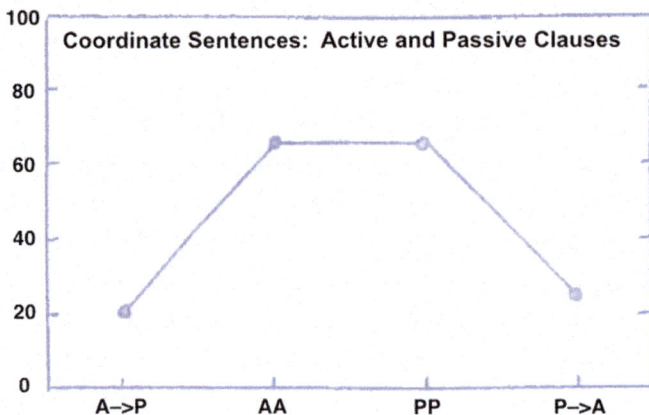

FIGURE 2.4 The percentage of correct acting-out responses produced by the 20 children in the US study, classified according to the four types of coordinate structures (see Table 2.2 for description of sentences).

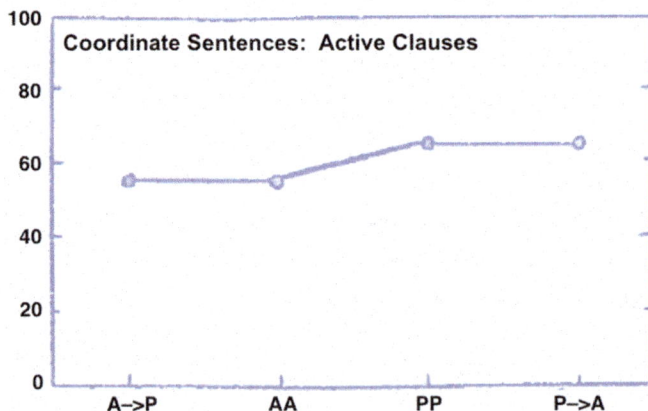

FIGURE 2.5 The percentage of correct acting-out responses produced by the 20 children in the US study, classified according to the four types of coordinate structures (see Table 2.2 for description of sentences).

Similarly, in the coordinate sentences with active and passive clauses, eliminating ties, we found no differences between the two sentences without role-switch (2 vs. 3, p = .745, N = 6) and no differences between the two sentences with role-switch (1 vs. 4, p = .50, N = 5). And again, we found a significant difference between the sentences with and without role-switch (2 vs. 1, p = .005, N = 9; 2 vs. 4, p = .004, N = 8; 3 vs. 1, p = .006, N = 11; 3 vs. 4, p = .011, N = 10).

    To summarize, we found that when children responded differently to the four types of sentences within a category (i.e., to the four types of relative clause sentences, and to the four types of coordinate sentences with active and passive clauses), linguistic structure did not predict those differences in comprehension.

Instead, semantic structure – whether an entity plays the same role or switches roles in both clauses – determined sentence comprehension. An obvious follow-up question to ask is whether the US children, as they continued to develop, mastered the relative clause construction and the coordinate clause construction with passives at the same moment in development. If so, the bottleneck to comprehension would appear to be primarily semantic. If not (and this is the more likely outcome, with coordinate sentences containing passive clauses mastered first), the bottleneck to comprehension is likely to involve both semantic and syntactic structure. Overall, our findings underscore how important a sentence's message is to the comprehension of that sentence. This seems like an obvious point, but it can easily be overlooked. If a message is beyond a child's cognitive grasp, we can't even begin to probe the child's comprehension of linguistic structures conveying that message.

Support from the subsequent literature on children and adults

In 1974, Sheldon explored relative clause comprehension in 33 English-speaking children, ages 3.8 to 5.5 years, and manipulated precisely the same three factors that Annette and I had included in our study in Geneva (see Table 2.1): (1) The position of the relative clause, whether the RC was embedded or not. (2) Word order in the relative clause, whether the RC followed SVO order (the order in our RSC sentences) or OSV order (the order in our ROC sentences). (3) Parallel function, whether the noun phrase that appeared in both clauses had the same grammatical function or different grammatical functions (same role or different roles in our terms). Using an acting-out paradigm, Sheldon found that the third factor, parallel function, accounted for the children's responses better than the other two factors. She cites Brown (1971) and Maratsos (1973) as providing additional evidence for the importance of parallel function in young children's language.

If we are correct that children's difficulty with certain types of relative clause sentences stems from their inability to envision the semantic element that is shared across clauses in two different roles, then this problem should be short-lived, resolved once children are able to easily conceptualize two facets of an object at the same time. Interestingly, Sheldon (1977) explored the same three factors in a sentence processing study in adults and found strong evidence for the effects of the first two factors (RC position, and word order in the RC), but not for the third factor (parallel function), providing support for the hypothesis that this factor is a product of the cognitive constraints of childhood.

The long reach of the relative clause

In hindsight I can see that the project I did with Annette in 1970 set the stage for the research program that I began in graduate school and am still pursuing. Another way to explore the impact, or lack of impact, that linguistic structure has on children's communicative development is to observe children who have not been exposed to linguistic input. My dissertation took this path by studying deaf children whose hearing losses prevented them from making use of the spoken language input that surrounded them, and whose hearing parents had not yet exposed them to sign language. We might guess that children in this situation would fail to communicate simply because they have no model for language. But this guess would be wrong – deaf children not exposed to a usable linguistic model communicate with their hearing family members, and use gesture to do so. Moreover, these gestures – called *homesigns* – display many of the structural properties found in natural language (Goldin-Meadow, 2003).

For example, homesign contains lexical markers that modulate the meanings of sentences (negation and questions, Franklin, Giannakidou, & Goldin-Meadow, 2011), grammatical categories such as nouns, verbs (Goldin-Meadow, Butcher, Mylander, & Dodge, 1994) and subjects (Coppola & Newport, 2005), and devices that refer to non-present events (i.e., that allow displacement, Butcher, Mylander, & Goldin-Meadow, 1991; Morford & Goldin-Meadow, 1997). Homesign gestures are also composed of parts, akin to a morphological system (Goldin-Meadow, Mylander, & Franklin, 2007), and those gestures combine to form structured sentences, akin to a syntactic system (Feldman, Goldin-Meadow, & Gleitman, 1978). Homesign is thus characterized by levels of structure, and those levels are organized hierarchically. For example, homesigners use multi-gesture combinations – a demonstrative gesture plus a noun gesture – to serve the same semantic and syntactic functions as either the demonstrative gesture or the noun gesture used on its own. The larger unit can thus substitute for the smaller units and, in this way, functions as a complex nominal constituent embedded within a sentence (i.e., a sentence with hierarchical structure [[[that] [bird]] [pedals]], rather than flat structure [[that] [bird] [pedals]] (Hunsicker & Goldin-Meadow, 2012; Flaherty, Hunsicker & Goldin-Meadow 2021).

The structures found in homesign have not been copied from a conventional language model, nor can they be traced back to the co-speech gestures that the homesigners' hearing family members used when interacting with them (Goldin-Meadow & Mylander, 1983, 1984, 1998; Goldin-Meadow et al., 2007; Flaherty, Hunsicker, & Goldin-Meadow, 2019). Moreover, there is no evidence that the way in which the hearing family members respond to their homesigner's gestures plays a role in shaping the structure of those gestures (Goldin-Meadow & Mylander, 1983; 1984). These structures, therefore, come as close as we can currently envision to revealing the human child's predispositions to communicate in a structured way. Because they have not been influenced by an established language model, the structures in homesign may be a relatively straightforward reflection of the cognitive structures children bring with them to language-learning.

Returning to the relative clause, if role-switch is cognitively difficult for children learning language from a model, it is likely to be just as difficult (maybe more so) for children who are creating their own linguistic structures. We would then expect that, when homesigners produce sentences with two clauses and a semantic element shared across those clauses, they should be biased toward producing sentences in which the shared element plays the same semantic role in the two clauses. This prediction is, in fact, supported by the homesign data.

The shared semantic element in a relative clause sentence is marked by the relative pronoun (*that* in the study Annette and I did together), which is either the subject or object of the subordinate clause. But as the examples in Table 2.2 illustrate, sentences that have coordinately conjoined clauses (as opposed to subordinately conjoined clauses) can have shared elements too. Sometimes the shared element is repeated in the surface structure of the sentence (e.g., The PIG pushes the mouse and the PIG licks the bear) and sometimes it is omitted (e.g., The PIG pushes the mouse and licks the bear).

We analysed all of the gesture sentences containing more than one clause produced by a US homesigner between the ages of 2 years;10 months and 4 years; 10 months. The homesigner produced 267 two-clause sentences (with either coordinate or subordinate sentence structure), and 80% of these sentences contained shared elements. The *horse* is the shared element in example (1) and the *village* is the shared element in example (2) (Goldin-Meadow, 1982). Iconic gestures are in CAPS in the examples, pointing gestures are in lower case; a gloss of the sentence is in parentheses.

1. CLIMB–SLEEP–horse (horse climbs and then sleeps, describing a picture of a horse sleeping on top of a house)
2. toy1–village–toy2–village (you put toy1 in village and toy2 in village, requesting the experimenter to put the toys in the village).
   A shared semantic element can play the same role in the two clauses, as in these examples. In (1) the *horse* is an actor in both clauses, and in (2) the *village* is a locative recipient in both clauses. But a shared element can also switch roles, as in example (3).
3. PUSH–truck–CIRCLE–truck (I push the truck and then the truck circles, describing his own actions on the truck).

Here the shared element, the *truck*, is the patient of the pushing action and the actor of the circle action.

The interesting question given the theory that Annette and I constructed concerns the distribution of shared elements that play either the same or different roles in the two clauses of a complex sentence. Consistent with our theory, the homesigner produced significantly more sentences in which the shared semantic element played the same role in the two clauses (N = 112, 85% of complex sentences containing two action propositions) than sentences in which the shared element played different roles (N = 20, 15%) (p < .0001, binomial test, two-tailed).

A child who is not exposed to a language model can nevertheless produce two-clause sentences in which a semantic element appears in both clauses. More to the point for our question, the child seems to be biased against producing two-clause sentences in which the shared semantic element plays a different role in each of the two clauses. The linguistic structures that homesigners display in their sentences are good candidates for cognitive structures that exert an influence on language development, consistent with Piagetian theory (see Sinclair, 1967). The homesign findings, thus, lend support to the hypothesis that at least one of the difficulties children face in comprehending relative clauses is cognitive and not purely linguistic.

Annette and I set out in 1969 to determine whether Piaget's theory had anything say about children's acquisition of the relative clause, and found that it did. But the truly important aspect of the experience for me was that I got to work with Annette, who was (even then) a gifted researcher. And I got to watch first hand as Annette managed being a young mother and a student, and did it with her characteristic excellence – the beginning of one of the important themes in Annette's life – achieving a sensible work-life balance.

As I look back on my work, I now see many places where Annette's work had an important influence on me. For example, her belief in the value of the microgenetic method was instrumental in getting me to do my own microgenetic study of the role that spontaneous gestures can play in learning math (Alibali & Goldin-Meadow, 1993), and her ideas about *representational redescription* provided a framework for the theory my students and I are building on the role that gesture plays in the transition from concrete action to abstraction (Novack, Congdon, Hemani-Lopez, & Goldin-Meadow, 2014; Goldin-Meadow, 2015; Wakefield, Hall, James, & Goldin-Meadow, 2018) – gesture turns out to be a viable vehicle for the redescription that propels developmental change (Goldin-Meadow & Alibali, 1994). And then there is my work on homesign, which addresses questions that Annette and I asked together. Much to my surprise, the homesign data corroborate the theory that Annette and I proposed back in 1970, a realization that I came to in writing this chapter, and am saying out loud for the first time. I have come full circle and only wish that I could discuss it all with Annette, who would know just how to put the insight into perspective and creatively build on it.

Note

1 The graphs in this paper are taken from my 1971 honors thesis and thus were constructed by hand (as graphs were at that time), which accounts for their home-grown feel.

References

Alibali, M. W., & Goldin-Meadow, S. (1993). Gesture-speech mismatch and mechanisms of learning: What the hands reveal about a child's state of mind. *Cognitive Psychology*, 25, 468–523.

Brown, H. D. (1971). Children's comprehension of relativized English sentences. *Child Development*, 42, 1923–1926.

Butcher, C., Mylander, C., & Goldin-Meadow, S. (1991). Displaced communication in a self-styled gesture system: Pointing at the non-present. *Cognitive Development*, 6, 315–342.

Coppola, M., & Newport, E. (2005). Grammatical Subjects in homesign: Abstract linguistic structure in adult primary gesture systems without linguistic input. *Proceedings of the National Academy of Sciences, 102*, 19249–19253.

Dick, F., Bates, E., Wulfeck, B., Aydelott, J., Dronkers, N., & Gernsbacher, M. (2001). Language deficits, localization, and grammar: Evidence for a distributive model of language breakdown in aphasic patients and neurologically intact individuals. *Psychological Review, 108*(3), 759–788.

Dick, F., Wulfeck, B., Krupa-Kwiatkowski, M., & Bates, E. (2004). The development of complex sentence interpretation in typically developing children compared with children with specific language impairments or early unilateral focal lesions. *Developmental Science, 7*(3), 360–377.

Feldman, H., Goldin-Meadow, S., & Gleitman, L. (1978). Beyond Herodotus: The creation of language by linguistically deprived deaf children. In A. Lock (Ed.), *Action, symbol, and gesture: The emergence of language* (pp. 351–414). New York: Academic Press

Flaherty, M., Hunsicker, D., & Goldin-Meadow, S. (2021). Structural biases that children bring to language learning: A cross-cultural look at gestural input to homesign. Revision under review.

Franklin, A., Giannakidou, A., & Goldin-Meadow, S. (2011). Negation, questions, and structure building in a homesign system. *Cognition, 118*, 398–416.

Goldin-Meadow, S. (1982). The resilience of recursion: A study of a communication system developed without a conventional language model. In E. Wanner & L. R. Gleitman (Eds.), *Language acquisition: The state of the art* (pp. 51–77). New York: Cambridge University Press.

Goldin-Meadow, S. (2003). *The resilience of language: What gesture creation in deaf children can tell us about how all children learn language*. New York: Psychology Press.

Goldin-Meadow, S. (2015). From action to abstraction: Gesture as a mechanism of change. *Developmental Review, 38*, 167–184, doi: 10.1016/j.dr.2015.07.007.

Goldin-Meadow, S., & Alibali, M. W. (1994). Do you have to be right to redescribe? *Behavioral and Brain Sciences, 17*, 718–719.

Goldin-Meadow, S., Butcher, C., Mylander, C., & Dodge, M. (1994). Nouns and verbs in a self-styled gesture system: What's in a name? *Cognitive Psychology, 27*, 259–319.

Goldin-Meadow, S., & Mylander, C. (1983). Gestural communication in deaf children: The non-effects of parental input on language development. *Science, 221*, 372–374.

Goldin-Meadow, S., & Mylander, C. (1984). Gestural communication in deaf children: The effects and non-effects of parental input on early language development. *Monographs of the Society for Research in Child Development, 49*, 1–121.

Goldin-Meadow, S., & Mylander, C. (1998). Spontaneous sign systems created by deaf children in two cultures. *Nature, 391*, 279–281.

Goldin-Meadow, S., Mylander, C., & Franklin, A. (2007). How children make language out of gesture: Morphological structure in gesture systems developed by American and Chinese deaf children. *Cognitive Psychology, 55*, 87–135.

Hunsicker, D., & Goldin-Meadow, S. (2012). Hierarchical structure in a self-created communication system: Building nominal constituents in homesign, *Language, 88*(4), 732–763.

Limber, J. (1973). The genesis of complex sentences. In T. E. Moore (Ed.), *Cognitive development and the acquisition of language*. New York: Academic Press.

Maratsos, M. (1973). The effects of stress on the understanding of pronominal coreference in children. *Journal of Psycholinguistics, 1*, 1–8.

Menyuk, P. (1969). *Sentences Children Use*. Research Monographs No. 52. Cambridge, MA: MIT Press.

Morford, J. P., & Goldin-Meadow, S. (1997). From here to there and now to then: The development of displaced reference in homesign and English. *Child Development, 68,* 420–435.

Novack, M. A., Congdon, E. L., Hemani-Lopez, N., & Goldin-Meadow, S. (2014). From action to abstraction: Using the hands to learn math. *Psychological Science, 25*(4), 903–910. doi: 10.1177/0956797613518351

O'Grady, W. (2011). Relative clauses: Processing and acquisition. In E. Kidd (Ed.), *The acquisition of relative clauses: Processing, typology and fFunction* (pp. 13–38). Amsterdam: John Benjamins.

Piaget, J. (1967). *La Psychologie de l'Intelligence.* Paris: Librare Armand Colin.

Piaget, J., & Inherlder, B. (1966). *La Psychologie de l'Enfant.* Paris: Pressess Universitaires de France.

Sheldon, A. (1974). The role of parallel function in the acquisition of relative clauses in English. *Journal of Verbal Learning and Verbal Behavior, 13,* 272–281.

Sheldon, A. (1977). On strategies for processing relative clauses: A comparison of children and adults. *Journal of Psycholinguistic Research, 6*(4), 305–318.

Sinclair, H. (1967). *Acquisition du langage et développement de la pensée: Sous-systèmes linguistiques et opérations concrètes.* Paris: Dunod.

Slobin, D. I. (1971). Developmental psycholinguistics. In W. O. Dingwall (Ed.), *A survey of linguistic science.* Linguistics Program, University of Maryland.

Wakefield, E. M., Hall, C., James, K. H., & Goldin-Meadow, S. (2018). Gesture for generalization: Gesture facilitates flexible learning of words for actions on objects, *Developmental Science, 21*(5), DOI: 10.1111/desc.12656

3

ON THE CONSTRUCTION OF THE DEVELOPMENTAL PROBLEM IN KARMILOFF-SMITH'S THEORY

Jean-Paul Bronckart

Annette Karmiloff-Smith developed a neo-constructivist theory of psychological development that is distinguished by the breadth and relevance of the empirical data that feed it, as well as the accuracy and depth of the epistemological positioning that supports it. Having had the privilege to witness the setting up of her theoretical project in the first half of the 1970s at the University of Geneva, I will try, by consulting the available documents and my still vivid memories, to report on the conditions under which this project took shape. This retrospective analysis will be done in two stages. The first will propose a chronology of scientific, organizational and relational events marked by the very peculiar configuration of a research institution endowed with a very great master, a few strict chamberlains and a small crowd of devoted craftsmen. The second will consist in an examination of the problems encountered by Piagetian constructivism as early as the 1950s and an analysis of the various attempts to solve them, which will enable us to propose an interpretation of the type of relationship that Annette Karmiloff-Smith had with Piaget and his close collaborators.

Annette Karmiloff-Smith's first steps in psychology

Annette Karmiloff-Smith (hereinafter AKS) arrived in Geneva in 1965 to work as a conference interpreter for various international institutions, including UNESCO, BIT and the *Bureau International de l'Éducation*. It is in this last institution that she heard of Jean Piaget, who had been its director for a long time, but who had also developed at the University of Geneva a new and particularly creative theory of the child cognitive development. Wishing to train with this scientist, she enrolled in September 1967 in the *École de Psychologie et des Sciences de l'Éducation* (*EPSE*), from which she graduated, in late 1970, with a *Licence en psychologie de l'enfant*.

After spending almost 2 years in Beirut, doing interventions and research in various institutions[1], she applied for an assistant position with Piaget and Bärbel

Inhelder, both of whom had greatly appreciated the intellectual qualities she had shown during her training[2]. "Le patron[3]" immediately invited her to participate in the research being carried out at the *Centre international d'épistémologie génétique*[4] (hereafter *CIEG*) and suggested that, in order to complete her job she should speak to Professor Hermina Sinclair-De Zwart, who had just created a unit of developmental psycholinguistics and was looking for collaborators with linguistic knowledge or abilities. AKS then met Sinclair, who hired her immediately as a research assistant in this unit, and whose other collaborators at the time were Emilia Ferreiro, Ioanna Berthoud-Papandropoulou, Harold Chipman and the author of this chapter.

AKS carried out research at *CIEG* from 1972 to 1975 and held an assistant position in the psycholinguistic unit until 1979. But from 1974 onwards, she also participated, initially informally and then officially, in Inhelder's research program, who then became her mentor and soon gave her the status of leader of her group of collaborators. In short, when she came back to Geneva in 1972, AKS wanted to work exclusively with Piaget. However, to be involved in the work of the *CIEG*, she was also forced, in order to get a full-time job, to become part of Sinclair's psycholinguistic unit. In fact, it is actually through her interactions with Inhelder, the third role of this unforeseen scenario, that she found her psychological path, set up her theoretical framework and her first research projects.

Three "seasons" of research at CIEG

AKS participated in the work of the *CIEG* from 1972 to 1975. During the first 2 years, she carried out an empirical research project with me. In the third year, she completed two research projects alone.

The *CIEG* had a mode of operation in accordance with the principles and rules established since the creation of this institution. For one academic year, Piaget invited colleagues, mathematicians, logicians or physicists, asking them to collaborate with psychologists in the laboratory of experimental psychology. The common effort was to delimit the questions in such a way that they can give rise at the same time to experimental research, the results of which are of an epistemological interest, and to a theoretical elaboration, which in turn, can shed light on the experiments on development or to provoke new ones.

As I had shown elsewhere (Bronckart, 1980), the research themes of the *CIEG* can be grouped into three periods: first (from 1955 to 1960) the examination of aspects of the relationship between language, logic and cognition, carried out from a perspective of contestation of logical empiricism (in particular of Carnap's theses, 1934); then (from 1960 to 1969) a historical and genetic re-examination of the construction of the major "Kantian" cognitive categories (space, time, identity, causality); finally (from 1969 to 1980), a focus on the processes and/or factors of cognitive development (awareness, contradiction, generalization, abstraction, etc.). The *CIEG* had two sub-groups of participants, renewed each year: on the one hand, experienced psychologists and a few scholars from other disciplines[5], and, on the other hand, assistants[6] who designed and carried out developmental psychology experiments on the theme of the year. The members of the *CIEG* met for 2 hours each Monday morning under the direction of "le patron", with 1 hour of theoretical presentation or discussion from a scientist, then 1 hour of description of the state of empirical research by one or several assistants.

Piaget had refocused the work of the *CIEG* on development processes because previous research on *causality* (conducted for 4 years, from 1965 to 1969) had shown that causality must be interpreted as an *attribution* of operational structures to the object, and can, therefore, clearly be differentiated from *legality*, which is obtained from observation; but this work had also shown that this causality implies another step, that is the search for connections for non-observable deductive links. It was, therefore, necessary to identify the role that, in this establishment of links, could be played by processes such as consciousness, reflective abstraction, generalization, morphisms, correspondences, etc.

For her first 2 years of contribution to *CIEG*, AKS was invited to conduct a research project with a collaborator who already had experience of how the *Centre* functioned, in this case it was me[7]. The theme of the 1972/1973 academic year was that of *generalization* processes. We built an experimental device that consisted in placing or dropping objects of different sizes and weights on surfaces of different stiffness and resistance, and in interviewing about 100 children (from 5 to 12 years old) to know by what type of generalization these children, having initially maintained that there are forces only in motion, come to accept that a still weight continues to exert a thrust on a flat surface. The following year, on the theme of *correspondences*, we conducted, on the same sample of children, a research project on the types of similarities and differences existing between cartons, having a set of randomly distributed properties (colour, size, geometric shape), to understand the role played by classes and relationships in mapping processes. The following year, AKS alone produced two particularly ingenious studies on the subject of *morphisms and categories* that extended previous work on correspondences. Although her research was particularly appreciated by Piaget, at the end of this work she decided to put an end to her participation in the *Centre*[8].

Like all the experimentalists of the *CIEG*, at the end of the year we had to present to Piaget the results of our research, by establishing a progression of the cognitive levels of the children in the form of stages and sub-stages. We also had to transcribe samples of children's arguments that particularly illustrated each of the levels identified. We also had the opportunity to share any difficulties encountered, suggestions for interpretation, or even methodological criticism. On this methodological level, AKS and I made some remarks, notably on the difficulties of interpreting certain children's statements, by suggesting that they did not necessarily reflect the real level of their cognitive representations. Piaget took no notice of this; his approach was to rescan the verbal interactions between the child and the experimenter, which had been handed to him, and to rewrite almost entirely the reports that accompanied them. Nevertheless, Piaget established the testers as co-authors of the chapters evoking their research, and this was an arrangement that the experimenters did not generally complain about. In this context, for AKS as for myself, the participation in the *Centre* had been an exceptional opportunity to interact with Piaget and with many renowned scientists, as well as an opportunity also to train seriously in the "Piagetian clinical" methodology, although ultimately it made us understand that it was now necessary to follow other paths in the field of psychology.

A long and difficult path into psycholinguistics

In 1972, AKS got a research-assistant job in the developmental psycholinguistics unit, whose program, inspired by the initial works of Sinclair (1967) and Ferreiro (1971), consisted largely of identifying and conceptualizing the stages of acquisition of certain linguistic properties, and to highlight the relations existing between this linguistic development and the stages of cognitive development described by Piaget. Each assistant had to choose a dimension of the language, to design and implement a research subject aimed at highlighting the conditions and stages of the mastery of this dimension by the children, and to compare this progression with that of their cognitive development. Moreover, in principle this research work should lead to the completion of a doctoral thesis. When AKS joined this unit, the research in progress dealt with many topics, including the relationship between the properties of the signifiers and those of their referent, the mastery of the temporal marks, of the pronoun values and of various syntactic structures or operations. It was, therefore, necessary to find another theme of research that remained unexplored. It is without enthusiasm that she finally agreed to take on the theme of the mastery of French language determinants.

It is, therefore, on this theme that AKS began the research path that was to give rise to her doctoral thesis, but this path proved problematic in several respects. First of all, AKS took a firm theoretical position on the type of linguistics to be considered. Whereas Sinclair generally adhered to the theory of *Generative Grammar,* and I adhered to the newly emerging theories of discourse, AKS wanted to develop an original functional approach distinguishing two key aspects of the language; she clearly expressed these ideas as text prepared for a symposium that I organized in 1975[9]:

> Une discussion des relations entre acquisition du langage et développement cognitif appelle, en premier lieu, la distinction fondamentale entre ce que l'on pourrait appeler « outil langagier » et « système langagier ». Par « outil langagier », nous entendons le langage dans sa fonction représentative.

> C'est l'outil langagier que Piaget a pris en considération et c'est à ce niveau qu'il est pertinent de se poser les questions concernant les rapports entre langage et développement cognitif.

> Par contre, lorsqu'on considère le « système langagier », c'est-à-dire le système de règles internes dont le langage est composé, on peut se demander s'il est pertinent de poser la question en termes de « relations entre langage et cognition ».
>
> *(A. Karmiloff-Smith, 1977, p. 169)*[10]

In addition to the question of the linguistic orientations on which the debate centred, there was a second problem, more delicate because it could endanger the very program of research of the unit; namely, the choice of the kind of cognitive development that should be related to one aspect of language development. Piagetian theory, in fact, does not propose a single or general timing of the stages of the operative development, but instead, shows sometimes very different cognitive progression timings (called *décalages*), according to the complexity of the domain to be treated (seriation, quantification, volume, weight, etc.). Consequently, the question that arose was what specific level of progress needed to be taken as the baseline level of performance when evaluating the possible impact of cognitive development on language development. This question was at the very heart of the unit's research program and was, in fact, unresolved. While other assistants generally tried to avoid it, AKS regularly addressed it, indirectly in the preceding quote, but more firmly in the discussions within the unit. Beyond the relative inadequacy of the terms used ("system" vs "tool"), her position was that, on the one hand, language is an element of the environment with which the child interacts, and the stages of the *knowledge* of this object necessarily have a kinship with those evidenced by Piaget for other objects; but, on the other hand, the capacities for *functional implementation* of the various domains of language have a development which is, in its very principle, unrelated to cognitive development.

It is from this perspective that she conducted her research on the mastery, in production and comprehension, of the determinants of the noun in French, showing that if the meta activities (classification of terminal affixes in disjoined classes, or consciousness of the plurifunctionality of certain determinants) are likely to be dependent on general cognitive development, the strategies adopted for their adequate implementation in production, are not linked to a particular level of cognitive development. Finally co-directed by Sinclair and Inhelder, this thesis, titled *Little Words Means a Lot*[11] was brilliantly defended in March 1977 before a jury including also Jerry Bruner and John Lyons.

This brilliant thesis defence almost coincided with my resignation from the post of *Chef de travaux* that I occupied in the psycholinguistic unit. AKS quite legitimately envisaged being appointed to this vacant position. But Sinclair preferred to appoint a different colleague, because she disagreed, as did Inhelder, with the firmness of the theoretical and methodological positions that AKS had advocated for years, in particular her regular criticism of the "limitations" of Piaget's operative theory. AKS, thus, remained as an assistant in a painful context that also made her move away from Inhelder. This remained the case until she was hired by the University of Sussex in 1979.

Looking for the processes of development with Inhelder

While Piaget devoted all of his work to genetic epistemology, Inhelder conducted a research program that was in line with those she had conducted with Piaget in the late

1930s but with a clearer focus on the *psychological subject* (rather than the epistemic subject of Piaget), and on the *developmental processes* (rather than their products).

As we have indicated, Inhelder had invited AKS to collaborate on her own research program, and it is within this context that AKS had, in fact, developed her own scientific project, which aimed to identify the processes that children use to progress in their cognitive development. Her program had three major characteristics. The first was to consider that psycholinguistic problems, or at least a large part of them, must be integrated into the general problem of development, insofar as, as Piaget argued, there are psychological mechanisms able to apply to any type of object, but also to the extent that, contrary to what Piaget seemed to think, the different types of development (cognitive, verbal, social) are in permanent interaction and influence each other:

> Piaget's genetic epistemology [...] has always posited the constructive interaction of the child and his environment. Whilst Piaget has explicitly stressed the effects of cognition on language growth, he has left implicit the effects the latter might have on cognitive growth. Yet, language acquisition is a dialectical form of problem-solving and the child's constructive interaction with his linguistic environment suggests that language must be a crucial problem area for the child in its own right.
>
> *(Karmiloff-Smith, 1979, p. 228)*

The second specific feature was to pay more attention to the linguistic and logical properties of children's statements. AKS argued that these statements testified to the ability of children, even at a very young age, to develop *"theories"* that may be inadequate from a scientific point of view but have their own logic and properties that reflect their abilities of self-overtaking:

> Our present analysis is focused less on particular, explicit notions than on the gradual unification of ideas as observed in action sequences. There is no doubt that the generalized application of a theory will ultimately lead to discoveries which in turn serve to create new or broader theories. However, it seems possible for the child to experience surprise and to question his theory only if the prediction he makes emanates from an already powerful theory expressed in action. Our observations indicate that children hold on to their initial theory for as long as they can. Even when they finally do take counter-examples into consideration, they first prefer to create a new theory, quite independent of the first one, before finally attempting to unify all events under a single, broader theory.
>
> *(Karmiloff-Smith & Inhelder, 1975, p. 209)*

> [...] a shift of emphasis is suggested from conservation-attainment to the psychological function of conservation-seeking [....] from the logical necessity of final levels to the psychological necessity of stages leading to

them. The typical « errors » of the nonconserver should be analyzed in terms of whether they represent powerful heuristics in development or merely shortcomings to be surmounted later.

(Karmiloff-Smith, 1978, p. 189)

Perhaps the most important feature of this project, which Inhelder shared to a large extent, was to identify and conceptualize the *dynamic processes* under which psychological development takes place, or to identify the *strategies* implemented by learners, or *developmental micro-genesis*, in their general properties and in their diversity/specificity:

> Although Genevan research has provided a detailed analysis of cognitive structures, our knowledge of cognitive processes remains fragmentary. The focus is now not only on macro-development but also on changes occurring in children's spontaneous action sequences in micro-formation.
>
> *(Karmiloff-Smith & Inhelder, 1975, p. 195)*

Goodbye Geneva!

After Inhelder's retirement in 1978, Howard Gruber continued with her program and maintained her research team, but at the end of this transition phase, the Faculty decided to give a new direction to this professorial position and abandon almost all the Piagetian-oriented projects[12], which implied in fact that AKS could not succeed Inhelder to the professorial position, as some colleagues had wished.

AKS, therefore, left Geneva to accomplish the impressive scientific body of work that we all know. She mentions much later, certainly indirectly, the difficulties she encountered in Geneva, which may have been the main reason for her departure from the City of Calvin. In a critical review of the Inhelder tribute book, *Working with Piaget*, AKS rebelled against the title and against the fact that, in many chapters, the writers spoke mostly about Piaget, sometimes ignoring the specific contributions of the supposedly honored researcher. The coda of this collective article, undoubtedly of her pen, said a lot on her concern not to suffer the same fate as Inhelder:

> Let us end with a very different kind of musing. Are the editors of this volume really guilty for letting their contributors ignore Inhelder to varying degrees? Or, rather, are they guilty of failing to bring to the fore in their introduction the following possibility: that working with Piaget – a formidable man who tried to dominate all his collaborators – represented, despite its incredible richness, a constant struggle to assert one's own theoretical ideas? If Inhelder was party to this constant struggle, remaining in Geneva as she did for the vast majority of her admirable career, perhaps the contributors can be forgiven after all!
>
> *(Scerif, Paterson, & Karmiloff-Smith, 2001, p. 120)*

The paradoxical pitfall of Piagetian constructivism

In this second part, we will propose an analysis and an interpretation of the reasons why, in the 1970s, the question of the developmental processes (as Piaget sometimes said, the question of the nature of the "fuel" that powers the "engine" that is the cognitive system) was central, as much for Piaget himself as for Inhelder and some young psychologists, including AKS and I. But in fact this question had guided some initial research in the *CIEG* and had undoubtedly emerged from the very formulation of the operative theory in the 1940s.

Explicitly inspired by Bergson's *L'évolution créatrice* (1907), which emphasized the continuity of the mechanisms giving rise to the evolution of the living, Piaget's initial project was to develop a scientific approach to the conditions of the emergence and the development of these "special forms of life" that constitute human knowledge. After two decades of trial and error, this project succeeded in the late 1930s, under the joint effect of two innovations. The first was the creation, by Alina Szeminska and Bärbel Inhelder, of a methodology (later described as the *Piagetian clinical method*), consisting of presenting children with clever problems of object manipulation, and inferring their level of cognitive development on the basis of their objective behavior and their verbal justifications. The second was the research carried out by Piaget himself in the fields of logic and mathematics, leading to the identification of systems capable of formalizing the successive logical structures underlying psychological functioning. This conjunction gave rise to the constructivist model of cognitive development, which comprises, as we know, three general stages: the *sensorimotor stage* characterized by practical coordination of actions, the stage of *concrete operations* characterized by thought operations that are still dependent on the properties of the objects and the problems to which they apply (a level that can be formalized by the Boolean algebra), and the stage of the *formal operations* characterized by operations of thought detaching themselves from the properties of the objects to transform themselves in capacities of reasoning *in abstracto*.

In this way, Piagetian genetic psychology helped to identify and describe cognitive stages of development whose content and organization were subsequently validated by multiple studies; however, it has not been possible to identify and describe the conditions for the emergence of these stages. It, therefore, remained to analyze the processes involved in the important transformations characterizing the transition from one structure to another, or to try to go from the description-formalization of the *"developed"* (the *stages*) to the description-explanation of the *development itself as a process*.

A first approach, formal and "algebraic"

The first attempt in this direction was undertaken in the *CIEG*'s 1962 research program, introduced by Piaget as follows:

> A la filiation « naturelle » des structures, on pourrait peut-être faire correspondre une filiation abstraite ou théorique, fournissant un modèle algébrique d'un tel

développement. [...] Du point de vue épistémologique, la question est centrale, car on ne saurait espérer résoudre le problème de la nature des structures logico-mathématiques sans dominer la question préalable de leurs relations avec la pensée « naturelle » des sujets. Or, cette question ne saurait être posée en termes significatifs que sur le terrain génétique, car ce n'est pas avec la « conscience » du sujet (intuition, évidence, etc.) que ces structures peuvent être mises en rapport valable, mais seulement avec leur mode de construction donc avec leur genèse. A cet égard il va alors de soi qu'une correspondance possible entre les filiations réelles et la généalogie abstraite et théorique des structures constituerait une donnée décisive.[13]

(Piaget, 1963, p. 4)

As this excerpt shows, Piaget argued that in order to identify the processes of filiation or generation of successive stages, it is quite useless to solicit and analyze the conscious representations of children (their opinions, their feelings or their intuitions), but that it is better to elaborate an algebraic model formalizing the intermediate states between these structures. Such a model would therefore have for him the ability to account for what happens in the "natural thinking" of children during the long transition periods between one stage and the next. In a study entitled *Des groupements à l'algèbre de Boole* (1963), Grize tried to construct this algebraic model of how to move from sensorimotor stage to the operation stages. His work led to the formalization of six intermediate structural levels, each of which had at least two subsystems. Piaget seemed to be satisfied with this result, but this was not the case for Grize who pointed out that the identified sub-structures could intersect and combine in a largely random manner while nevertheless leading to a same type of result. For the author, the model he had constructed could not claim to constitute a formalization of the psychological processes of children. Moreover, on a strictly formal level, the transitional relations between the formalized structures could not include the lowest dimension of vection (or "oriented progression"):

[Nous ne prétendons] ni présenter une nouvelle formalisation de la logique, ni une formalisation, au sens propre du terme, de données psychologiques. [...] Nous ne pouvons guère songer à faire davantage, une structure réelle pouvant bien contenir la nécessité de son dépassement, mais jamais celle d'un dépassement vers telle autre structure fixée d'avance.

(Grize, 1963, p. 29)[14]

In another study of the same work (entitled *Structure et genèse*), Apostel undertook a piece of work centered, not on the formalization of hypothetical intermediate structures, but on the question, explicitly functional, of the genesis of these same structures:

[...] toute genèse a une structure et toute structure a une genèse. Nier qu'une structure ait une genèse signifie ou bien la poser comme perma-nente, ou bien la faire apparaître brusquement, sans que rien ni ne la prépare ni ne l'explique. Nier qu'une genèse ait une structure signifie qu'on abandonne la naissance des ensembles fermés et organisés en faveur d'une intervention de la contingence pure. Les deux négations sont d'ailleurs solidaires, comme les deux affirmations le sont.

(Apostel, 1963, p. 66)[15]

Stepping away from Piaget's command, Apostel considered that if algebra was a relevant instrument for the study of structures, it was not relevant for the study of transitions or development. For him, the instrument adapted to the study of these breaks of equilibrium, or of these transitions between forms of organization, was "the analysis", which he conceived of as a comparative deepening of the properties of the structures of different levels: "to understand the form of the passages (as changes) from the comparison of static sections, the passages themselves being extremely difficult to observe" (*ibid.*, 67). Although different from that of Grize, Apostel's approach nevertheless arrived at the same conclusion: the comparative deepening of the properties of the structures and the conditions of their possible transformations revealed a multitude of modalities of passage from one structure to another, which led the author to declare that it was necessary to abandon any idea of a single model capable of accounting for the genesis of the stadium-structures (*ibid.*, p.65).

The initial attempt to develop a logical model of the filiation of the structures had therefore resulted in a real failure, and it was not until the 1970s that this issue was reintroduced, in the new form mentioned above, into the *CIEG*'s program.

The two psychological attempts of Inhelder

Having admitted that the algebraic or logical pathways were not relevant to addressing the issue of developmental processes, Inhelder undertook, a decade later, to explore the purely psychological path of development, and in this case attempted to resolve this problem through the implementation of empirical work directly addressing the *natural thinking* of children.

With Sinclair and Bovet, she first conducted a series of research projects aimed at providing children with learning activities designed to facilitate their access to a higher stage or sub-stage of development:

L'un des deux buts essentiels des travaux décrits en ce volume était de compléter nos informations sur le développement lui-même des fonctions cognitives en cherchant à atteindre les mécanismes formateurs assurant le passage d'un niveau au suivant, sur lesquels nous étions assurément bien trop peu renseignés.

(Inhelder, Sinclair, & Bovet, 1974, p. 5)[16]

These activities were of a "crypto-Vygotskian" nature, insofar as they consisted in proposing to the children a standard test of the clinical method (for example, matter conservation), to make them act and express their reasoning when solving the problem, then to create *contradiction* by introducing concrete counter-examples or verbal counterarguments. According to the authors, these conceptual disturbances produced three types of effects: (1) a null effect (neither progress nor regression), (2) a positive effect of acceleration with regards to what would have been the spontaneous developmental timeline, and (3) a momentary negative effect. But in their interpretation of these results, they considered that all of these effects were only consequences of the current state of the natural cognitive capacities of the child, capacities which they requalified of *competences*:

> A l'analyse, le premier de ces cas (effet nul) se produit lorsque l'enfant est trop jeune […]. Le second cas (effet positif) se présente lorsque le facteur introduit constitue d'emblée un instrument d'assimilation […], mais cette propriété dépend naturellement elle aussi du niveau, donc de la "compétence" du sujet. Enfin le troisième cas est celui où le facteur introduit constitue une perturbation et nécessite une accommodation compensatrice entre schèmes d'abord hétérogènes, d'où échec si cette régulation est encore impossible, ou dépassement du conflit si une construction nouvelle est accessible au sujet selon son niveau.
>
> *(ibid., pp. 8–9)*[17]

The general conclusion of this study was, again, clearly negative: for the author's learning does not constitute a factor of development; it is, on the contrary, only a consequence of development. In other words, it is the structural state of the child's cognitive abilities that makes certain given forms of learning possible or not.

Inhelder's second attempt was the one that gave rise a few years later to the research program already mentioned, in which AKS was heavily involved. In the *Introduction* to the book that belatedly reported on this research (see above, note 12), Inhelder indicated that this new approach was intended to identify and conceptualize the heuristics of "real" or *psychological subjects*:

> L'originalité profonde de Piaget a été d'orienter d'emblée son œuvre vers l'étude des catégories fondamentales de la connaissance, sans lesquelles aucune adaptation à la réalité et aucune pensée cohérente ne seraient possibles […] C'est en ce sens que le sujet épistémique apparaît surtout comme le sujet d'une connaissance normative. […] Par contraste, le *sujet psychologique individuel* est étudié par un observateur qui s'attache à déceler la dynamique des conduites du sujet, leurs buts, le choix des moyens et les contrôles, les heuristiques propres au sujet et pouvant aboutir à un même résultat par des chemins différents, afin que l'on puisse pénétrer dans le fonctionnement psychologique et dégager les caractéristiques générales des procédures ou enchaînements finalisés et organisés d'actions.
>
> *(Inhelder & de Caprona 1992, pp. 20–21)*[18]

Breaking in fact with the strictly cognitive orientation of the Piagetian approach (i.e., extracting data only from the logical properties of the physical interactions between an individual subject and the environment), Inhelder and her collaborators attempted to take into account the affective, social and semiotic dimensions of child's behavior when confronted with an intellectual problem. However, if the various empirical results presented highlighted, for each of the experimental situations constructed, an impressive number of subtle transition mechanisms from one type of cognitive reasoning to another, the interpretation of the general significance of these results proved to be extremely difficult, as AKS acknowledged in the *Preface* that we asked her to write:

> A l'époque, nous avions énormément de peine à transmettre toute la richesse de nos découvertes à d'autres si ce n'est en les prenant par la main et en les installant devant notre écran. J'en vins même à douter qu'il fût jamais possible de communiquer ces travaux par écrit.
>
> *(Karmiloff-Smith, 1992, p. 11)*[19]

Despite the richness and precision of the empirical results it produced, this line of research has remained unfulfilled, because of the institutional reasons already mentioned, but also probably because of the ambivalence of the epistemological framework that it was intended to underpin. On the one hand, Inhelder's focus on the real operations of real psychological topics called for a focus on familiar patterns as individual and variable discovery procedures, and emphasized the decisive role that semiotic (or figurative) instruments played in the constitution of these same schemas:

> Les procédures du savoir-faire de l'enfant, dont nous suivons le déroulement sous forme d'une microgenèse, connaissent toutes une dynamique. [...] L'unité du fonctionnement s'est révélée être [...] le « schème familier », haut placé dans la hiérarchie d'accessibilité, tantôt producteur de découverte tantôt obstacle provisoire. L'aspect instrumental des diverses représentations de nature symbolique (gestes, figurations spatiales, langage) joue un rôle particulièrement important en tant qu'« objets à penser »
>
> *(Inhelder, 1992, p. 15)*[20]

But on the other hand, the new theoretical framework, called *psychological constructivism*, of which Cellérier tried to lay the foundations in the two final chapters of the book, remained in fact inspired by the embryological model ("an epistemology based on biology through psychology") and for this reason proved to be incapable of really taking into consideration the role played in development by the conditions of use of semiotic instruments and more broadly by the various forms of social interaction.

Three ways to reframing Piaget and Inhelder's inquiries

In this last part we will briefly discuss three directions of research taken by psychologists who were initially trained in Geneva and whose approach is characterized both by an adherence to the general achievements of the Piagetian work and by the research of new theoretical and methodological ways of clarifying the question of processes generating psychological development.

In a differential psychology approach, De Ribaupierre and Lautrey completed an impressive research program exploiting the method of the *"épreuves piagétiennes"*, and highlighted, beyond the horizontal shifts (*décalages*: differences of cognitive level in function of complexity of the domains to be treated: quantity, weight, volume, etc.) that Piaget had already identified, that the cognitive progression of the children is characterized not only by an important inter-individual variability but also and especially by a regular intra-individual variability:

> Traduite sur un plan développemental, la variabilité intra-individuelle observée, soit sous forme de décalage soit sous forme de facteurs de groupe, implique [...] qu'il y aurait plusieurs voies développementales différentes, certains sujets avançant d'abord sur la voie logico-mathématique, d'autres sur une voie infra-logique, d'autres en parallèle sur ces deux voies, toutes les combinaisons étant ensuite possibles.
>
> *(de Ribaupierre, 1998, p. 540)*[21]

These authors consider that their own data do not invalidate the Piagetian model as such, which remains valid as long as the averages of behaviors and argumentation of a substantial sample of children of the same age are taken into account. But they also consider that the importance of the two kinds of variability highlights a plurality of developmental processes interacting with each other. The results of this work are important in that they provide psychological data that correspond, and even explain to a certain extent, Grize and Apostel's research into logic. Although it is not difficult to identify and formally describe stages of development, the conditions and modalities of passage from one stage to another are, in subjects and subjects, so diverse that they are hardly definable.

For my part, after having been trained in what was called at the time "Soviet psychology" (Bronckart, 1970), I carried out various research programs inspired by Vygotsky, Saussure's real work and the linguistics of discourses-texts (Bronckart, 1997). I have, in particular, re-examined the question of the conditions for the emergence of conscious thinking in young children, and the role of language in the subsequent development of cognitive abilities (Bronckart, 2012, 2013). On the first point, based on the Saussurian analysis of verbal signs (see Saussure, 2011[22]), which highlights their properties of radical arbitrariness, linearity and psychic splitting, I have shown that it is the internalization of these three properties that can explain the emergence of a conscious thought alone, or again, to use the terms of Piaget (1974), which can explain the transformation of a psychological ability initially organized into *causal rules*, into a psychism organized into a system of *significant implications*. Regarding the further development of thought, I have tried to show that this is the internalization of the three fundamental levels of language structuring (signs, predicative relations, and discourse types) that is literally *constitutive* of the three fundamental levels of thought identified in the *Logique de Port-Royal* (Arnauld & Nicole, 1965), namely ideas or concepts, propositions (or operations), and reasoning. But these analyses also point to the fact that the object of internalization are signs, predicative relations, and discourse types *as they are shaped and organized in a given natural language*. Consequently, the initial thought of the child carries the imprint of the specific properties of the language spoken by his or her social group and is thus *verbal thought*, in the strict sense of the term. The initial thought of the child is thus, by and large, determined, on the one hand, by the semantic and syntactic properties of the language spoken in his or her environment and, on the other hand, by the systems of purposeful activity implemented in the same environment. But subsequently, alongside this "practical" verbal thought, another kind of thought develops, which is liberated from these initial constraints and which leads to abstract reasoning, founded on formal operations. The results that I have obtained, like those resulting from all Piagetian and post-Piagetian research concerning developmental processes, lead to consider that *the factors and processes of cognitive development are not themselves cognitive*; these factors are psychological interactions with the world and the others in which the affective, social and linguistic dimensions play a determining role, and the processes of this development, while they certainly mobilize cognitive operations (abstraction, differentiation, generalization, etc.) are so variable that they cannot be organized into a properly cognitive system.

As for the post-Geneva work of AKS, the other contributors to this volume are eminently more competent than me to comment on them. I will note, however, that unlike the two approaches mentioned above, AKS's approach is part of the mainstream contemporary Anglo-Saxon debate, particularly the discussion of the thesis that neural equipment plays a key, even exclusive, role in cognitive development. As she clearly stated in the book *After Piaget*[23], in her own work she adopted a neuroconstructivist position in which, according to the first component of this appellation, on the one hand, she gives a central role to the dimension "neuro":

> The brain starts out with a number of basic-level biases, each of which is somewhat more relevant to the processing of certain kinds of input over other and which become domain-specific over time through neuronal competition and a process of gradual modularization.
>
> *(Karmiloff-Smith, 2012, pp. 1–2)*

But according to the second component of this appellation, she stressed that it is necessary "with Piaget, to take development seriously" (*ibid.*, p. 8), because infants are active participants in their own learning, and cognitive structures are emergent and not innately specified.

And more generally, her reflections lead to a position that both closes her chapter of *After Piaget*, and also closes the present contribution, a position to which Inhelder, as well as Vygotski, would have undoubtedly adhered:

> It is clear that development […] is fundamentally characterized by plasticity for learning, with the infant brain dynamically structuring itself over the course of ontogeny. The infant brain is not a collection of static, built-in modules handed down by the evolution. Rather, the infant brain is the emergent property of dynamic multidirectional interactions between biological, physical, and social constraints.
>
> *(ibid., p. 11)*

Notes

1 In particular, she worked in refugee camps and obtained the status of *Associate in Psychology* at the *American University of Beirut* from February 1971 to February 1972.

2 This appreciation is testified to in a letter that Piaget had sent her directly (a rare thing) on May 13, 1968, and a letter from Inhelder dated September 23, 1971, bitterly regretting her "flight" to Beirut.

3 Term used by collaborators and colleagues to address or to mention Jean Piaget, but that AKS challenged and tried never to use in public.

4 *International Center of Genetic Epistemology* – Piaget had ceased teaching at the University since 1968, and devoted all of his activities to the direction and animation of the *CIEG* he had created in 1955.

5 Some 30 scientists actively participated in the work of *CIEG*, including – philosophers or logicians: L. Apostel, J.-B. Grize, T.S. Kuhn, J. Ladrière, W. Mays, W. Quine; – physicists: O. Costa de Beauregard, B. Mandelbrot, I. Prigogine, L. Rosenfeld; – mathematicians: R.B. Deal, G. Granger, P. Libois; – psychologists: F. Bresson, P. Greco, H. Gruber, S. Papert, G. N. Seagrim, etc.

6 From 1955 to 1980, about a hundred psychologists served as assistant-experimenters.

7 I had been hired at *CIEG* in 1969, and before the duo's constitution with AKS, I had conducted three research projects on the themes of consciousness, contradiction and reflective abstraction.

8 The *CIEG* was then in decline; its research objectives were beginning to be unclear and the *Presses Universitaires de France* had just given up publication of the serie *Etudes d'Epistémologie Génétique*. After the death of Piaget, Pierre Mounoud and I retrieved the manuscript *Morphismes et catégories* and published it in 1990 in the collection that we directed at Delachaux et Niestlé.

9 This text was written in 1974 for a symposium planned in Spain in 1975 but cancelled for political reasons. The texts prepared were published in 1977 in Bronckart *et al.*

10 A discussion of the relationship between language acquisition and cognitive development firstly calls for the fundamental distinction between what we might call "language tool" and "language system". By "language tool" we mean language in its representative function. It is the language tool that Piaget took into consideration and it is at this level that it is pertinent to pose the problem of the relationship between language and cognitive development. On the other hand, when we consider the "language system", that is to say the system of internal rules whose language is composed, we can

ask ourselves whether it is relevant to pose the problem in terms of "relations between language and cognition". (AKS, 1977, 169; our translation)

11 *A functional approach to child language*, published in 1979, constitutes a reworked version of this original thesis.

12 The results of Bärbel Inhelder's extensive research program had not given rise to any substantial publication, and it was Pierre Mounoud and myself who published, in 1992, the book of Inhelder & Cellérier, entitled *Le cheminement des découvertes de l'enfant...* with a preface by AKS.

13 To the "natural" filiation of structures, one could perhaps match an abstract or theoretical filiation, providing an algebraic model of such a development. [...] From the epistemological point of view, the question is central, because one cannot hope to solve the problem of the nature of the logico-mathematical structures without dominating the preliminary question of their relations with the "natural" thought of the subjects. However, this question can be posed in significant terms only in the genetic field, because it is not with the "consciousness" of the subject (intuition, evidence, etc.) that these structures can be put into valid relation, but only with their mode of construction so with their genesis. In this respect, it goes without saying that a possible correspondence between real filiations and the abstract and theoretical genealogy of structures would constitute a decisive point. (Piaget, 1963, p.4)

14 [We do not claim] to present a new formalization of logic, nor a formalization, in the true sense of the term, of psychological data. [...] We can hardly think of doing more, a real structure may well contain the need for its overtaking, but never that of an overtaking to another structure set in advance. (Grize 1963, 29; our translation)

15 [...] every genesis has a structure and every structure has a genesis. To deny that a structure has a genesis means either to pose it as permanent, or to make it appear suddenly, without anything either preparing or explaining it. Denying that a genesis has a structure means abandoning the birth of closed and organized sets in favor of an intervention of pure contingency. The two negations are, moreover, in solidarity, as the two affirmations are. (Apostel 1963, 66; our translation)

16 One of the two main aims of the work described in this volume was to supplement our information on the development of cognitive functions itself by seeking to reach the training mechanisms ensuring the passage from one level to the next, about which we certainly had too little information. (Inhelder, Sinclair, & Bovet, 1974, 5; our translation)

17 On analysis, the first of these cases (null effect) occurs when the child is too young [...]. The second case (positive effect) arises when the introduced factor constitutes from the outset an instrument of assimilation [...], but this property also naturally depends on the level, therefore on the "competence" of the subject. Finally, the third case is where the introduced factor constitutes a disturbance and requires a compensatory accommodation between initially heterogeneous schemas, hence failure if this regulation is still impossible, or if the new construction is accessible to the subject according to its own level. (*ibid.*, pp. 8–9; our translation)

18 The profound originality of Piaget was to direct his work from the beginning towards the study of the fundamental categories of knowledge, without which no adaptation to reality and no coherent thought would be possible [...] It is in this sense that the epistemic subject appears above all as the subject of a normative knowledge. [...] In contrast, the individual psychological subject is studied by an observer who seeks to detect the dynamics of the subject's behaviors, their goals, the choice of means and controls, the heuristics of the subject and can lead to the same result by different paths, so that one can penetrate into the psychological functioning and clarify the general characteristics of the procedures or finalized and organized sequences of actions. (Inhelder & de Caprona, 1992, pp. 20–21; our translation)

19 At the time, we had a lot of trouble transmitting all the wealth of our discoveries to others if not by taking them by the hand and by installing them in front of our screen.

I even doubted that it was ever possible to communicate this work in writing. (Karmiloff-Smith, 1992, 11; our translation)

20 The procedures of the child's know-how, which we follow in the form of microgenesis, are all dynamic. [...] The unity of operation has proved to be the [...] "familiar schema", high in the hierarchy of accessibility, sometimes a producer of discovery, sometimes a temporary obstacle. The instrumental aspect of the various representations of symbolic nature (gestures, spatial figurations, language) plays a particularly important role as "objects of thought". (Inhelder, 1992: 15); our translation)

21 Translated on a developmental level, the observed intra-individual variability, either in the form of an offset or in the form of group factors, [...] implies that there would be several different developmental pathways, some subjects advancing first on the logico-mathematical path, others on an infra-logical path, others in parallel on these two paths, all combinations being then possible. (de Ribaupierre, 1998, 540; our translation)

22 The true Saussurian analysis of the status of signs is presented in *La double essence du langage* (2011).

23 This book is composed of chapters written by psychologists who worked at the CIEG in the years 1960–1970, and it is with good reason that the first chapter was entrusted to AKS.

References

Apostel, L. (1963). Structure et genèse. In L. Apostel, J.-B. Grize, S. Papert & J. Piaget (Eds.), *La filiation des structures* (pp. 65–106). Paris: PUF.

Arnauld, A., & Nicole, P. (1965). *La logique ou l'art de penser*. Paris: Vrin [original edition: 1662]

Bergson, H. (1907). *L'évolution créatrice*. Paris: Alcan.

Bronckart, J.-P. (1970). Le rôle régulateur du langage: critique expérimentale des travaux d'A.R. Luria. *Neuropsychologia*, *8*, 451–463.

Bronckart, J.-P. (1980). The International Center of Genetic Epistemology (Geneva). *French-Language Psychology*, *1*, 241–252.

Bronckart, J.-P. (1997). *Activité langagière, textes et discours. Pour un interactionisme sociodiscursif*. Lausanne: Delachaux et Niestlé.

Bronckart, J.-P. (2012). Contributions of Piagetian constructivism to social interactionism. In E. Martí & C. Rodríguez (Éd.), *After Piaget* (pp. 43–58). London: Transaction Publishers.

Bronckart, J.-P. (2013). Qu'est-ce que le développement humain? Interrogations, impasses et perspectives de clarification. In J. Friedrich, R. Hofstetter & B. Schneuwly (Éd.), *Une science du développement est-elle possible? Controverses du début du XXe siècle* (pp. 207–226). Rennes: PUR.

Bronckart, J.-P., Malrieu, P., Siguan, M., Sinclair De Zwart, H., Slama-Cazacu, T., & Tabouret-Keller, A. (1977). *La genèse de la parole*. Paris: PUF.

Carnap, R. (1934). *Logische Syntax der Sprache*. Vienne: Springer Verlag.

Ferreiro, E. (1971). *Les relations temporelles dans le langage de l'enfant*. Genève: Droz.

Grize, J.-B. (1963). Des groupements à l'algèbre de Boole: essai de filiation des structures logiques. In L. Apostel, J.-B. Grize, S. Papert & J. Piaget (Eds.), *La filiation des structures* (pp. 25–63). Paris: PUF.

Inhelder, B. (1992). Avant-propos. In B. Inhelder & G. Cellérier, G. (Ed.), *Le cheminement des découvertes de l'enfant. Recherche sur les microgenèses cognitives* (pp. 13–16). Lausanne: Delachaux et Niestlé.

Inhelder, B., & Cellérier, G. (1992). *Le cheminement des découvertes de l'enfant. Recherche sur les microgenèses cognitives*. Lausanne: Delachaux et Niestlé.

Inhelder, B., & de Caprona, D. (1992). Vers le constructivisme psychologique: structures? Procédures? Les deux indissociables. In B. Inhelder & G. Cellérier, G. (Ed.), *Le cheminement des découvertes de l'enfant. Recherche sur les microgenèses cognitives* (pp. 19–50). Lausanne: Delachaux et Niestlé.

Inhelder, B., Sinclair, H., & Bovet, M. (1974). *Apprentissage et structures de la connaissance.* Paris: PUF.

Karmiloff-Smith, A. (1976). *Little Words Mean a Lot. The plurifunctionality of Determiners in Chiuld Language.* Thèse de doctorat, Université de Genève, Faculté de Psychologie et des Sciences de l'Education.

Karmiloff-Smith, A. (1977). Développement cognitif et acquisition de la plurifonctionnalité des déterminants. In J.-P. Bronckart *et al.*, (Éd.), *La genèse de la parole* (pp. 169–177). Paris: PUF.

Karmiloff-Smith, A. (1978). On stage: The importance of being a non-conserver. *Behavioral and Brain Sciences, 2,* 188–190.

Karmiloff-Smith, A. (1979). *A functional approach to child language. A study of determiners and reference.* Cambridge: Cambridge University Press.

Karmiloff-Smith, A. (1992). Préface: « Il est difficile d'imaginer Piaget sans Inhelder la réelle ». In B. Inhelder & G. Cellérier, G. (Ed.), *Le cheminement des découvertes de l'enfant. Recherche sur les microgenèses cognitives* (pp. 9–11). Lausanne: Delachaux et Niestlé.

Karmiloff-Smith, A. (2012). From constructivism to neuroconstructivism: The activity-dependent structuring of the human brain. In E. Martí & C. Rodríguez (Éd.), *After Piaget* (pp. 1–14). London: Transaction Publishers.

Karmiloff-Smith, A., & Inhelder, B. (1975). If you want to get ahead, get a theory. *Cognition, 3,* 195–212.

Piaget, J. (1963). Le problème de la filiation des structures. In L. Apostel, J.-B. Grize, S. Papert & J. Piaget (Eds.), *La filiation des structures* (pp. 1–23). Paris: PUF.

Piaget, J. (1974). L'explication en psychologie et le parallélisme psychophysiologique. In P. Fraisse et & J. Piaget (Ed.), *Traité de psychologie expérimentale, Vol. I* (pp. 137–184). Paris: PUF.

Piaget, J., Karmiloff-Smith, A., & Bronckart, J.-P. (1978). Généralisations relatives à la pression et à la réaction. In J. Piaget, *Recherches sur la généralisation* (pp. 169–191). Paris: PUF.

Piaget, J., Karmiloff-Smith, A., & Bronckart, J.-P. (1980). Correspondances et relations. In J. Piaget (Éd.), *Recherches sur les correspondances* (pp. 133–151). Paris: PUF.

Piaget, J., & Karmiloff-Smith, A. (1990). Un cas particulier de symétrie inférentielle (Carte routière à lire à l'envers). In J. Piaget (Éd.), *Morphismes et catégories* (pp. 117–127). Lausanne: Delachaux et Niestlé.

Piaget, J., & Karmiloff-Smith, A. (1990). Conflits entre symétries. In J. Piaget (Éd.), *Morphismes et catégories* (pp. 129–140). Lausanne: Delachaux et Niestlé.

de Ribaupierre, A. (1998). Développement cognitif et différences individuelles/Individual differences in cognitive development. In M. Sabourin, F. I. M. Craik & M. Robert (Eds.), *Advances in psychological science, Vol. 2: Biological and cognitive aspects* (pp. 531–555). London: Taylor & Francis.

Saussure, F. (2011). *De la double essence du langage.* Edition établie par R. Amacker. Genève: Droz.

Scerif, G., Paterson, S., & Karmiloff-Smith, A., (2001) What Piaget could have learnt from working with Inhelder: A review of *Working with Piaget. Archives de Psychologie, 69,* 115–120.

Sinclair, H. (1967). *Acquisition du langage et développement de la pensée.* Paris: Dunod.

4

INTELLIGENCE: TAKING THE DYNAMICS OF DEVELOPMENT SERIOUSLY

Frank D. Baughman and Mike Anderson

If you've already had a look at the introduction to this volume, you will have seen the impressive number of awards and honours that were bestowed upon Annette Karmiloff-Smith over the course of her career – testament to some of the achievements and contributions that she made to developmental psychology. Annette was, of course, a great intellect but she was also an extraordinarily generous and good person. And, like many others, we feel privileged to have known her.

Given the influence Annette has had within the field of developmental psychology, it is natural that the reader may wonder (just as we did, initially!) – why *us*? What warrants *our* contribution to this volume? Though we have important personal histories and treasured memories of Annette, the answer is that Annette's thinking continues to be of pivotal importance to our own research goals. Specifically, these are to understand (1) the causes of individual differences in thinking, reasoning and problem solving and (2) the mechanisms that explain the development of those abilities over time. Our aim in this chapter is to detail Annette's influence on our attempts to shed light on key theoretical questions concerning the nature of intelligence.

The origin of Annette's influence on us

Though it might appear facile, the essence of Annette's enduring influence on us can be traced back to a single idea. Indeed, it is the same idea around which this volume is organised:

To understand development, it is necessary to take development seriously.

Within the field of developmental psychology, this was something that Annette became well known for promoting wherever the opportunity arose. Readers may also recognise this was the idea that was fundamental in her book, Beyond Modularity (Karmiloff-Smith, 1995), and that permeated Annette's subsequent research. However, over the years it has become apparent that for some (even prominent) scholars within psychology, precisely what was meant by 'taking development seriously' has not been obvious. A great amount of research, some undoubtedly inspired by Annette's work, has produced a great amount of data evidencing differences in the capabilities of children at various ages. However, Annette was crystal-clear on this point: taking development seriously does not mean simply that one must study children – it is not about cataloguing what children can and cannot do at different ages. Annette vehemently challenged those who offered such snapshots of younger versus older children's competencies, without providing a cognitive explanation for how those changes took place. She did not compromise – that was not taking development seriously!

While perhaps a seemingly simple idea, the practicalities of devising a research approach that truly takes development seriously is complex. For Annette, this came to mean studying multiple, nested influences across time. This included examining the effects of highly dynamic interactions between genes, the environment, as well as various social, physiological, physical, and cognitive factors (see e.g., Karmiloff-Smith, 1998; Karmiloff-Smith, 2009). For our own part, taking development seriously has meant attempting to isolate a research programme that allows us to study the precise contribution of the mechanisms theorised to underlie between-age and within-age differences in intelligence (see also Baughman, Thomas, Anderson, & Reid, 2016; Davis & Anderson, 2001). Importantly, most current theories of intelligence happen to be theories of *adult* intelligence. That is, they attempt to provide an explanation for the differences seen in abilities between individuals, but they neglect to provide an account of where the processes responsible for differences come from, or how these processes are shaped by development. To this end, we suggest a codicil to Annette's aphorism:

> To understand intelligence, it is necessary to take development seriously.

Before we describe our attempt to take development seriously in the study of intelligence, we shall address another question that the reader may have in mind – *why study intelligence in any case?*

Why intelligence?

Our answer to this is that questions concerning intelligence are fundamental to an understanding of the mind. Consider, what exactly is intelligence? What *property* differentiates a more intelligent person from a less intelligent person? Where does this property come from? And, what influences, or modulates it over time? We find such questions inherently interesting, and the answers have potentially far-reaching consequences. Admittedly, not everyone shares our enthusiasm for the topic. Indeed, within academic psychology the subject of intelligence appears to have been somewhat unpopular – perhaps not surprising when the research has often been marred by linking contentious scientific claims about intelligence, race, social class and genetics with socio-political outcomes (e.g., Jensen, 1998, 2006; Rushton, Jensen, Philippe Rushton, & Jensen, 2003; for discussion, see Anderson, 2007). Though certainly a troubled history, the study of intelligence has nevertheless revealed important empirical facts that beg the development of better theories. Next, we briefly outline these facts and describe why the time has come for the discipline of psychology to finally engage with the core construct of intelligence – g.

The empirical regularities in intelligence

1. A single factor underlies performance on tests of intelligence. Anyone reading this volume will almost certainly have heard of the term 'g', and will also likely know that it refers to 'general intelligence'. It was Charles Spearman who first made the discovery of g, following the observation that for a population, the subtest scores from an intelligence test (he used the Binet-Simon test; 1905) correlated strongly, and in a positive direction. Put simply, this meant that an individual scoring well on one test was likely to score well on the others. Conversely, an individual scoring poorly on one test was likely to also score poorly on the others. Through factor analysis of these positive correlations, Spearman found that a single higher-order factor accounted for a large proportion of variance on test performance. This finding led Spearman to propose his two-factor theory of intelligence: that an individual's performance on tests was influenced most by a single, general factor that he called the *general factor* of intelligence (g) and then by secondary 'specific factors' that he considered unique to the particular test (Spearman, 1904). The presence of the g-factor, that is the statistical extraction of a single higher-order factor from test scores, has not diminished since its discovery and has remained thus despite the development of ever newer and more sophisticated statistical techniques. Analogous to this, within the context of the study of psychopathology, recent work has also emerged supporting the presence of a single, general dimension termed the g-factor (see Caspi et al., 2014), and this too appears to have a substantial genetic component (e.g., Allegrini et al., 2019).

  2. The shape of developmental change is highly similar across cognitive domains. Figure 4.1 plots the average ability scores for typically developing children between 6 and 18 years of age, on six core scales comprised within the British Abilities Scales (BAS-II; Elliot, Smith, & McCulloch, 1997). From this, one can see, for example, that the typical ability score on the Word Definition subtest for a 6-year-old is approximately 70, and for a 9-year-old is close to 100. The tests depicted in Figure 4.1 are referred to as core scales because they are assumed to tap core cognitive domains, and because the scores on these six subtests are used to compute overall measures of intelligence (i.e., one's intelligence quotient, or IQ score). What is important about this figure is that it shows quite clearly that over time, on each of the subtests, the typically developing child can be expected to

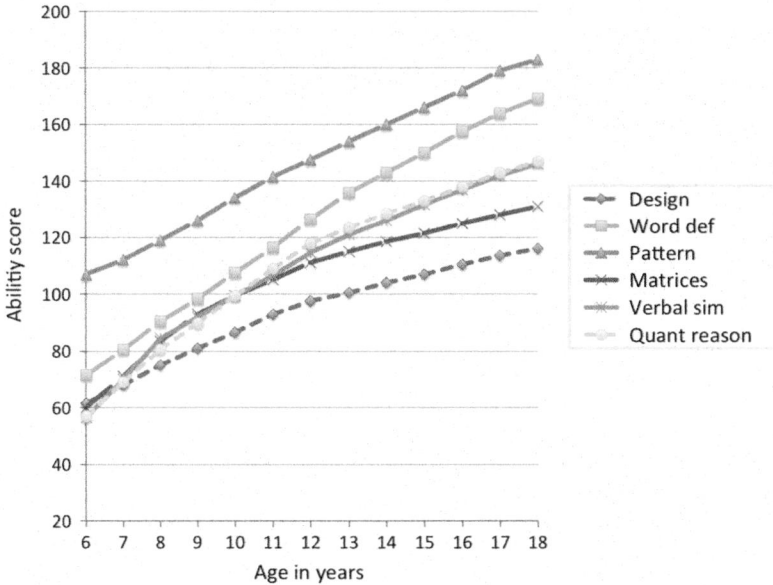

FIGURE 4.1 Increase in abilities with age. Data are of typical age equivalent scores from the core scales of the British Abilities Scales (BAS-II)

demonstrate very similar increases in their underlying abilities. That is, though the ability scores on the subtests vary, the trajectories of those abilities appear to take a common developmental path.

3. IQ scores show constancy across time. This is not to say that IQ scores do not change, nor that there are no changes in absolute cognitive performance or one's ranked position relative to age. Indeed, in typical development, and at least up until early adulthood, one's absolute cognitive performance must necessarily increase, whereas one's ranked position within a population (e.g., compared to others of the same age) may vary (i.e., may go up or down, or stay the same). Rather, the notion of constancy refers to the fact that one's IQ in early childhood predicts to a relatively high degree one's later IQ. Figure 4.2 shows data reported by Hindley and Owen (1978) of the correlations in IQ scores over childhood. The figure shows a very high degree of correlation between IQ scores of children when they were 3 years old compared to when they were 5 years old (depicted by the bar furthest left of Figure 4.2; $r = .78$). In plain language, this means that approximately 60% of the variation in IQ scores of children aged 5 years old are predicted by the IQ scores obtained from those same children aged 3 years old. There are two important observations we wish to highlight from these longitudinal data. First, even at the largest separation between ages that Hindley and Owen tested (14 years separate the test scores taken at 3 years old and then later at 17 years old) the correlation between IQ scores is not trivial ($r = .53$). Second, for each age bracket, the correlations between IQ scores are largest between adjacent ages (e.g., between 3–5, 5–8, 8–11, 11–14, and 14–17). While the actual test items

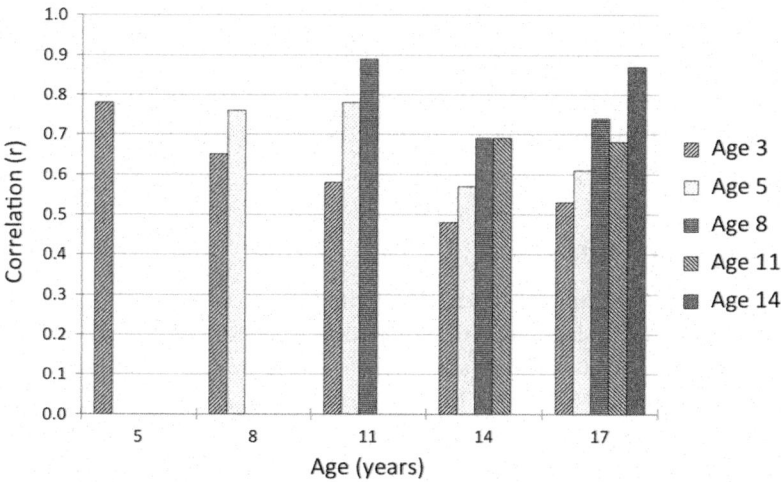

FIGURE 4.2 Correlations reported by Hindley and Owen's (1978) of IQ scores at 3, 5, 8, 11, 14 and 17 years of age

that the children received when they were 3-year-olds were quite different to the test items they had at age 17, the conversion of those test scores to IQ scores shows a high degree of stability across time.

Further evidence of the constancy of IQ comes from a study by Deary and colleagues who followed up 101 adults in Scotland, 66 years after they were first administered an IQ test (Deary, Whalley, Lemmon, Crawford, & Starr, 2000). Deary et al. showed that the relation between IQ scores obtained of these participants as children (aged 11 years old) and IQ scores collected as adults (aged 77 years old) was substantial (r = ~.7).

Given Hindley and Owen's (1978) and then Deary et al.'s (2000) findings, it might be tempting to think that the case for the constancy of IQ is won. Alas, the case is complicated when one looks at what evidence exists for constancy in the scores that comprise one's overall IQ score; most often overall IQ is comprised by separate scores relating to domains of spatial, verbal, and non-verbal reasoning. Here, rather than finding similarly high levels of constancy in ability scores on these domains, recent work has shown that constancy in overall IQ scores across development can mask quite large (i.e., up to 20 IQ points) differences in scores tapping each domain (see Ramsden et al., 2011). So, while a high degree of constancy may exist in overall IQ scores, there may be much less constancy in the relative contributions offered by each cognitive domain.

There remain significant challenges in reconciling such findings within a single theoretical framework. Nevertheless, at a broader level, the facts pertaining to the regularities in development offer key constraints for theories of intelligence concerned with understanding the possible architecture of the mind. And, clearly, any adequate theory must account for these facts. But, for understanding the mechanisms of individual differences and also developmental change it can often be the exceptions to the "rules" that developmental psychologists have looked to for what these might indicate of the cognitive architecture of intelligence. We now turn to some examples of exceptions to the rule of typical development, which can be identified from different patterns of unevenness in the observed profiles of cognitive abilities.

Exceptions in development

Developmental delay in childhood is the term often used to refer to instances whereby at one age a child shows significantly poorer abilities in just one cognitive domain but is then within the typical ability range at a later age. For example, an 8-year-old child might be described as having reading delay, if their reading ability is at the level of the average 5-year-old but then at age 12 is within the typical range for children of that age. We have depicted this pattern in Figure 4.3; Tile 1, where the typical developmental pattern (solid line) is contrasted to this exception (dotted line). The notion of delay is therefore somewhat complicated by the fact that it is only with time (i.e., development) that one can say whether a child's lower-than-typical score is best characterised as a delay, or as a persisting (specific) deficit.

Specific deficit refers to cases whereby an individual appears to show a persisting deficit (i.e., it does not resolve in time, as per delay) that, again, is isolated to a single cognitive domain. That is, there appears to be no system-wide impacts that are due to atypical genetic or environmental constraints on developmental processes. Two forms of specific deficit may be distinguished: *developmental* and *acquired*. The former refers to a type of deficit that is often observed earlier in

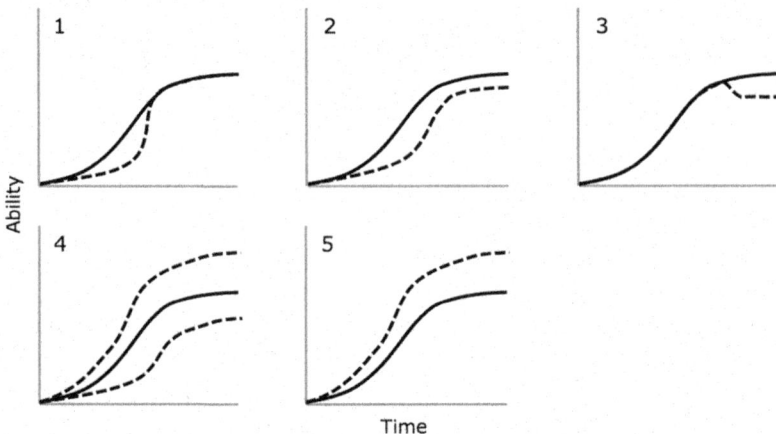

FIGURE 4.3 Uneven profiles of ability in typical (solid lines) and atypical (dashed lines) development. Tiles depict (1) developmental delay, (2) developmental specific deficit, (3) acquired specific deficit (4) savant abilities and (5) individuals with prodigious abilities

development, and which is potentially caused by genetic or environmental factors (see Figure 4.3; Tile 2). The latter refers to a type of deficit that occurs most often as a result of brain injury (e.g., stroke, or disease. See Figure 4.3; Tile 3).

Autistic savants provide a different profile of abilities in that these individuals are usually severely impaired across a range of cognitive or language domains, but yet show remarkable abilities in one or more other areas (e.g., with musical abilities, numbers, or calendrical operations. Figure 4.3; Tile 4).

Child prodigy is the term often used in relation to children under the age of 10 years, who show exceptional abilities compared to their typically developing peers. For example, they may demonstrate talents and skills in mathematics or music that resemble adult experts (see Figure 4.3; Tile 5).

The case for taking g seriously

If IQ scores were nothing more than the arithmetic mean, or sum total, of a set of unrelated cognitive abilities, then the analysis of test scores would show no g-factor, and IQ scores would (at least from a psychological stance) be of little relevance. However, the reality is that diverse abilities *are* related and so therefore the real question of theoretical interest is whether the g-factor relates to a substantive property of the mind or whether it is simply some epiphenomenon of the dynamics of the system. The challenge for any complete theory of intelligence is thus to provide an explanation for the g-factor and for the regularities and exceptions that are observed in development.

A quick sketch of the theory of the Minimal Cognitive Architecture

Anderson's solution was to propose the Minimal Cognitive Architecture (Anderson 1992, 2007). Within this framework the mind is comprised of two different types of information processing routes: route one (thinking), involving a slow, central process of reasoning that is subject to constraints imposed by variability within a single mechanism, termed the 'basic processing mechanism'. And, route two, involving computationally complex 'modules' that afford representations of the world that could not be achieved by the mechanisms of thought. Under route 1, it is individual differences in the *speed* of the basic processing mechanism that gives rise to a 'g' factor in all operations that require thought and reasoning. And, because the speed of the basic processing mechanism is held to be unchanging in development, the MCA also accounts for the observed phenomenon of the stability of IQ over time. Under route two, it is the coming on–line of modules through development that introduces substantial changes in cognitive ability, as children grow older. Consequently, it is largely due to delay in their coming on–line, or differences in the functioning of these modules that give rise to the exceptions in development, including those listed above. It is noteworthy that this architecture generates two dimensions to *g* itself: one related to

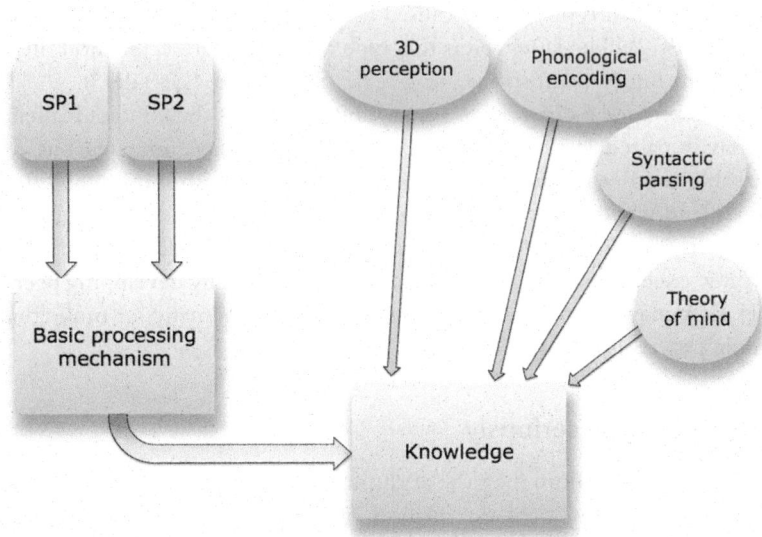

FIGURE 4.4   The Minimal Cognitive Architecture (from Anderson, 1992)

individual differences in speed and the fount of IQ-differences – and the other related to modular development (another example would be the maturation of systems underlying executive functioning) and developmental change (Anderson, 2017). A schematic of the MCA is given in Figure 4.4 and fuller accounts can be found in Anderson (1992, 2017).

A radically different interpretation of g

Though the MCA provides an account for the regularities and many of the exceptions that occur in development, it has a number of shortcomings, particularly when it comes to the developmental dimension of intelligence. In terms of its timing, the emergence of the MCA coincided with changes in key theoretical influences within developmental psychology. One such change relates to increasing dissatisfaction among theorists with traditional box-and-arrow models relating to information processing – i.e., models that neglected to provide the specifics of the mechanisms or processes of change. Another important change taking place at that time relates to a growing rift in views concerning the functional organisation of the mind. It is clear that the MCA is distinctly modular in design. However, at the time of its emergence, the view that cognition is underpinned by a number of domain-specific, fast, automatic and inherited modules – the sort of view exemplified in the seminal writings of Jerry Fodor (e.g., Fodor, 1983) – was being increasingly challenged, on the basis of mounting evidence that the cognitive system is far more interconnected, dynamic and distributed. And this, it appears, is especially true during early development. Annette was particularly involved in this debate – she argued that to the extent that the cognitive system is comprised of modules, that this is a *product* rather than the starting point of development. It was largely to examine fully modular accounts of development that Annette took a very active interest in the computational modelling of cognitive systems, that also inspires one of us (Baughman). And, a closer look at how computational approaches can be used to explore questions of the development of intelligence, reveals just how much of a radical departure they might offer to more conventional approaches such as the one offered by the MCA (see also, Thomas, Baughman, Karaminis, & Addyman, 2012).

The scope of our research program

In homage to Annette, we decided to begin a research program to explore how much a computational approach that is based in dynamical systems theory (DST; see e.g., Thelen & Smith, 2007) could emulate (if not surpass) the more conventional cognitive architecture embodied in MCA. In so doing we hoped to introduce a more fully developmental approach to understanding *g* and the regularities and exceptions in the development of intelligence – in short, we decided to take development seriously. Our approach extends on a dynamical systems model developed by van der Maas et al. (2006). Therefore, before we turn to the details of our approach let us first set out their base model.

A dynamical model of g

van der Maas et al. (2006) developed a mathematical model based on the Lotka-Volterra equation, comprising a fully connected dynamical system that simulates cognitive development via growth functions, which are themselves assumed to be the outcomes of developmental mechanisms within relevant cognitive components, in a number of processes (depicted in Figure 4.5c). Within the model, a limited number of parameters effect development for each individual process. However, the development of the model as a whole is influenced dynamically over time by all connected processes within the system. Two key features of the model may be highlighted. The first is that the model is neutral on the extent to which the parameters that effect development within each process are reliant on experience-dependent change. The second key feature is that the processes that are connected to each other interact with each other in a mutually beneficial way. Hence, the model is called the 'mutualism' model. Equation 4.1 gives the dynamics of the mutualism model (van der Maas et al., 2006).

$$\frac{dx_i}{dt} = a_i x_i \left(\frac{1 - x_i}{K_i} \right) + a_i \sum_{\substack{j=1 \\ j \neq i}}^{w} M_{ij} \, x_j x_i / K_i \qquad 4.1$$

The equation states that at each point in time (*t*) the change in the performance level *x* of a given process *i* (dx_i) is a product of the sum of the interaction weights

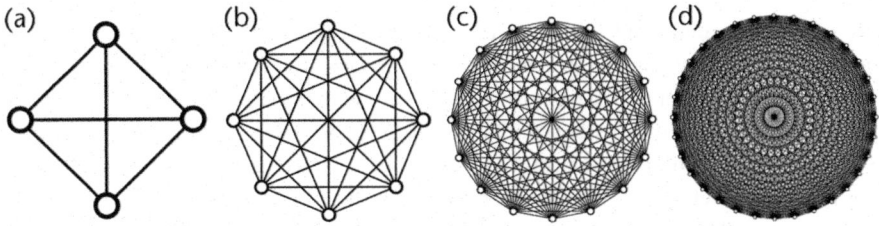

FIGURE 4.5 Four functional architectures comprised of (a) 4 processes, (b) 8 processes, (c) 16 processes, and (d) 32 processes. Connections show the functional influence of the development of a given process on the development of others

of each process j to which it is functionally connected ($M_{ij}x_jx_i$), multiplied by the rate of growth of process i (a_i), multiplied by the current level of performance of process x_i, and divided by the asymptote level for that process (K_i). Changes in x_i at each time step are thereby constrained by the performance (and thus the individual properties) of all other processes to which it is connected. The approach taken by van der Maas et al is noteworthy because it captures several features concerning the development of intelligence. For instance, within a population of models van der Maas showed that while the individual performance levels of processes were uncorrelated in early development, through the process of interactions, their performance became highly correlated later in development (i.e., when the model reached asymptote). Importantly, factor analysis of these data revealed a single higher-order factor, resembling the g-factor. However, crucially, there was no single parameter in their model that was the basis of this. In other words, the g-factor represented a statistical artefact of the dynamics of many interconnected processes.

The mutualism model has since been extended to explore cognitive accounts of the mechanisms underlying variability in development. For example, within the context of examining the case of developmental language disorders, Baughman and Thomas (2008) showed that the extent to which cognitive systems can buffer the effects of atypical variability applied to a single domain, and the extent to which unevenness in that domain affects the development of other processes are dependent on the functional architecture, and where in the system the locus of atypical variability was induced. More recently, Baughman, Baughman, and Mills (2012) adapted the mutualism model to investigate accounts relating to cognitive damage combined with general cognitive decline in older adulthood. They examined several functional architectures, and for each architecture model variants were created such that endstate performance varied from low to high – thereby simulating lower versus higher overall levels of IQ. Ageing was simulated by applying a fixed level of decay to all processes late in development, before a single cognitive process was selected for complete removal/destruction. Baughman et al. found that within a given architecture, the effects of damage were proportionate across level of IQ. However, the effects on the overall cognitive profile of models were different depending on the functional architecture. As these examples show, the precise functional architecture, that is the extent to which cognitive processes influence each other's development, can have profound effects on lifespan development.

Extending the base model

In extending the work of van der Maas et al. (2006), we start with a very simple question. If g does not correspond to a single process in a DST system and is simply an emergent property of interaction between processes in a dynamic system, then an obvious question is how does the number of processes affect the generation of a g-factor? And, second, because we are taking development seriously, might architectures of varying numbers of processes generate different profiles of g over development? To explore these questions, we decided to compare architectures comprised of different numbers of processes, across a range of developmental timepoints.

Simulations

Population dataset

We used the model parameters reported by van der Maas et al. (2006) to generate distributions of values for 50 processes, for a total of 0.5 million pseudo-subjects[1]. Distributions for growth rate (a), carrying capacity (K), and starting values (x) were normally distributed around means and standard deviations as follows: a mean=6, sd=.5; k mean=3, sd=.5; and, x mean=0.05, sd=.01. For each process within a pseudo-subject, parameters were randomly assigned so that the values of a, k, and x were uncorrelated at the first timepoint. The parameters for the same 1000 randomly selected pseudo-subjects were used in each model run, to ensure that any differences between models could be attributed to the effect of the model architecture, and not to differences in sampling. The four functional architectures we tested comprised of 4, 8, 16 and 32 processes each (see Figure 4.5).

Measures

Within each model, we sampled the performance levels of each process at 5 timepoints during development in order to assess the degree to which each functional architecture yields a single, higher-order factor. However, because asymptote levels vary as a consequence of the number processes, we computed the sampling timepoints relative to the endstate performance for that model. These timepoints, indicated by triangles in Figure 4.6, loosely correspond in each instance to birth (0% of endstate), early childhood (25% of endstate), middle childhood (50% of endstate), late childhood (75% of endstate), and adulthood (end-state)[2].

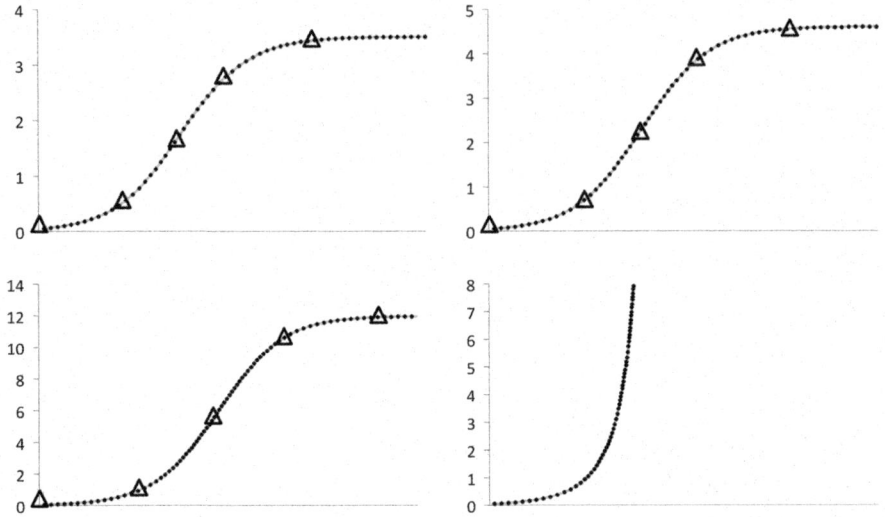

FIGURE 4.6 Developmental trajectories in the 4-process (top left), 8-process (top right), 16-process (bottom left) and 32-process (bottom right) architectures. Triangles indicate approximate points that the performances for each process were sampled for factor analyses. See text for further detail, and for an explanation for the absence of time-samples for the 32-process architecture

A key feature of dynamical systems is that manipulations to model parameters can, at times, have highly unpredictable consequences. For instance, in the models we tested increasing the number of processes had the effect of increasing the overall endstate achieved by the model. Put simply, models comprised of a greater number of processes achieved higher levels of ability, compared to models with fewer processes. This is because with greater numbers of processes connected to one another, there was greater opportunity for each process to influence each other process. However, varying the number of processes did not result in simple scaling differences in the endstate achieved by each model. That is, the endstate reached for example, in the 4-process architecture was not simply 50% lower than the endstate reached in the 8-process architecture. Another feature, captured by the results from the 32-process model, relate to massively overwhelming influences of many processes. As Figure 4.6 (bottom right) shows, the effect of this many processes had the dramatic effect within the model, with the model unable to reach a stable endstate. We return to discuss what this might mean shortly, however, because the data from the 32-process architecture cannot be analysed, in the next section we focus instead on the results of the other three models.

Results

For each model, we first tested the correlations between timepoints to see the extent to which they simulated the constancy of IQ. Figure 4.7 shows these results

(a)

(b)

(c)

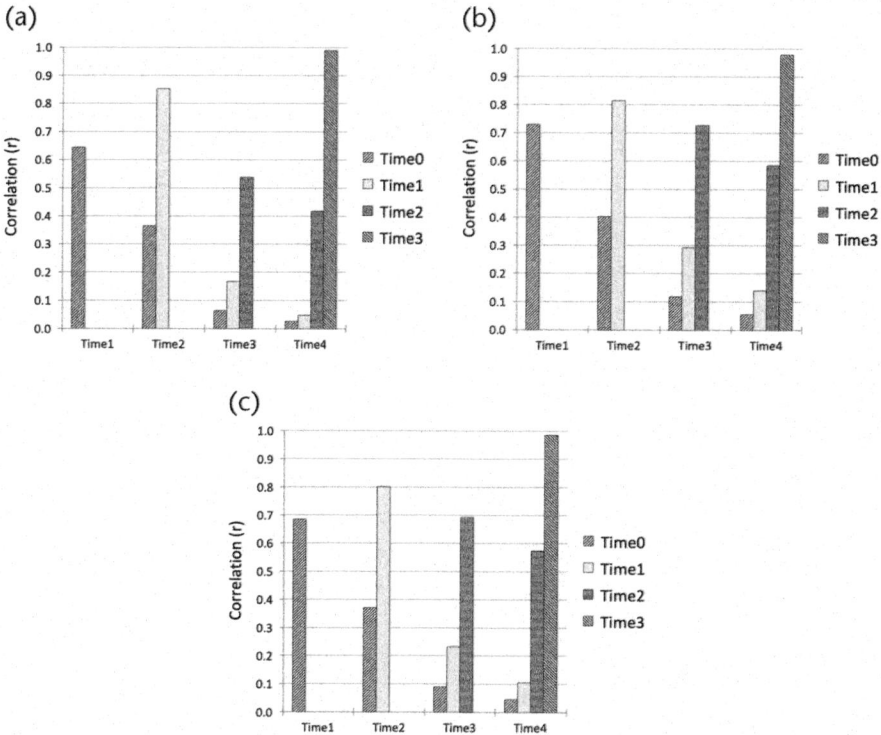

FIGURE 4.7   Examining the constancy of IQ in the 4-process, 8-process and 16-process architectures

for the 4-process (left), 8-process (middle) and 16-process (right) architectures. In each instance, the results show (1) correlations are highest between adjacent timepoints, (2) the strength of correlation reduces over greater time intervals, and (3) the lowest overall correlation is observed between furthest timepoints (Time0 – Time4). In comparing these results to those of Hindley and Owen (1978), one can see that while the differences between adjacent timepoints are more exaggerated in our models, the results simulate the pattern they reported.

Next, we factor analysed the data for each model, and at each timepoint, to determine how the factor structure might differ as a consequence of the functional architecture (i.e., the number of processes), and development. The results of these analyses are presented as scree plots, in Figure 4.8. Scree plots are useful for displaying the size and number of factors that explain variance in scores, within a larger set of observed variables. Here, the variance to be accounted for relates to performance scores of processes within each model. The scree plots in Figure 4.8 show as many potential factors (x-axes) as there are observed variables. Thus, the plots pertaining to the 4, 8 and 16-process models each show a total of 4, 8 and 16 potential factors, respectively. If scores for each process were unrelated then there would be no clear way to reduce the number of factors, and each observed variable

4-module · 8-module · 16-module

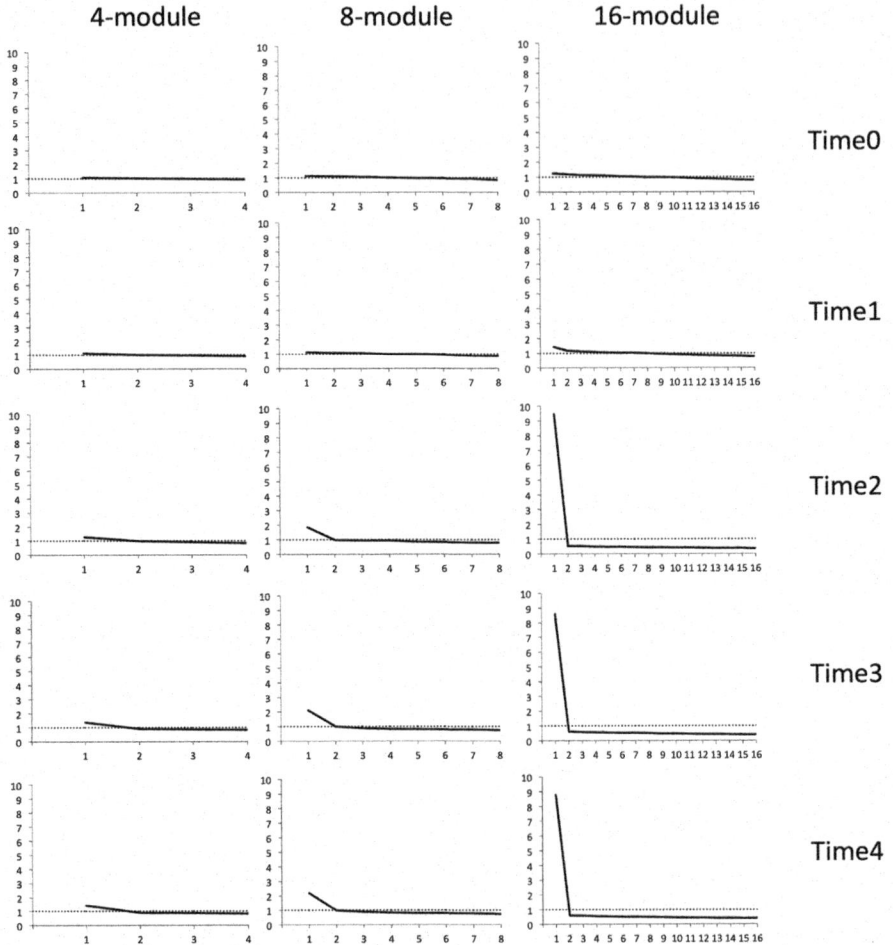

FIGURE 4.8 Table of scree plots showing eigenvalues for the 4-process (left column), 8-process (middle column) and 16-process (right column) architectures, over time (rows). The dotted line represents an eigenvalue of 1

would constitute a factor. The reduction of variables to fewer factors relies on calculations of the amount of variance across the observed variables that can be accounted for. These are represented as eigenvalues, and typically factors with eigenvalues greater than 1.00 explain more variance than a single observed variable. Thus, the critical feature of a scree plot is that the number of factors that may be deemed to underlie the larger set of observed variables, can be determined by the number of factors that appear in the plot up until the 'elbow' – the point on the figure where the size of eigenvalues drop below 1.00, and where they appear to plateau. From the scree plots in Figure 4.8, we can see that during early developmental timepoints (Time0 and Time1), and in each architecture, no single higher-order factor accounts for model performance. However, by Time2 (that is, midway through development) we find evidence of a single factor emerging. Although this is most clearly represented in the 16-process model, the same pattern can also be discerned in the figures relating to the 4 and 8-process models. Noteworthy is the fact that irrespective of timepoint, the size of the first factor is larger in those models with greater numbers of processes. In our simulations, this difference in size of the first factor is due to more limited variability in performance scores in models with fewer numbers of processes.

Discussion

So, what is to be made of the results from these models? The first point to note is that van der Maas and colleagues are right – one does not need to posit a single property of a cognitive system as the basis of *g*. The computational approach that we have taken here shows that a single higher-order g-factor can occur as an emergent property via interactions between processes within a developmental architecture. But in attempting to take the development of intelligence seriously, we have learned a few additional things. Firstly, without modifications to the parameters underlying these DST models, we can see that as a framework such DST models do not permit scaling beyond a number between 16–32 processes. Somewhere in that range, the models fail to reach asymptote – an outcome caused by massively interactive influences of each process on each other process. This of course makes such models unrealistic simulations of development. But it could also be important in that it might suggest an upper limit to the number of processes that are viable in a cognitive system. Secondly, for the models that we tested, and which were viable (4, 8, and 16-process), we found a single factor emerged midway during development (Time2 being roughly equivalent to 9 years of age). This is potentially important, as it mirrors empirical phenomenon showing that abilities are relatively uncorrelated early in development, but these become correlated during development and remain correlated in adulthood. Thirdly, and perhaps counter-intuitively, the more processes there are (perhaps up to the limiting factor discussed in point 1), then the greater is the g-factor. This difference between models in the size of the first eigenvalue may prove useful in isolating plausible architectures, based on what is known of the size of the g-factor in real populations.

The approach we have taken represents our first steps towards a research program using dynamical systems to explore questions of the development of *g*. But, we wonder what would Annette have to say about these results? Or, what she might make of our approach more generally? First, in order to focus on exploring the dynamics of change over time, our models necessarily had to abstract out details of development. That is, they did not specify the actual mechanisms responsible for age-related changes in abilities. Instead, the parameters for each process offered growth curves aimed at capturing the outcome of those processes but the processes themselves were not realised within the models. Thus, Annette would undoubtedly challenge us to expand our approach to include possible mechanisms of development within the component processes. Second, Annette did not rule out the possibility that modules, of a sort, might exist in the cognitive system. She argued that modules could emerge as a product of development, as parts of the cognitive system became functionally specialised. In our models, however, the functional architecture did not change. Within each model, processes were fully connected from the very first timepoint, and they remained fully connected throughout development. Thus, we think another criticism Annette would have of our approach is that it is still too static. More plausible would be a system that allowed for functional changes to the architecture – moving from a system characterised by higher levels of connectivity to one more specialised (modular) in its architecture. Lastly, we are sure Annette would take issue with the fact that within each of our models we assumed equal strength of connectivity between processes in the system. And, these too did not change over development. A more plausible alternative would be that the strength of connectivity would change over time, both as a product of externally driven (e.g., greater exposure to resources) and internally driven (e.g., increased levels of abilities within other connected processes) influences. Likewise, other parameters of the models, such as the rate of change each process exhibited, and the asymptote levels (a loose measure of resources) within a model were also fixed and did not change. We think Annette would insist that we examine what kinds of differences arise from a fuller program examining variability in each of these parameters on developmental outcomes. Thanks to Annette's continued influence on our thinking, this is something we are keen to pursue.

Notes

1 This was to allow for future work exploring scalability in these architectures.
2 In each model, these would be roughly analogous to ages 0 years, 4.5 years, 9 years, 13.5 years and 18 years old, respectively.

References

Allegrini, A. G., Cheesman, R., Rimfeld, K., Selzam, S., Pingault, J., Eley, T. C., & Plomin, R. (2019). The p factor: genetic analyses support a general dimension of psychopathology in childhood and adolescence. *Journal of Child Psychology and Psychiatry.* https://doi.org/10.1111/jcpp.13113

Anderson, M. (2007). Biology and intelligence-the race/IQ controversy. In *Tall tales about the mind and brain: separating fact from fiction* (p. 123). Oxford University Press.

Anderson, Mike. (1992). Intelligence and development: A cognitive theory. In *Cognitive Development.* Retrieved from http://search.ebscohost.com/login.aspx?direct=true&db=psyh&AN=1992–97999-000&site=ehost-live

Anderson, Mike. (2017). Binet's Error: Developmental change and individual differences in intelligence are related to different mechanisms. *Journal of Intelligence, 5*(24). https://doi.org/10.3390/jintelligence5020024

Baughman, F. D., Baughman, N., & Mills, S. A. (2012). Cognitive reserve and intelligence: Modulating the effects of damage in ageing dynamical systems. *Cognitive science, 1*(2), 1314–1319. Washington, USA.

Baughman, F. D., & Thomas, M. S. C. (2008). Specific impairments in cognitive development: A dynamical systems approach. *Cognitive Science*. In B. C. Love, K. McRae, & V. M. Sloutsky (Eds.), Proceedings of the 30th Annual Conference of the Cognitive Science Society (pp. 1819–1824). Austin, TX: Cognitive Science Society.

Baughman, F. D., Thomas, M. S. C., Anderson, M., & Reid, C. (2016). Common mechanisms in intelligence and development: A study of ability profiles in mental age-matched primary school children. *Intelligence, 56*. https://doi.org/10.1016/j.intell.2016.01.010

Binet, A., & Simon, T. (1905). The development of intelligence in children. *L'Année Psychologique, 12*, 191–244.

Caspi, A., Houts, R. M., Belsky, D. W., Goldman-Mellor, S. J., Harrington, H., Israel, S., & ... Moffitt, T. E. (2014). The p factor: One general psychopathology factor in the structure of psychiatric disorders? *Clinical Psychological Science, 2*(2), 119–137. https://doi.org/10.1177/2167702613497473

Davis, H., & Anderson, M. (2001). Developmental and individual differences in fluid intelligence: Evidence against the unidimensional hypothesis. *British Journal of Developmental Psychology, 19*(2), 181–206.

Deary, I. J., Whalley, L. J., Lemmon, H., Crawford, J. R., & Starr, J. M. (2000). The stability of individual differences in mental ability from childhood to old age: Follow-up of the 1932 Scottish mental survey. *Intelligence*. https://doi.org/10.1016/S0160–2896(99)00031-8

Elliot, C., Smith, P., & McCulloch, K. (1997). *The British Ability Scales II*. London: nferNelson.

Fodor, J. A. (1983). *The Modularity of Mind*. Cambridge, MA: MIT Press. Fodor, J.A (pp. 73–77). Retrieved from http://search.ebscohost.com/login.aspx?direct=true&db=psyh&AN=1988-98490-005&site=ehost-live

Hindley, C. B., & Owen, C. F. (1978). The extent of individual changes in IQ for ages between 6 months and 17 years, in a British longitudinal sample. *Journal of Child Psychology and Psychiatry, 19*(4), 329–350.

Jensen, A. R. (1998). *The G Factor: The Science of Mental Ability*. Westport: Praeger. Retrieved from http://search.ebscohost.com/login.aspx?direct=true&db=psyh&AN=1998-07257-000&site=ehost-live

Jensen, A. R. (2006). *Correlated Chronometric and Psychometric Variables*. Retrieved from http://www.sciencedirect.com/science/article/B84BV-4MT6KH7-P/2/9ff6324c7de6ceaf1a34eb89c6ffff10

Karmiloff-smith, A. (1998). *Neuroconstructivism, 3*(2), 99–102.

Karmiloff-Smith, A. (1995). *Beyond Modularity: A Developmental Perspective on Cognitive Science*. MIT Press.

Karmiloff-Smith, A. (2009). Nativism versus neuroconstructivism: Rethinking the study of developmental disorders. *Developmental Psychology, 45*(1), 56.

Ramsden, S., Richardson, F. M., Josse, G., Thomas, M. S. C., Ellis, C., Shakeshaft, C., & ... Price, C. J. (2011). Verbal and non-verbal intelligence changes in the teenage brain. *Nature, 479*(7371), 113–116. https://doi.org/10.1038/nature10514

Rushton, J. P., Jensen, A. R., Philippe Rushton, J., & Jensen, A. R. (2003). African-White IQ differences from Zimbabwe on the Wechsler Intelligence Scale for Children-Revised are mainly on the g factor. *Personality and Individual Differences, 34*(1), 177–183. Retrieved from http://www.sciencedirect.com/science/article/B6V9F-47425XM-B/2/d0f49af399271d761e67411f06e5eec3

Spearman, C. (1904). General intelligence, objectively determined and measured. *American Journal of Psychology, 15*(2), 201–293. Retrieved from http://psychclassics.yorku.ca/Spearman/

Thelen, E., & Smith, L. B. (2007). Dynamic Systems Theories. In *Handbook of Child Psychology* (pp. 258–312). https://doi.org/10.1002/9780470147658.chpsy0106

Thomas, M. S. C., Baughman, F. D., Karaminis, T., & Addyman, C. J. M. (2012). Modelling developmental disorders. In *Current Issues in Developmental Disorders*. https://doi.org/10.4324/9780203100288

Thomas, M. S. C., Purser, H. R. M., & van Herwegen, J. (2011). *The developmental trajectories approach to cognition*. Neurodevelopmental disorders across the lifespan: A Neuroconstructivist approach. Oxford: Oxford University Press.

van der Maas, H. L. J., Dolan, C. V., Grasman, R. P. P. P., Wicherts, J. M., Huizenga, H. M., Raijmakers, M. E. J., & … Raijmakers, M. E. J. (2006). A dynamical model of general intelligence: The positive manifold of intelligence by mutualism. *Psychological Review, 113*(4), 842–861. https://doi.org/10.1037/0033-295X.113.4.842

5

BEING A MENTOR

Kang Lee

Kang Lee grew up in China. He moved to the UK in 1989, initially as a visiting research fellow at the MRC Cognitive Development Unit under Dr. Annette Karmiloff-Smith's supervision. He then moved to Canada to obtain a Ph.D. degree and a faculty position at Queen's University at Kingston in 1994. Then, he returned to London to work with Annette as a postdoctoral fellow in 1996. He has been a full professor at the Erick Jackman Institute of Child Study, the University of Toronto and a senior scientist at the Department of Psychology, the University of California, San Diego since 2004. His research initially focused on notational development and subsequently on the development of lying. In this tribute, he describes his initial experiences of working with Annette, and particularly focuses on her role as a mentor.
(This chapter is based on the transcript of a presentation made in honour of Annette at her Festschrift)

Many of us work within universities. We have three primary responsibilities as professors: to do research, to teach and to mentor the next generation of scientists. To prepare us, most of the doctoral programs devote a lot of effort to teaching us how to do research, some effort to teaching us how to teach, and hardly any effort to teaching us how to mentor. It is my observation that most of us learn to mentor by observing and imitating our own mentors, similar to how we learn parenting from our own parents.

It's also my observation that there are at least three types of mentors: the first are those who care very little about their students, the second terrorize their students and the third treat their students with care, respect and unselfishly passing along their wisdom through either words or deeds. I was very lucky to have Annette as my mentor, who epitomizes the third type of mentors and much more.

Here I am going to share with you a few stories about how I was mentored by Annette and the lessons I learned from my experience as her student.

Annette accepted me to work with her in 1989 without knowing anything about me beyond what James Greeno wrote in his letter of recommendation. The year 1989 was when the Tiananmen Square massacre took place. Because I had some association with the student movement, the Chinese government was hesitant to allow me to leave the country. Luckily, because I had a letter from Annette, I was allowed to leave. Nevertheless, before I set out for London, I had to endure a few weeks of brainwashing sessions in Beijing. When I arrived in London, the brainwashing sessions continued in the Chinese embassy for one more week. By the end of that week, the embassy people told me that I could leave only if my mentor could come to claim me. Otherwise, I would need to continue the brainwashing sessions for one more week. So I immediately called Annette and told her exactly that. As soon as she heard about the situation, Annette immediately rushed over, and whisked me out of the Chinese embassy as fast as she could. That was our first introduction.

Though meeting me for the first time, Annette welcomed me into her home for months until I found my footing. Annette's house was the first place I ever lived in the West and it's the first place I felt the meaning of freedom. It was also at her house I met my future wife. So Annette changed my life. I once joked to Annette that without her, I would likely be brainless, wifeless and careerless.

So the lesson I learned from Annette here is:

Being a mentor is a solemn responsibility that we must take very seriously, because the seemingly routine act of accepting students into our labs may completely transform their lives.

At that time in London I was very poor. Each week after meals and rent, I only had 5 pounds left. So to save money I regularly skipped fares when taking the Tube. One of Annette's friends saw me doing this and reported it to Annette.

A few days later Annette called me to her office. She said, "Kang, I have this survey project with an Australian visiting scholar about the night life of gays and lesbians in London. I need an RA to do data entry. I know it's a low-level job but at least you'll get paid." I jumped at the job offer.

This job turned out to be a great experience. It opened my eyes to a world I did not know ever existed and realized all the fun the gays and lesbians in London were having. More importantly, my English vocabulary about sexual orientations and sexual positions increased exponentially. Sometimes, I couldn't understand the meaning of certain words the participants used in their responses. I would ask Annette and Mark, Annette's husband, what they meant. Even to this day I can still see the vivid image of Annette and Mark standing in their kitchen struggling not to laugh and patiently explaining to me what a particular sexual word really meant.

So the lesson I learned from Annette here is:

We should not be afraid of giving our students seemingly mundane jobs. You never know when they may turn out to be mind expanding learning experiences.

After I had settled in London, I went to see Annette about what research topic to study. She said she could supervise me on a topic she knew and cared about, and she gave me three questions that fit these categories. She told me not to decide right away, but rather to think them through carefully and choose the one I was most passionate about. She also said that I could choose whatever I was passionate about, except, for God's sake, not theory of mind! At the time I did not fully appreciate the implications of this choice. Only after years of experiencing a few ups and many downs did I realize it was this initial passion that had sustained me throughout the hard times and allowed me to pursue the answers to the same research questions for decades.

So the lesson I learned from Annette here is:

We must let our students choose a research topic they love because only this love will give them joy in doing research and allow them to endure many hardships ahead in their career.

I met with Annette regularly to discuss research ideas or papers. When I met with her she had a 5-minute rule. This rule was that I needed to give her an overview of a study in no more than 5 minutes. She said that if I needed more time, it meant that there was something wrong with the study or my understanding of the study was wrong. This was an incredible exercise for sharpening my mind and my communication skills. I have been using this rule ever since.

So the lesson I learned from Annette here is:

A good study should not take forever to explain.

Annette was a very passionate person. It's only natural that occasionally she became angry, in particular, if you failed to meet her expectations or showed disrespect to others. She was angry at me only once. It was when Daniel Ansari just joined the lab as a graduate student. Having met him only briefly, I called him Dan. Annette pulled me aside and told me: "You cannot change people's names without their permission. It is a sign of lack of respect." This incident made a very strong impression on me. Since that day, whenever I meet a David, an Elizabeth, a Michael, and of course, a Daniel, I would not call them Dave, Liz, Mike, and Dan unless I receive permission to do so.

The lesson I learned from Annette here is:

The act of respecting your fellow lab mates can be as small as calling their names properly.

Another time I saw Annette become angry was when Child Development took 6 months to review and reject outright one of our papers. We both thought the reviewers were wrong and the action editor made a terrible decision to reject us. So we sat down and wrote a long and angry letter. In the concluding sentence we said we would never submit our papers to Child Development ever again! After we finished the letter, Annette said, "let's come back tomorrow, polish up the letter, and send it." The next day Annette called me to her office and said: "Since we already expressed our emotion yesterday, let's not send the letter." That was such a wise decision because we have published quite a few papers in Child Development ever since.

So the lesson I learned from Annette here is:

Sometimes the best strategy to go forward is holding back.

Annette and I worked on development of notations for a while, but we had a difficult time getting our papers accepted into top-tier journals and there was simply no traction. Nobody seemed to care about the topic or our papers (Lee, Karmiloff-Smith, Cameron, & Dodsworth, 1998; Lee & Karmiloff-Smith, 1996a, 1996b). So Annette suggested that we should abandon this line of research and move on. I am not quite sure whether that was the time when Annette moved on to study neurodevelopment and Williams syndrome. I moved on to study the development of lying, which turned out to be career changing for me.

So the lesson I learned from Annette here is:

Sometimes the best strategy to go forward is to cut your losses and start over.

I have been studying how children learn to tell lies for more than 20 years. We discovered that lying is a common and universal phenomenon in childhood and beyond. Children begin to tell lies at two and half years of age. But most 2-year-olds are honest. Only about one-third will lie. At 3 years of age, 50% lie. However, after their fourth birthday, most children lie. We searched for decades to identify the key factors that might predict young children's tendency to lie, with little success. We found that children's moral understanding of lying and honesty, IQ, and temperament do not predict whether they will lie. We also found no relations between children's tendency to lie and parenting styles, parental religiosity, and socioeconomic status. Many researchers had suggested me to look at theory of mind. Initially, with Annette's advice in mind, I resisted this suggestion. After more than a decade's failure to find any significant predictors of lying in young children, I relented and decided not to heed Annette's advice. We did a study around 2006 (Talwar & Lee, 2008). We found theory of mind to be uniquely associated with children's lying. More specifically, those children with better theory-of-mind understanding tell lies earlier, and lie better, than those with poorer theory-of-mind understanding (Ding, Wellman, Wang, Fu, & Lee, 2015; Ding, Heyman, Fu, Zhu, & Lee, 2017; Ding et al., 2018; Lee, 2013). This finding has been consistently replicated all over the world. In fact, the crucial role of theory of mind turns out to be an important discovery about the development of lying in children.

So the lesson I learned from Annette here is:

On occasion, you can deviate from your mentor's advice.

One of the qualities I admired about Annette the most was her insatiable desire to learn and the unrelenting ways to learn whatever she wanted to learn. When she published her first Science paper, I asked her how she did it. She said: I just read all the relevant psychology papers published in Science and learned how to write in a way that the Science editors will like. When Annette's Williams syndrome work began to make a major impact on the field, I ask her how she managed to move into a previously unfamiliar field and make a major impact on it. She said: I spent about 2 years reading all the relevant textbooks and papers on genetics, genetic disorders and on Williams syndrome.

I was so inspired by her that about 5 years ago, I began to explore whether I also have what it takes to move into an unfamiliar field and make an impact. I decided to study affective artificial intelligence. Let me share with you some very preliminary results.

The artificial intelligence field in the last few years has produced many highly intellectually smart systems, such as Watson and Siri. However, they have very limited emotional intelligence. I think we psychologists can help change that. Modelling after Annette, I went ahead and read many papers in signal processing, optics, human physiology, and machine learning. We invented a new imaging method that we believe can change how affective artificial intelligence is done. This method is called Transdermal Optical Imaging (Barszczyk and Lee, 2019; Luo et al., 2019; Wei et al., 2018; Liu, Luo, Zheng, Wu, & Lee, 2018). This technology uses a conventional digital camera that is now ubiquitous in smart-phones to extract facial blood flows from the video images. Using advanced machine learning and signal processing, we are now able to measure, with high accuracy and reliability, various important health indexes such as heart rate, breathing, blood pressure, vascular stiffness and even body mass index. We are also able to measure stress, emotional valences, and even emotions when the facial expression is neutral. At this point, our technology is still in its infancy. To improve it for actual applications in the real world, much is still needed to be learned. I feel so lucky to have Annette as a role model to inspire me.

So the lesson I learned from Annette here is:

To inspire students to learn new things, you need to be an inspired life-long learner yourself.

I want to end with another story about Annette to yet again illustrate what kind of role model she was. Sometime ago, I went to London to celebrate her 70th birthday and stayed with her and Mark. After a long day's work at the lab, we went back to their apartment. Annette said: "Let's go out for dinner tonight. But I need to work a little more." I assumed the dinner time would be around 7 and wrapped up my own work around that time. But Annette kept working until around 10 when we finally went out for dinner. The next morning, when I got up at 7, I discovered that Annette had already exercised and worked for two hours! I had never met anyone who worked this hard. It made me realize how much she loved what she was doing.

This is another, and perhaps the most important lesson I have learned from Annette:

When you love what you do, it is just so much fun to work hard.

It also reminded me that it is largely due to Annette's mentorship that I have a career doing work that I love. Thank you, Annette, for the wisdom, friendship and inspiration that you have generously given to me for nearly three decades. I miss you.

References

Barszczyk, A., & Lee, K. (2019). Measuring blood pressure: From cuff to smartphone. *Current Hypertension Reports*, *21*(11), 84.

Ding, X. P., Wellman, H. M., Wang, Y., Fu, G., & Lee, K. (2015). Theory of mind training causes honest young children to lie. *Psychological Science*, *26*, 1812–1821. doi: 10.1177/0956797615604628.

Ding, X. P., Heyman, G. D., Fu, G., Zhu, B., & Lee, K. (2018). Young children discover how to deceive in 10 days: A microgenetic study. *Developmental Science*, *21*(3), e12566. DOI: 10.1111/desc.12566.

Ding, X. P., Heyman, G. D., Sai, L., Yuan, F., Winkielman, P., Fu, G., & Lee, K. (2018). Learning to deceive has cognitive benefits. *Journal of Experimental Child Psychology*, *176*, 26–38. Doi: 10.1016/j.jecp.2018.07.008

Lee, K., Karmiloff-Smith, A., Cameron, C. A., & Dodsworth, P. (1998). Notational adaptation in children. *Canadian Journal of Behavioural Science/Revue canadienne des Sciences du comportement*, *30*, 159–171. doi: 10.1037/h0087059

Lee, K., & Karmiloff-Smith, A. (1996a). The development of cognitive constraints on notations. *Archieves de Psychologie*, *64*, 3–26. Retrieved from http://psycnet.apa.org/index.cfm?fa=search.displayRecord&uid=1996-03503-001

Lee, K., & Karmiloff-Smith, A. (1996b). Notational development: The use of symbols. In E. C. Carterette & M. P. Friedman (Eds.), *Handbook of perception*, Vol. 13. R. Gelman & T. Au, (Volume Eds.), *Perceptual and cognitive development* (pp.185–211). New York: Academic Press.

Lee, K. (2013). Little liars: Development of verbal deception in children. *Child Development Perspectives*, *7*, 91–96. doi: 10.111/cdep.12023

Liu, J., Luo, H., Zheng, P., Wu, S. J., & Lee, K. (2018). Transdermal optical imaging revealed different spatiotemporal patterns of facial cardiovascular activities. *Nature Scientific Reports* 8:10588 doi:10.1038/s41598-018-28804-0

Luo, H., Yang, D., Barszczyk, A., Vempala, N., Wei, J., Wu, S. J., Zheng, P. P., Fu, G., Lee, K., & Feng, Z. (2019). Smartphone-based blood pressure measurement using transdermal optical imaging technology. *Circulation: Cardiovascular Imaging*. https://doi.org/10.1161/CIRCIMAGING.119.008857

Talwar, V., & Lee, K. (2008). Social and cognitive correlates of children's lying behavior. *Child Development, 79,* 866–881. doi: 10.1111/j.14678624.2008.01164.x

Wei, J., Luo, H., Wu, S. J., Zheng, P. P., Fu, G., & Lee, K. (2018). Transdermal Optical Imaging reveal basal stress via heart rate variability analysis: A novel methodology comparable to electrocardiography. *Frontiers in Psychology.* https://doi.org/10.3389/fpsyg.2018.00098

6

BIOLOGICAL EVOLUTION'S USE OF REPRESENTATIONAL REDESCRIPTION

Aaron Sloman

Introduction

Annette Karmiloff-Smith and I nearly became close colleagues when she was offered a chair at Sussex University. I wonder how different this tribute would have been if we had interacted more closely over a longer period. I'll try to summarise some of the similarities and differences between our interests and our proposed explanatory theories. We were both deeply influenced by Jean Piaget, but in different ways. In particular some of Piaget's thinking, like much of my work, was closely related to Immanuel Kant's ideas about the nature of mathematical discovery, an interest Annette did not share, though I'll try to show that some of her work on child development was relevant to mathematical cognition. We both drew attention to largely unnoticed complexities of gene expression – but they were different complexities. There are implications for psychology, neuroscience, philosophy, artificial intelligence (AI) and theories of biological evolution.

After she moved to London, we met a few times at workshops and conferences, including an occasion when she invited me to London, to talk about representation. I think we may have discussed some of the ideas in (Karmiloff-Smith, 1990). We also both contributed to a book on "Forms of Representation" (Peterson, 1996), and in 2007, she was one of the guest speakers at a cognitive robotics project meeting in Paris in 2007, which I helped to organise.[1] But our interactions were rare and limited. We last met when I gave a departmental colloquium at Birkbeck in 2012 on "Toddler Theorems, Representational Redescription and Meta-Morphogenesis", and we talked after the seminar. She seemed to be disappointed that although I specified some of what needed to be explained, criticising over-simple alternative views, I was not able to demonstrate or report on a working AI system with the right explanatory powers. I think she had hoped that I would demonstrate or at least describe a neural net based solution to the problem.

Despite the overlap in our ideas about cognitive development, described below, we had different centres of interest, and although I found her ideas and theories relevant to what I was trying to do, I do not think the interest was reciprocated, partly, I suspect, because I did not share her confidence in neural networks. We also had different but overlapping views of the relationship between the genome and development, a difference that I think I understand better as a result of writing this paper. In particular, I think Annette viewed epigenesis as a process of richly interacting branching processes, whereas the meta-configured genome theory, explained later, treats the genome as a multi-layered structure, in which some of the more recently evolved higher-level layers are expressed later, using information acquired from interactions between the earlier layers and the environment as parameters. This form of epigenesis allows dramatic differences between results of gene expression in different environments, which could not result from any known type of uniform learning process. In particular, it allows a genome to specify discrete (parametrised) developmental stages, possibly related to Annette's examples of "representational redescription" mentioned below.

The rest of this chapter provides some (Kantian) background, then attempts to present some ideas about gene expression (the meta-configured genome) that seem to be capable of explaining Kant's observations, e.g., about mathematical discovery, which I'll relate briefly to Piaget's ideas, Annette's theories of re-presentational redescription and more generally to current theories in psychology, neuroscience and AI/Computational cognitive science.

Background: Immanuel Kant on mathematical knowledge

A major theme in my research is trying to explain and extend what was correct in Kant's philosophy of mathematics in (1781). Kant argued that mathematical knowledge, including knowledge of Euclidean geometry and elementary ar-ithmetic, had features that distinguish it from both of the two kinds of knowledge identified by Hume, namely *analytic* knowledge ("relations of ideas"), based only on definitions of terms plus purely logical reasoning, and *empirical* knowledge ("matters of fact and real existence"), based on observation and experiment. Kant pointed out that besides being non-analytic (synthetic) and non-empirical (*apriori*), mathematical knowledge also had a *modal* feature: it was concerned with what is *possible* (e.g., polygons with N sides are possible for any N greater than 2), or *impossible* (e.g., two straight lines cannot completely enclose a finite portion of a plane) or *necessarily* the case: for example, angles of a planar triangle necessarily sum to half a rotation, i.e., 180 degrees[2]. Many such mathematical discoveries, some too trivial to be included in formal courses on mathematics, are unwittingly made and used in everyday intelligent actions by young children, without recognising their nature. I call them "toddler theorems".[3]

My doctoral thesis (Sloman, 1962) defended Kant against 20th Century claims that his theories had been refuted by the success of Einstein's General Theory of Relativity, which treated physical space as non-Euclidean. However, my defence of Kant did not propose *mechanisms* explaining ancient human mathematical abilities in geometry, topology and arithmetic. Later, after being introduced to artificial intelligence by Max Clowes, in 1969, and learning to program, I hoped to substantiate Kant's claims by building working models of the relevant ancient geometrical discovery processes. Since mathematical competences of a 6-year-old child are clearly not present at birth we need to find mechanisms that can produce mathematical knowledge of the kind Kant discussed, that is not available from birth and does not merely express generalisations from experience. I hoped to show how to design a "baby" robot that could learn about structures, processes and causation in something like the way a human child seems to, including going through processes of empirical discovery in various kinds of play and exploration, and then somehow coming to grasp deep non-empirical necessary truths, often without realising that that has happened.

Jean Piaget, Annette's early mentor, had read Kant and Frege and by 1952 had made interesting discoveries about the development of mathematical cognition, including, for example, evidence suggesting that before the fifth year children could not grasp one of the facts that are crucial to understanding the natural numbers, namely that one-to-one correspondence is necessarily a transitive relation (a fact that is often distorted using the label "conservation"). However, Piaget was unable to propose mechanisms capable of explaining such mathematical insights.

He tried to find ways of modelling cognitive transitions using ideas from Boolean algebra and group theory, but they were not really relevant, and he apparently did not encounter computer programming and artificial intelligence until near the end of his life, when it was too late for him. If he had learnt to use a collection of important programming techniques half a century before they were developed in AI and software engineering, the history of developmental psychology (along with cognitive science and neuroscience) might have been very different. He commented on that in a talk he gave (in his wheelchair) at a conference in Geneva on Genetic Epistemology and AI in 1980, to which I had been invited.

Re-representing the precocial-altricial distinction

Informal observation of young children, squirrels and magpies in our garden, and reports and video recordings of other animals led me, during the 1990s, to conjecture that Kantian mathematical reasoning competences were related to the distinction between species with adequate competences available soon after birth or hatching (e.g., chicks able to peck for food and follow a hen, or deer able to walk to the mother's nipple shortly after birth) and species that started off helpless but later achieved deeper and richer forms of intelligence, including humans, hunting mammals, squirrels, elephants and some birds. In the early 1990s, I learnt that I was talking about the precocial/altricial distinction, long known to ethologists! Around that time, I also heard Manfred Spitzer give a talk emphasising the importance of delayed development of frontal lobes in humans – suggesting that evolution had discovered that development of some parts of brains in altricial species need to be delayed until they have access to enough information provided by older parts of brains interacting with the environment. That clearly had the benefit of allowing genetic mechanisms producing sophisticated competences to tailor their products to features of the environment, allowing such species to live and learn in environments with different kinds of complexity, as proposed in the meta-configured genome theory, below.

My thinking was disrupted when Jackie Chappell joined the school of Biosciences in Birmingham in 2004. She was one of the authors of a headline-grabbing paper (Weir, Chappell, & Kacelnik, 2002) reporting on Betty, a New Caledonian crow who repeatedly made hooks from straight pieces of wire, and used them to extract a bucket of food from a vertical glass tube. I later discovered, from the online videos at the Oxford ecology lab, that in 10 trials Betty had made usable hooks in at least four significantly different ways, suggesting that something more was going on than merely acquiring food.

Jackie had earlier studied some of Betty's other forms of problem-solving behaviours, all involving spatial reasoning. Could Betty's desire to explore alternative solutions to an already solved problem be related to some of the phenomena Annette found in children's drawings? Certainly, in both cases fairly deep knowledge of alternative possible spatial arrangements and rearrangements of objects was used. But there are many unanswered questions. How do brains represent sets of possible alternatives? How do they identify some combinations as impossible (e.g., *A inside B, B inside C, and C inside A*)? Recognising and eliminating impossibilities from a search for solutions is an important aspect of intelligence – saving enormous amounts of wasted effort (unlike a wasp repeatedly trying to get through a pane of glass).

In response to my vague conjecture about altricial species, Jackie suggested that the basic distinction was not between *species*, or even individuals, but between *competences*. For example, humans are labelled an altricial species because newborn infants are helpless and incompetent. Nevertheless, they are born with very important competences, including the ability to obtain nourishment by sucking and swallowing, which requires complex coordination of a collection of muscles – a precocial competence. So we recommended replacing the biologists' distinction between altricial and precocial *species* with a distinction between altricial and precocial *competences* (Sloman & Chappell, 2005, Sloman & Chappell, 2007, Chappell & Sloman, 2007b). Later we changed our terminology and labelled the competences "pre-configured" and "meta-configured".

Our claim was that some genetic specifications are *parametrised*, with information gaps that can be filled (given parameters) during gene expression. Obvious examples include genetically specified physiological structures, such as bones or muscles, that change in size, weight, strength and other features during individual development, in coordination with changes in associated mechanisms, including neural control mechanisms. For example, larger bones will usually require larger, stronger muscles. We extended this kind of variability from genetically specified *physical* structures to genetically specified *competences*, including sucking, chewing, swallowing, grasping and various forms of sensory information processing.

Preconfigured competences are genetically specified, and any changing parameters they need are either intrinsically generated or continuously derived from the changing developmental environment, including other parts of the developing organism. I think that description fits many of the cases of neural development described by Annette, e.g., in Karmiloff-Smith (2006), where multiple processes of development occur in parallel, with mutual influences. Later capabilities are results of such parallel, interacting developmental processes, in which information from the environment can play an important role, e.g., determining which language is spoken, in what accent, and various other features.

In contrast, *meta-configured* competences result from abstract, parametrised, *genetic* specifications that may be activated at different stages of development, using parameters that depend on information acquired during earlier gene expression. For example, if some gene expression produces behaviours whose results are recorded and stored, and a later phase of gene expression makes significant use of those recorded results, then later genes whose expression uses the previously acquired information are meta-configured, in our sense. Information acquired at an earlier stage is recorded for use as parameters for meta-configured competences developed later. For example, development of spoken words can use previously acquired phoneme production competences, partly derived from surrounding speakers. The differences between the two accounts of development can be understood by comparing the Meta-configured genome idea with the ideas expressed in Karmiloff-Smith (2006), which states: "...modules could be the result of ontogenesis over developmental time, not its starting point".

In contrast, competences associated with meta-configured genes are essentially partly abstract *meta-competences* specified in the genome, though not expressed in the earliest phases of development. Instead, the abstractions are instantiated at appropriate later stages, using as parameters information previously gained through use of *pre-configured competences* developed earlier, or using results of earlier phases of meta-configured gene expression. Human language development is an example of such multi-layered meta-configured gene expression, including spectacular cases of later developments substantially re-organising the information gained in earlier phases of language development, including coping with exceptions to previously learnt syntactic regularities. All this allows genetically specified meta-competences, competence- *patterns*, to be instantiated in widely varying ways in different individuals, in different physical, social, linguistic, or cultural contexts, as illustrated by historical changes in children's playthings across centuries. A spectacular example is the recent development of new ontologies and extended concepts of causation required for use of wirelessly connected electronic devices, now taken for granted in many cultures.

The key idea is that meta-configured competence-patterns are instantiated using (not necessarily numeric) *parameters*, that are not specified by the genome but instead extracted from results of earlier interactions with the environment during development of simpler competences while the individual is younger and, in many cases, less physically mature. Simpler forms of parameter-based genomic flexibility (e.g., required for control of body parts changing in size, weight and strength, and possibly other features) may have been an evolutionary precursor of meta-configured genomes that take account of far more variability than that produced simply by individual growth and development in a standard environment.

Meta-configured genome expression

A meta-configured genome can produce meta-configured competences that are parametrised by information acquired using results of earlier gene expression influenced by the environment.

The diagram in Figure 6.1 crudely summarises processes associated with a meta-configured genome. (A video explaining how to read the diagram is available at https://www.youtube.com/watch?v=G8jNdBCAxVQ.) Downward arrows in-dicate multiple processes of gene expression, starting with early, relatively direct,

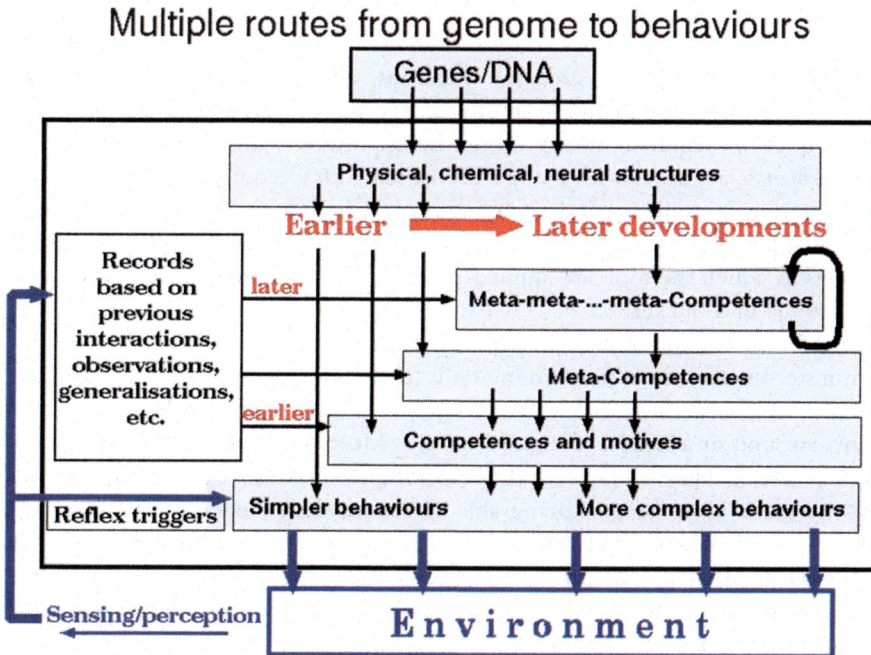

FIGURE 6.1   The meta-configured Genome

gene expression (down arrows to the left) and later gene expression (down arrows to the right), where later genes have parameters/gaps that are filled using information derived (in increasingly complex ways) from records produced by interactions with the environment during earlier gene expression (represented by down arrows more to the left). The diagram does not indicate the important fact that some of the records of gene expression will be in the environment rather than in the brain. Because of that, gene expression processes in different individuals can be mutually enhancing.

In general, there is no requirement for genetically specified mechanisms or behaviours to be produced at birth or soon after: the relevant genes may be expressed at a much later stage. Educational systems that ignore this are likely to be highly sub-optimal. A teaching strategy that produces high scores on a particular test could seriously interfere with later developments that build on unnoticed side effects of earlier teaching and learning processes. For instance, emphasising phonics-based early reading may produce higher scores in reading-aloud tests, while interfering with development of deeper forms of text comprehension and creative thinking, whose effects cannot be measured until much later.

Human language development provides spectacular examples of epigenetic variation based on a common genome. Several thousand different languages, including signed languages, have emerged, produced by communities sharing roughly the same language creation genetic mechanisms. The common human genome machinery enables production of thousands of different languages without specifying any of them! Similar comments apply to products of human technology at different times and locations. Even in some non-human species, the variety of structured percepts, intentions, intentional behaviours, information acquired (e.g., about locations of various resources and dangers, and techniques for achieving goals), would be impossible without the use of structured *internal* languages with compositional semantics, required, for example, for representing the unique, and changing, configuration of 3D obstacles, routes, potential perches, types of tool, control of body parts manipulating tools and the objects to which the tools are applied. Betty first used a newly made hook to lift the handle of a bucket.

A hive or colony of insects without this sort of meta-configured intelligence can instead make use of pheromone trails left by foraging individuals, but that is not available to animals that do not cooperate in such large numbers and which move around on complex extended terrain. Moreover, chemical trails may not survive strong winds and rain, and can only be used transiently by flying animals. So there are advantages in being able to reduce dependence on continuously available external cues and instead use enduring *internal* brain structures storing information about those surfaces. There are also enormous hidden costs in the complexity of mechanisms required for acquiring, transforming, combining, storing and using such information. That is one reason why brains are so complex.

Use of internal languages

All of this suggests that use of *internal* structured languages with compositional semantics evolved long before the use of language for communication (Sloman, 1978b). It is hard to explain some of the abilities of pre-verbal humans and other intelligent animals without postulating mechanisms for creating and manipulating internal information structures with compositional semantics, including structures that change both when perceived structures and relationships in the environment change, and also during consideration of *possible* actions or events and their consequences.

If such internal language use evolved first, then genetically specified mechanisms that originally made rich *internal* languages possible in a variety of species might later have been copied, modified and used to provide *external* languages. After some mechanism is copied for a new use, the new version can undergo changes that produce entirely new features. I suggest that originally communication between humans used a form of *sign* language, growing out of movements involved in cooperative actions (e.g., indicating direction of movement when cooperatively lifting and moving a heavy object), later extended to include *sounds* when hands were occupied or for communication to someone out of sight.

Such internal languages could support causal reasoning based on understanding of structural relationships rather than mere observed correlations – another product of a meta-configured genome – for example, understanding why a meshed pair of toothed wheels *must* rotate in opposite directions. As Piaget understood, this contradicts the widely shared (Humean) hypothesis that all causal reasoning is based on empirically learned generalisations, weighted by probabilities derived from recorded frequencies. Creative causal problem solving by crows, squirrels, pre-verbal humans and other intelligent animals renders statistical inference unnecessary for understanding some causal relationships.

After reading *Beyond Modularity* (Karmiloff-Smith, 1992), I felt that its claims about "Representational Redescription" were closely related to our work on the meta-configured genome, and also some ideas about how results of exploration and experiment in a rich environment could feed into mathematical discoveries about the nature of the environment, illustrating Kant's claims in (1781) that the mathematical knowledge of necessary truths and impossibilities is (1) synthetic (not based on logical deductions from definitions), (2) *a priori* (non-empirical – not *derived from* nor *justified by* but *awakened by* experience) and (3) involves *necessary truths* and *impossibilities*, rather than degrees of probability.

Piaget on possibility and necessity

Annette did not share Piaget's interest in the Kantian questions that drove my research, including how humans could discover and make use of necessary connections and impossibilities, the topics of his last two (posthumously published) books (Piaget, 1981; Piaget, 1983). Annette had not read them, but it was she who informed me of their existence in 2010, after I asked her, by email, if she knew of any developmental or neural research on recognition of necessity or impossibility. She wrote back: *"Piaget devoted a whole year of the Centre d'Epistemologie Genetique to the child's becoming aware of necessity and not just empirical likelihood – 'it has to be so' – if you do a google there should be a symposium book on necessity"*. I followed her advice and acquired the two books arising out of that work, one on possibility and one on necessity. I found the experimental data fascinating but Piaget's explanatory theories too obscure and imprecise – probably because he had not learnt to produce computational explanations. Piaget and his collaborators in that research could not have been aware of Annette's work reported some time later, on children exploring possibilities by modifying drawings, and she seems to have been unaware of his work on the possibility in which children were asked to specify new possibilities by modifying configurations of objects rather than by producing new drawings. I will not try to speculate about how each might have viewed the relationships.

My own work in AI was originally triggered by being challenged in 1971 to write a critical response to the defence of logic-based AI in (McCarthy & Hayes, 1969). On that approach, all complexity of representation comes from the application of functions to arguments, as in logic and algebra, which I labelled use of "Fregean" representations, contrasted with "analogical representations", where properties and relations represent (not necessarily identical) properties and relations, as in 2D projections of 3D scenes (Sloman, 1971). Analogical forms of representation play important roles in some of Annette's examples of "representational redescription", but her interest in those examples was different from mine. Perhaps our deepest disagreement was about mechanisms: I seek any AI techniques or mechanisms capable of explaining ancient mathematical discoveries and spatial reasoning in pre-verbal children, whereas she was particularly interested in general features of neural development, and how they contradicted some popular beliefs about how genes work, illustrated, for example in Williams syndrome. As a result, she seemed to me to be interested only in neurally inspired AI. I do not know whether she ever considered the possibility that much neural computation uses changing sub-neural chemical structures, with a mixture of continuous and discrete interactions rather than being restricted to changes in signal strengths and synaptic weights. Her emphasis on roles of brain chemistry in brain development might eventually have led her in that direction.

Overlaps

Annette and I shared some deep partly overlapping interests concerning what needs to be explained. In particular, we have both thought about *competence domains* but in different ways. I originally encountered this idea with the label "microworld" used by AI researchers, including Minsky and Papert (1971), and I used it in teaching programming as well as characterising research problems (Sloman, 1984). Later, I switched to using "domain" partly because of the overlap with ideas in *Beyond Modularity*, and partly because "micro-" suggests a restriction of scope. However, much of my thinking, unlike Annette's, has been concerned with trying to understand Kantian aspects of development of mathematical competences: how can learners acquire and use knowledge of non-definitional necessary truths and impossibilities, rather than having to rely only on empirical generalisations, where nothing is known unless based on observation and experiment.

Mathematical domains are implicit in some of Annette's work, for example research on *balancing* tasks discussed in (Karmiloff-Smith & Inhelder, 1974–1975). But as far as I know, like most developmental psychologists studying number cognition or spatial cognition, apart from Piaget, she never had the Kantian goal of trying to understand reasoning about necessary truths and impossibilities.

We were both interested in relationships between genetic mechanisms and development, but we disagreed about likely mechanisms insofar as she hoped that neural net models could explain all the phenomena (when we last talked in 2012), whereas I think that probabilistic mechanisms cannot explain mathematical understanding involving necessity and impossibility. Those are not extreme points on a probability scale. Despite such differences, I learnt a great deal from her work; I also found some of her recorded lectures on the complexity and diversity of developmental processes after fertilization inspiring, for example her 2011 talk at the National Academy of Sciences Sackler Colloquium on "Biological Embedding of Early Social Adversity: From Fruit Flies to Kindergartners".

Representational redescription

Annette's work on "Representational Redescription" turned out to have interesting overlaps with some of my work on linguistic and mathematical development, including abilities to discover necessary connections and impossibilities, and the ideas concerning the meta-configured genome, mentioned above. However, as far as I know, Annette did not explicitly emphasise, as Jackie Chappell and I did, that new types of learning competence can arise out of the interplay between the "meta-configured" genetic mechanisms expressed relatively late in development, and what has previously been learnt. For example, a 9-year-old is capable of learning things a toddler cannot, and some of that may be dependent on having learnt ways of learning that are relevant to the current environment. Another example would be 21st Century 5-year-olds learning how to use a computational system with a graphical interface, which none of their ancestors ever had a chance to do.

Such examples illustrate the fact that cultural evolution can feed information into the products of biological evolution in ways that affect the learning and developmental competences produced by the genome, but without altering the genome. This depends on the later-developing learning mechanisms produced by evolution not being fully specified. In effect, they are parametrised and the parameters can be filled by products of earlier learning. This seems to be consistent with the general thesis of *Beyond Modularity*. The discussion of language learning in *Beyond Modularity*, which involves differences between cultures and between learning a spoken language and a sign language, implicitly makes the point. Such learning depends on the parameter-based meta-configured genome, which allows parameters acquired at an earlier stage of development to feed into a template "turned on" by the genome at a later stage, as illustrated by the meta-configured genome diagram above. For example, a template relevant to grammatical structures may obtain inputs acquired from spoken, written or sign languages.

I do not know whether any of Annette's ideas changed after we last met or whether she would have been interested in my attempts to fill gaps in current explanations of ancient mathematical competences. I know of nobody with good explanations of how ancient mathematical brains made possible the amazing discoveries of Archimedes, Euclid, Zeno and many others. I suspect they used mechanisms that had previously enabled more general discoveries about spatial impossibilities and necessities made much earlier, using mechanisms shared with many other species with deep spatial intelligence, though without human reflective (meta-cognitive) abilities.

Neither current symbolic AI, nor artificial neural nets, nor known brain mechanisms seem to have the capabilities required to explain mathematical discoveries of the sorts described by Kant, including representing "alethic" modalities, such as *impossibility* and *necessity*. These have nothing to do with probabilities or statistical evidence. Moreover, the ancient examples are not cases of logical/definitional impossibility or necessity, but involve discovery of necessary features of spatial structures, such as transitivity of containment, or the impossibility of intersection of two convex 2D or 3D shapes producing a non-convex shape. At the time of writing I do not know of any good account of the workings of ancient mathematical minds, or even the spatial reasoning in squirrels, weaver birds, elephants, octopuses, apes, pre-verbal toddlers and other animals that are good (but not perfect) at spatial reasoning. Neither are there working robots with those abilities. I do not think these abilities can be innate (neither did Kant). Rather their discovery involves something like what Annette called representational redescription, and Kant called synthetic necessary truths discovered nonempirically: not derived from experience but "awakened" by experience.

Much work in artificial intelligence attempts to design machines that start with very little world knowledge but include a powerful learning mechanism that enables such machines, possibly helped by teachers who set challenges and indicate whether responses to questions and challenges are correct or not. The hope of such research is that over time the learning mechanism will discover statistical relationships between features in its sensory and motor records that allow discoveries to be made concerning how to predict what is likely to follow some collection of actions and sensed features, and on that basis will reliably allow intelligent selection of actions to achieve desired results. Like Kant, I do not believe that such statistical competences can explain mathematical discoveries in geometry and topology – because noticing statistical regularities cannot lead to the ancient mathematical discoveries about *necessary* relationships. Those discoveries require special mechanisms. Perhaps they grew out of mechanisms that evolved because they were powerful tools for spatial reasoning, like the use of diagrams (rather than logical formulae) in constructing geometric proofs, since ancient times?

Numbers

I think much of what psychologists and neuroscientists have written about development of number competences is wrong, or at least seriously incomplete, because they are based on an incorrect analysis of what numbers (of various kinds) are. That is not surprising – philosophy of mathematics is a complex subject and is not normally taught in those disciplines.

One use of numbers by scientists and engineers is in measurements of length, area, volume, weight, force and a host of other properties of physical objects, such as tension, torsion, momentum and energy. Some animals have behaviours that seem to include very precise numerical measures, for example spider monkeys swinging on, bouncing on, and leaping between structures, including performing several such actions in quick succession, implying both very rapid and accurate measurements and very rapid and accurate control of muscular tension and compression, producing forces that accurately launch them through space onto a rigid or flexible target object. Presumably, a combination of an extended period of evolution, combined with learning processes during development, produces a tightly integrated collection of "compiled" skills, but with very little reflective understanding of what those skills are and how they interact.

Such competences, including all the acquired measures, may be available for immediate use during performance of actions (e.g., in spider monkeys), without ever becoming objects of reflective thought, for instance thoughts about relationships between the elasticity of a launching support, the distance to a target branch, the muscular forces needed to produce an accurate launch and many other cases. In contrast, a human with vastly inferior versions of those skills (like me) can notice them, think about them, and ask questions about them without being able to acquire and use them (trapeze artists are rare exceptions.) That suggests that the brain mechanisms required for the different uses of "numerical" measures are very different and may take different forms in different evolutionary lineages. I suspect similar comments are applicable to tests for "numeracy" in various non-human animals and in young children.

This line of thought makes me deeply suspicious of research reports claiming to have identified numerical/mathematical competences in infants or toddlers and other animals when nobody knows exactly what sorts of brain functions are actually being used in the various cases. I suspect similar comments can be made regarding the discussion of "The Child as a Mathematician" in *Beyond Modularity*. If our natural number concepts correspond to equivalence classes based on 1–1 correspondence, as agreed by David Hume, Gottlob Frege and Bertrand Russell, then many of the behavioural tests alleged to identify onset of number competences may be irrelevant if they merely produce evidence for competences that produce the right results in a very limited set of contexts, just as the jumping precision of a spider monkey happens to correspond to use of the physicists concepts of length, force, mass, and momentum, without justifying the assumption that anything in the monkey's brain uses the same concepts.

Example mechanisms capable of detecting or creating one-to-one correspondences in various practical tasks were presented in (Sloman, 1978a), which assumed the existence of our current number and counting systems (based on one-to-one correspondence), with tentatively sketched mechanisms capable of supporting their use (I here refer not to physical (e.g., brain) mechanisms, but to computational (virtual machine) mechanisms that could be implemented in different physical mechanisms, including mechanisms capable of generating parallel, coordinated, discrete processes, such as pointing while counting). I do not think there is any empirical or neural evidence that human brains are *innately* programmed with abilities to think about and reason about 1–1 correspondences involving arbitrarily large sets. The relevant abilities must have developed through cultural processes, building on more general innate or learnt capabilities. For understanding numbers (or geometry), however, each individual has to develop insight into impossibilities and necessary connections.

Beyond Modularity also emphasises the fact that much learning that requires innate mechanisms can have features that are strongly dependent on what it is about the environment that is learnt; for example, learning about properties of physical structures and processes that vary widely across cultures – such as building materials and what they are used for. Such learning can occur at different stages of development, and still be heavily influenced or constrained by the genome, in addition to being influenced by portions of the environment acted on and perceived – for example, abilities to write, understand, test and debug computer programs, for which opportunities did not exist until very recently. It is likely that among known animals only humans have genes that have the ability to support learning to debug computer programs, and have had them for millennia. Some other species may have related but simpler abilities to debug action strategies.

Chapter 4 of *Beyond Modularity* "The Child as a Mathematician" requires critical discussion: firstly, because learning about numbers is a much more complex and diverse process than the chapter acknowledges; and secondly, because there is much more to mathematical development, including toddler mathematical development, than learning about numbers. For example, as Piaget understood, it should include learning about topological relationships, including containment and connectivity.

Nevertheless, *Beyond Modularity* (or an updated version) should be compulsory reading for all researchers working in the areas of intelligent robotics, cognitive robotics, AI learning systems, and more generally computational cognitive science. Reading the book will give such researchers important new insights into what needs to be modelled and explained as well as experience of debugging a complex theory!

I do not agree with everything in the book. In particular, I believe that the emphasis only on *representational* change, without discussing required changes in information-processing *architectures and mechanisms* during development, is a major gap. In her opposition to modularity theories that postulate specific innate competences, Annette claims that the processes that happen within a domain to produce observed developments use general mechanisms. This is not correct, because learning to learn, as discussed in (Chappell & Sloman, 2007b), can include learning to learn different kinds of skills, facts, and learning strategies relevant to different environments or different developmental stages. Towards the end of *Beyond Modularity*, Annette raises doubts about the adequacy of the concepts and theories presented. It is possible that later work stimulated by her ideas has made important progress about which I am still ignorant.

Evolution's use of representational redescription

I have been working on evolution's use of compositionality and its role in ancient mathematical discovery, which I think is at least loosely related to Annette's ideas. Roughly, evolution discovered increasingly powerful abstractions that can contribute to increasingly powerful modes of compositionality in its designs, by means that cannot be replicated in individual learning and development. Some of the results were the ancient mathematical discoveries that are now in constant use by scientists, engineers, architects etc., but which are not explicable by current neuroscience, nor replicated in AI systems. Those evolutionary discoveries, whose precise history remains to be charted, have some of the character of processes of representational redescription that Annette attributed to individuals. But I am attributing them to evolutionary processes. I cannot tell whether she would have approved.

By encoding abstracted versions of new discoveries in genomes, evolution left mechanisms of development in individuals the "task" of instantiating and combining the abstractions, not re-discovering them, though differences in individual trajectories resulting from layered, context-sensitive, stages of instantiation, may produce amazing individual creativity. If what is in the genome is appropriately abstract, then it can go on being applied to new products of cultural evolution instead of being restricted to the original applications. Linguistic examples would include late-expressed genes that deal with differences between singular and plural forms, or differences in tenses. If the information is not encoded in terms of the precise language in use when the new capability is added to the genome then it can be useful in a far greater variety of future environments. This can apply to semantic and pragmatic features of language, and to internal uses of language for reasoning, as well as to grammatical details.

There are many striking examples in the (presumably) common collection of parametrised genomic features that can be instantiated in different ways to produce thousands of different languages, differing in primitive sounds or signs used, in syntactic forms, in semantic contents, in pragmatic communicative functions, and in various social processes. For example, the ability to congratulate someone on achieving something admirable but difficult is unlikely to have direct genetic support yet can be expressed in multiple different languages by combining more primitive linguistic functions, a possibility that can, in principle be re-discovered in different cultures, or by different individuals. Moreover, that possibility must have been supported by the human genome long before any human ever made use of it.

Among the products of all those evolutionary mechanisms are results that cannot be produced by current models of deep learning or known neural mechanisms, nor by purely logical/algebraic reasoning, for example the achievements of ancient geometers mentioned above. A major remaining task is to specify *detailed requirements* for the mechanisms underlying ancient mathematical discoveries, and other aspects of natural intelligence, a task that I do not think has ever been done systematically, comprehensively and adequately, although *Beyond Modularity* contributes important relevant ideas. In particular, we need a theory about the internal languages required, not for communication with other individuals, but for encoding/representing information contents for *internal* use, including perceptual contents, emotional states, online action control, offline consideration of actual or possible actions, and reflection on these internal representations.

Conclusion

Originally inspired by Kant's philosophy of mathematics, I have been working for some time on evolution's use of compositionality and its role in ancient mathematical discovery, which I think is related to Annette's ideas. Roughly evolution discovered increasingly powerful abstractions that can contribute to increasingly powerful modes of compositionality by means that cannot be replicated in individual learning. So those abilities need to be supported genetically. In other words, powerful cognitive abilities depend on genetically determined modules, but some of the most important modules are not active from birth, and they are only partially specified abstractions, with important parameters acquired from the environment before the modules are activated.

By encoding abstracted versions of those discoveries in genomes, evolution provided opportunities for individuals to solve difficult problems, including engineering design problems, mathematical problems and problems of explaining new scientific observations, by instantiating powerful abstractions, without having to re-discover them, though with differences in individual trajectories resulting from layered, context-sensitive, results of such instantiation. This can produce amazing individual creativity, illustrated, for example, in Annette's work on children's drawings. Results based on understanding spatial necessities/impossibilities cannot be produced by current models of deep learning or known neural mechanisms, since those mechanisms cannot represent, let alone discover, impossibilities and necessities. Neither do powerful current logic-based AI reasoning systems capture the ancient mathematical discovery processes using spatial reasoning, or the closely related spatial reasoning capabilities of pre-verbal toddlers and other intelligent animals.

I felt very honoured, and very surprised, when I was invited to contribute to this book. However, reading or re-reading both things Annette wrote and work by others reporting, commenting on or criticising her work, has helped me to understand better the depth and breadth of her work, and some of the details of our agreement and disagreement. I hope some of the relationships summarised briefly here will trigger interest in new research that addresses important scientific and educational problems, as it has prompted me to clarify some aspects of the Meta-Configured Genome project. Thank you Annette!

Notes

1 http://www.cs.bham.ac.uk/research/projects/cosy/conferences/mofm-paris-07/latest.html
2 A summary of Kant's claims about mathematics is in (Sloman, 1965) and http://www.cs.bham.ac.uk/research/projects/cogaff/misc/kant-maths.html
3 For examples, see http://www.cs.bham.ac.uk/research/projects/cogaff/misc/toddler-theorems.html1

References

Chappell, J., & Sloman, A. (2007a). Contributions to WONAC: International Workshop on Natural and Artificial Cognition Two ways of understanding causation: Humean and Kantian. International Workshop on Natural and Artificial Cognition. Pembroke College, Oxford. June 25 and 26, 2007. http://www2.cs.arizona.edu/projects/wonac/
Chappell, J., & Sloman, A. (2007b). Natural and artificial meta-configured altricial information-processing systems. *International Journal of Unconventional Computing, 3*(3), 211–239.
Clowes, M. (1971). On seeing things. *Artificial Intelligence, 2*(1), 79–116.
Clowes, M. (1973). Man the creative machine: A perspective from Artificial Intelligence research. In J. Benthall (Ed.), *The limits of human nature.* London: Allen Lane.
Fuson, K. C. (1979). Review of The Child's Understanding of Number by R. Gelman and C. R. Gallistel. *Journal for Research in Mathematics Education, 10*(5), 383–387.
Gelman, R., & Gallistel, C. R. (1978). *The child's understanding of number.* Cambridge: Harvard University Press.
Hilbert, D. (1899). The Foundations of Geometry, Project Gutenberg, Salt Lake City. Translated 1902 by E.J. Townsend, from 1899 German edition. http://mate.dm.uba.ar/~glaroton/Hilbert.pdf
Huffman, D. A. (1971). Impossible objects as nonsense sentences, In B. Meltzer & D. Michie (Eds.), *Machine intelligence* (*Vol. 6*, pp. 295–323). Edinburgh: Edinburgh University Press.
Jablonka, E., & Lamb, M. J. (2005). *Evolution in four dimensions: Genetic, epigenetic, behavioral, and symbolic variation in the history of Life.* MIT Press, Cambridge MA.
Kant, I. (1781). *Critique of pure reason.* Macmillan, London. Translated (1929) by Norman Kemp Smith.
Karmiloff-Smith, A. (1990). Constraints on representational change: Evidence from children's drawing. *Cognition, 34,* 57–83.

Karmiloff-Smith, A. (1992). *Beyond modularity: A developmental perspective on cognitive science.* Cambridge, MA: MIT Press.

Karmiloff-Smith, A., & Inhelder, B. (1974-1975). If you want to get ahead, get a theory. *Cognition, 3,* 195–212. 3.

Karmiloff-Smith, A. (2006). The tortuous route from genes to behavior: A neuroconstructivist approach. *Cognitive, Affective & Behavioral Neuroscience, 6,* 9–17.

McCarthy, J. (2008). The well-designed child. *Artificial Intelligence, 172*(18), 2003–2014.

McCarthy, J., & Hayes, P. (1969). Some philosophical problems from the standpoint of AI. In B. Meltzer & D. Michie (Eds.), *Machine intelligence 4* (pp. 463–502). Edinburgh: Edinburgh University Press.

Minsky, M., & Papert, S. (1971). *Progress report on artificial intelligence.* MIT. AI Memo AIM-252.

Mumford, D. (2016). Grammar is not merely part of language, http://www.dam.brown.edu/people/mumford/blog/2016/grammar.html. Online Blog.

Neisser, U. (1976). *Cognition and reality,* W. H. Freeman, San Francisco.

Peters, L., Messer, D., Davey, N., & Smith, P. (1999). An investigation into Karmiloff-Smith's RR model: The effects of structured tuition. *British Journal of Developmental Psychology, 17,* 277–292.

Peterson, D. (Ed.), (1996). *Forms of representation: An interdisciplinary theme for cognitive science.* Exeter: Intellect Books.

Piaget, J. (1981). *Possibility and Necessity Vol 1. The role of possibility in cognitive development.* Minneapolis, MN: University of Minnesota Press, Tr. by Helga Feider from French in 1987.

Piaget, J. (1983). *Possibility and Necessity Vol 2. The role of necessity in cognitive development.* Minneapolis, MN: University of Minnesota Press, Tr. by Helga Feider from French in 1987.

Roberts, L. G. (1965), Machine perception of threedimensional solids. In J. T. et al., ed., *Optical and Electrooptical Information Processing.* Cambridge, MA: MIT Press.

Schrödinger, E. (1944). *What is life?* Cambridge: CUP.

Sloman, A. (1962). *Knowing and Understanding: Relations between meaning and truth, meaning and necessary truth, meaning and synthetic necessary truth (DPhil Thesis).* PhD thesis, Oxford University.

Sloman, A. (1965). 'Necessary', 'A Priori' and 'Analytic'. *Analysis, 26*(1), 12–16.

Sloman, A. (1971). Interactions between philosophy and AI: The role of intuition and non-logical reasoning in intelligence. In Proc 2nd IJCAI (pp. 209–226). London: William Kaufmann, Reprinted in Artificial Intelligence, vol 2, 3-4, pp. 209–225, 1971.

Sloman, A. (1978a). *The computer revolution in philosophy.* Hassocks: Harvester Press (and Humanities Press). http://www.cs.bham.ac.uk/research/cogaff/62-80.html#crp, Revised 2018.

Sloman, A. (1978b). What about their internal languages? Commentary on three articles by Premack, D., Woodruff, G., by Griffin, D.R., and by Savage-Rumbaugh, E.S., Rumbaugh, D.R., Boysen, S. in Behavioral and Brain Sciences Journal 1978, 1 (4). *Behavioral and Brain Sciences, 1*(4), 515.

Sloman, A. (1982). Computational epistemology. In *Cahiers De La Fondation Archives Jean Piaget, No 2–3 Geneva June 1982,* Archives Jean Piaget. (Proceedings Seminar on Genetic Epistemology and Artificial Intelligence, Geneva 1980), http://www.cs.bham.ac.uk/research/projects/cogaff/81–95.html#53.

Sloman, A. (1984). Beginners need powerful systems. In M. Yazdani (Ed.), *New horizons in educational computing* (pp. 220–234). Chichester: Ellis Horwood Series In Artificial Intelligence. http://www.cs.bham.ac.uk/research/projects/cogaff/81-95.html#45.

Sloman, A. (2016). Natural vision and mathematics: Seeing impossibilities. In U. Furbach & C. Schon (Eds.), *Second Workshop on: Bridging the Gap between Human and Automated Reasoning, at IJCAI 2016*, New York, pp. 86–101. http://ceur-ws.org/Vol-1651/.

Sloman, A., & Chappell, J. (2005). The altricial-precocial spectrum for robots. In *Proceedings IJCAI'05* (pp. 1187–1192). Edinburgh: IJCAI. http://www.cs.bham.ac.uk/research/cogaff/05.html#200502.

Sloman, A., & Chappell, J. (2007). Computational Cognitive Epigenetics (Commentary on (Jablonka & Lamb 2005)). *Behavioral and Brain Sciences, 30*(4), 375–376.

Smith, L. (1994). The development of modal understanding: Piaget's possibility and necessity. *New Ideas in Psychology, 12*(1), 73–87.

Trettenbrein, P. C. (2016). The Demise of the Synapse as the Locus of Memory: A looming paradigm shift?. *Frontiers in Systems Neuroscience*, Vol 10, No 88, ISSN 1662-5137.

Turing, A. M. (1938). Systems of logic based on ordinals. *Proc. London Mathematical Society*, pp. 161–228, doi: 10.1112/plms/s2-45.1.161

Turing, A. M. (1950). Computing machinery and intelligence. *Mind, 59*, 433–460.

Turing, A. M. (1952). The chemical basis of morphogenesis. *Philosophical Transactions of the Royal Society of London, 237*(237), 37–72.

Weir, A. A. S., Chappell, J., & Kacelnik, A. (2002). Shaping of hooks in New Caledonian crows. *Science, 297*(9 August 2002), 981.

7

REVISITING *RETHINKING INNATENESS*: 20 YEARS ON

Mark H. Johnson

Dedication: for Elizabeth Bates, Jeffrey Elman and Annette Karmiloff-Smith

In 1996 the multi-authored volume *Rethinking Innateness* (RI; Elman *et al.,* 1996) was published to both acclaim and critical comment. The impact of this book on the fields of developmental psychology and cognitive science was significant for at least the following decade, as evidenced by its impact (over 4,000 Google Scholar citations). In this contribution I reflect on the factors that drove the original writing of the book and its related spin-off publications, and the reasons for the impact it had at the time. I follow this by briefly assessing the current evidence for, and status of, some of the primary claims and ideas presented from the perspective of 20 years hindsight. This is an exercise I would have loved to have undertaken with the original books co-authors. Sadly, however, Jeffrey Elman, Elizabeth Bates and Annette Karmiloff-Smith are all now deceased, and I can only hope that they would have at least agreed with the spirit of what I have to say.

Rethinking Innateness: Origins

The book's roots began in the late 1980s when the MacArthur Foundation awarded a grant to Jeff Elman and colleagues at the Centre for Research in Language at UCSD to provide training for selected developmental psychologists in the tools and methods of connectionist computer modelling. After the program ran for several years, a workshop was held in 1991 where alumni presented the work they had done as a result of their training. At this point it became clear that a coherent perspective on development was emerging and that this had the potential to integrate information across different approaches – brain, cognitive and computational. Over some great conversations during that visit, the idea for the book was born, and I was fortunate enough to be invited to join my co-authors to work on it from the perspective of my background in developmental neuroscience and ethology. RI was very much Elman's baby that he nurtured from conception to publication, to subsequent post-publication dissemination events across the globe. Given that several of co-authors were passionately committed to their own particular views, developing a consensus statement required Jeff's full range of diplomatic and leadership skills! Subsequently, over the month in 1992 that we spent in San Diego developing the book, hardly a day went by without a heated debate ending with one or other of the authors walking out in protest on some key point or other. Jeff was always the one to bring the offended party back to the table, and remind everyone that the book would be all the more enriched for our different perspectives.

Rethinking Innateness: Impact

"Where does knowledge come from?.... The answer is not Nature or Nurture; its Nature and Nurture" (Elman et al., 1996; p.357). In the decades preceding publication of RI in 1996, the constructivist perspective that was brought in to human cognitive development by Piaget had suffered significantly from empirical attacks primarily focused on the details of his stage theory. As a result, Piagetian constructivism was largely replaced by a cognitivist approach in which a putative dissociation of nature and nurture was achieved through the underlying assumption that different cognitive units or structures could be dissociated by the extent of their malleability in the face of experience. For example, Spelke and others promoted the view that young children possess components of cognition termed "core knowledge", which are characterised by their presence at, or near, birth – and are thus separated from other cognitive units by their status as being "innate" (Spelke, 1994). This approach stands in sharp contrast to RI which addresses the basic question "Where does knowledge come from?" with the assumption that knowledge structures are always an emergent product of the interaction between nature and nurture (see also Oyama, 1985). From this latter view, the factors used to define cognitive units to explain and analyse the developing human mind are of a very different kind. In RI we sought to decompose the developing mind into units of cognition through examination of neural evidence and computational consideration, and not on their presence or absence in the months after birth. The knowledge/information (or representations in the strict sense used in the book) that emerged in these systems were the result of different levels of interactions including molecular, genetic, neural, and internal and external (to the organism) environments.

Another volume appearing around the same time was Thelen and Smith's A Dynamic Systems Approach to the Development of Cognition and Action (1993). While this book shared similarities in its focus on the processes of change, the view espoused was anti-representational in the sense that only accounts of processes of dynamic change were deemed to be necessary to explain a wide range of phenomena in human mental development. The advent of connectionist modelling allowed for an understanding of developmental change that also naturally includes representations, such as how partial or weak representations might generate different responses through different modes of action in infancy (Munakata, 2001).

Rethinking Innateness: The brain

One of the chapters in RI that would require substantive updating if the book were ever to be re-issued would be the one on Brain Development (Chapter 5). More than two decades is an age in the fast-moving field of neuroscience, and many nuances, twists and turns have been added as new data has come in. My intention here is not to summarise recent advances in human developmental neuroscience but rather to enquire whether any of the principle claims in RI need revision as a result.

The general idea that an understanding of the brain at a computational level requires the analysis, or modelling, of connections between nodes is still very much with us. Indeed, the whole sub-field of *Connectomics*, in which Graph Theory is commonly used to describe the topographical and hierarchical structure of both functional and structural brain networks, has thrived. In general, however, Connectomics has remained confined to descriptions of brain network architectures and their changes with age, pathology and the difference between individuals (human connectome project). Linking Graph Theory derived hubs and bottlenecks to computational models in ways that are informative for human cognition currently remains firmly on the "to do" list.

In the conclusion to Chapter 5 of RI we discussed the relevance of then-current developmental neuroscience to the key concepts (in RI) of *architectural* and *chronotopic constraints* on the emergence of representations. It was argued that while representational nativism (genetically encoded pre-wiring of detailed patterns of connectivity) was potentially a feature of some midbrain systems, the cerebral cortex is best viewed as an organ of plasticity in the sense that its intrinsic connectivity patterns were more malleable (albeit that the overall architecture of cortex provides constraints and limits on the nature and types of representations that are generated).

Despite much debate over the ensuing decades as to the extent to which the differentiation of cortex into distinct structural and functional regions is activity-dependent, the general view of the cortex as a machine evolved to generate and manipulate representations remains plausible today. The general models of cortical function available at the time of RI were largely based on feedforward networks in which greater abstraction away from sensory input was achieved through hierarchical stages of re-mapping (re-representation) (e.g., Shrager & Johnson, 1995, 1996). This hierarchical progression of abstraction of representations has a sequential temporal element (a *chronotopic* constraint according to RI) – with representations closer to the sensorium being required to be established before more abstract or integrative representations are formed (see pages 336–340 of RI). Of course, a crucial element of PDP/connectionist models was back-propagation of error signals, and RI contained hypotheses about the sources of this error information. Current models of cortical function still adhere to the notion that the cortex involves a modular design composed of canonical microcircuits that are replicated throughout the cortex. Thus, canonical neural computations are duplicated across sensory, motor and cognitive modalities (Heeger, 2017). In this sense, the architectural constraints discussed in RI remain as key factors – with the basic computational units of cortex imposing *architectural* constraints on the representations that can emerge.

Where modern theories of cortical function would require an updating of the views expressed in RI is evidence indicating that cortical processing can be modulated into different states in which different sources of processing get differentially weighted. For example, Heeger (2017) marshals evidence consistent with neural activity in each cortical area depending on combinations of feed forward drive (bottom-up from a previous stage in the processing hierarchy), feedback drive (top-down context from a subsequent processing stage) and prior drive (expectation). Heeger argues that the relative contributions of feedforward drive, feedback drive and prior drive are controlled by several state parameters implemented by neurotransmitters such as Acetyl Choline or frequencies of brain oscillation.

Rethinking Innateness: Acknowledged limitations

"Connectionism is still in its infancy, however, and development provides a rich set of phenomena which challenge existing technology" (Elman et al., 1996; p.359). In writing the quote above the authors of RI acknowledged that developments in the neural-inspired network computational models they engaged with could change future perspectives on the issues that motivated the book. The term "representation" was narrowly defined within RI as referring to the connection weights (assumed to model the fine-grained patterns of cortical connectivity) between nodes within a then-typical three-layered Parallel Distributed Processing network. These connection patterns encoded a representation as a specific patterned set of connection weights. In particular, a key argument of the book was that architectural (overall structure of the network) and timing constraints restricted the range of possible representations that could emerge in a given network. RI authors argued that there were few, if any, examples of representations (in the strict sense above) that were prespecified by direct genetic instruction. Rather, we argued, the emergence of representations (in this sense) is heavily constrained by the overall architecture of the brain, and by the sequential timing of development, with some representations by necessity preceding others.

In the subsequent two decades connectionism has moved on significantly. In the era of RI most connectionist models were composed of three-layer nets (with one being a "hidden layer") and they used variations on back-propagation learning algorithms (Rumelhart, McClelland, & PDP Research Group, 1986; McClelland, Rumelhart, & PDP Research Group, 1987). Over the past decade we have moved to multi-layered network systems that engage variations of genetic algorithms. Put simply, the networks essentially adapt their own architecture to best suit the nature and structure latent within the data that they are presented with ("deep learning", Bengio, 2009). In terms of RI this innovation clearly blurs the distinction between architectural constraints and representations, as the architectural constraints themselves can change in dynamic ways as progress is made by the network in discovering the structure of the data.

Viewed from a broader perspective, however, it could be argued that deep learning instantiates both the architectural and the timing constraints discussed in RI, in that the former depends on the state of representation acquisition ("developmental stage"). However, it must be acknowledged that the new field of deep learning/artificial neural networks is more distant from attempts to simulate the workings of the brain. Rather, the objective is to optimise learning within big datasets. Although unknown, it seems likely that architectural constraints imposed by brain anatomy are greater than those in most current artificial neural networks.

> *"Most models are passive. They exist in an environment over which they have no control and are spoon fed a preprogrammed diet of experiences. Babies, on the other hand are active. .... To a large extent, they select their environment by choosing what they will attend to and even where they will be."*
> *(Elman et al., 1996; p.393)*

While RI included some examples from Domenico Parisi's work in which he simulated simple organisms developing within simple virtual environments, the vast majority of connectionist modelling at the time required the programmer to create simple coded inputs that were intended to reflect the structure of sensory input from the environment. However, this approach entails a very passive view of learning systems in which they are essentially "spoon fed" information from the environment. Its abundantly clear that human infants are not like this in that they actively seek novel information through action on their environments. Indeed, I have more recently highlighted the importance of self-organisation in accounting for both typical and atypical human postnatal brain development (Johnson, 2015).

This limitation of the view presented in RI served as the springboard for a later volume published in 2007 called "Neuroconstructivism: How the brain constructs cognition" (Mareschal et al., 2007). In this volume, my co-authors and I emphasised the importance of embodiment (the brain being embedded within a body capable of acting on the world). Embodiment is critically important for at least two reasons. First, the representations that are constructed during development need only be sufficient to result in adaptive behaviour (action) in a given situation. Cognitively-relevant representations need not contain more information than is necessary for interfacing between perceptual inputs and motor outputs. As we described it in *Neuroconstructivism* "According to this view, the nervous system, the body and the environment are each complicated and highly structured dynamic systems, that are coupled together. Adaptive behaviour emerges from the interaction of all three systems" (p.70–71).

A second key consequence of embodiment is that having control over a body allows action on the external world in order to generate further interaction and learning. Mareschal et al. referred to this as "pro-activity" – a brain that actively initiates interactions with the environment. While these general ideas are not new and go back at least to Piaget's 'circular reactions' (in which children go through repeated cycles of acting on the world, and then observing the consequences of these actions, before modifying their subsequent actions based on their prior observation), Mareschal et al. identified key mechanistic features of Pro-activity. By their view, the internally generated activity of the brain initiates the feedback loop between the child and its environment, directing developmental trajectories along particular pathways. More recently, I have suggested that we should not just consider the role of pro-activity in generating typical neurocognitive trajectories, but also view at least some so-called developmental disorders as being ordered developmental responses to a different starting state (Johnson, 2017).

"From an early age, the infant's behaviours are highly social... In a similar vein, we would like to see models which have a more realistic social ecology" (Elman et al., 1996; p.394). To an extent, the social world can be seen as a sub-set of the environmental interaction issues raised above. However, there are a special – and arguably unique – features of the developing child's social world that necessitates additional explanation. The authors of RI recognised this gap in knowledge, and to a large extent this gap still remains today. While the proximal physical environment relevant for a child is ultimately predictable, the nature of interactions with other social beings is dependent on the other individual's moods, intentions and desires. One landmark in better predicting the social world is creating a predictive model of the other's mind – so-called theory of mind. Given the complexity of learning to create the necessary representations, there is continuing debate over the extent to which these functions are due to an innate cortical module (Thomas chapter, this volumeRefs). Potentially a more useful way to consider things from a neuroconstructivist perspective is to consider what is the minimal representation required for successful interaction between two individuals. Emerging evidence suggests the possibility that representations of the structure of interactions are constructed by the brain during development. Just as parts of our cortex specialise for faces, language and number, fragmentary evidence raises the possibility that parts of the social brain network become tuned up to the emergent product of interactions between individuals. For example, some EEG signatures of face perception in infants correlate with measures of the quality and quantity of social interaction between child and parent, and not with any measures of infant or parent behaviour independently during social interaction events (De Haan, Belsky, Reid, Volein, & Johnson, 2004).

Conclusion

I end with a quote from RI that I think encapsulates the spirit of the book, and was certainly imbued in the life and work of its now-deceased co-authors:

"It is with neither cynicism nor discouragement that we acknowledge what is usually true in science: At any given point in time it will be the case that we only know 5% of what we want to know, and 95% of that will eventually turn out to be wrong. What matters is that we might be on the right track, that we are willing to discard those ideas which are proven wrong, and that we continue searching for better ideas."

(Elman et al., 1996; p.392)

References

Bengio, Y. (2009). "Learning Deep Architectures for AI". *Foundations and Trends in Machine Learning. 2*(1), 1–127.

De Haan, M., Belsky, J., Reid, V., Volein, A., and Johnson, M. H. (2004). Maternal personality and infants neural and visual responsivity to facial expressions of emotion. *Journal of Child Psychology and Psychiatry, 45*, 1209–1218.

Elman, J., Bates, E. A., Johnson, M. H., Karmiloff-Smith, A., Parisi, D., & Plunkett, K. (1996). *Rethinking Innateness: A Connectionist Perspective on Development.* Cambridge, MA: MIT Press.

Heeger, D. J. (2017) Theory of cortical function. *Proceedings of the National Academy of Sciences USA, 114,* 1773–1782

Hellyer, P. J., Clopath, C., Kehagia, A. A., Turkheimer, F. E., & Leech, R. (2016). *Balanced activation in a simple embodied neural simulation,* arXiv preprint arXiv:1606.03592

Johnson, M. H. (2015). Neurobiological perspectives on developmental psychopathology. In Thapar, A., Pines, D. S., Leckman, J., Snowling, M. J., Scott, S., & Taylor, E. (Eds.), *Rutter's Child and Adolescent Psychiatry*, 6th Edition (pp. 107–118). John Wiley & Sons.

Johnson, M. H. (2017). Autism as an adaptive common variant pathway for human brain development. *Developmental Cognitive Neuroscience, 25,* 5–11. doi: 10.1016/j.dcn.2017.02.004

Mareschal, D., Johnson, M. H., Sirios, S., Spratling, M. W., Thomas, M. S. C., & Westermann, G. (2007). *Neuroconstructivism, Volume One. How the brain constructs cognition.* Oxford: OUP.

McClelland, J. L., Rumelhart, D. E., & PDP Research Group. (1987). *Parallel Distributed Processing Volume 2: Explorations in the Microstructure of Cognition: Psychological and Biological Models.* Cambridge, MA: MIT Press.

Munakata, Y. (2001). Graded representations in behavioural dissociations. *Trends in Cognitive Sciences, 5*(7), 309–315.

Oyama, S. (1985). *The ontogeny of information: Developmental systems and evolution.* Cambridge, England: Cambridge University Press.

Rumelhart, D. E., McClelland, J. L., & PDP Research Group. (1986). *Parallel Distributed Processing Volume 1: Explorations in the Microstructure of Cognition: Foundations.* Cambridge, MA: MIT Press.

Shrager, J., & Johnson, M. H. (1996). Dynamic plasticity influences the emergence of function in a simple cortical array. *Neural Networks, 9*(7), 1119–1129.

Spelke, E. (1994) Initial knowledge: Six suggestions. *Cognition, 50,* 431–445.

Thelen, E., & Smith, L. B. (1993). *A Dynamic Systems Approach to the Development of Cognition and Action.* Cambridge, MA: MIT Press.

8

REPRESENTATIONAL REDESCRIPTION: THE CASE OF THE EARLY MENTAL LEXICON

Kim Plunkett

> *I contend that the pre-requisite of fundamental macrodevelopmental change, in the form of progressive explicitation and restructuring of representational relationships, is "success" (i.e., positive feedback mechanisms) rather than failure (i.e., negative feedback mechanisms).*
>
> (Karmiloff-Smith, 1986, pg. 100)

Introduction

Recent statistical modelling estimates the average receptive vocabulary size of a typical 18-month-old at around 600 words (Mayor & Plunkett, 2011). Although many of these words may not be understood in an entirely adult-like fashion, the magnitude of the accomplishment highlights the extraordinary progress toddlers have made towards the construction of a mental lexicon before their second birthday. But mastery of a large repertoire of words does *not* necessarily imply possession of a *structured* mental lexicon similar to the adult system. For example, the young toddler's sensitivity to meaningful relations between words, such as those pertaining between 'dog' and 'cat' or between 'dog' and 'bone', or to the similarity in the sounds of words in their vocabulary, such as 'bus' and 'ball' or 'cat' and 'hat', has been relatively unexplored until recently.

I will argue that the toddler's success in learning words leads to a reorganisation of her mental lexicon, at the levels of both phonology and semantics. The consequences of re-organisation include both advantages and pitfalls: the advantages are the achievement of a more structured adult-like system, permitting greater speed, efficiency and accuracy of processing; pitfalls include interference effects in the recognition and processing of words, with its attendant slower processing and intrusion of errors. The reorganisation of the toddler's mental lexicon results in a temporary regression in performance. The regression is subtle and is revealed, sometimes, only by detailed experimental investigation.

Historically, perhaps the best-known example of linguistic reorganisation is in the domain of inflectional morphology and English children's acquisition of the past tense inflection (Berko, 1958). After successful learning of irregular past tense forms such as 'went', children sometimes produce erroneous inflections such as 'go-ed'. These errors are typically attributed to children's discovery of the regularities underlying past tense formation in English, as they are exposed to an increasing number of regularly inflected verbs in the language. Success leads to failure, albeit of a temporary and intermittent nature (Marcus et al., 1992). Children's erroneous behaviour is taken as an index of the reorganisation – or systematisation – of their knowledge of the past tense of English. A pre-requisite for reorganisation is successful learning of a representative sample of the past tenses of English verbs. The reorganisation of the toddler's mental lexicon follows a similar pattern: A sufficiently large repertoire of words must be learnt before patterns of similarity and regularity can be discovered. We shall see that a re-organisational process is set in train that leads to poorer performance in some tasks where language is needed to orient attention to a referent, such as a visual world task (Chow, Aimola Davies, & Plunkett, 2017).

Reorganisation of knowledge in the brain must be carried out in such a fashion that the organism can still function successfully in its environment, while re-structuring of the knowledge base takes place. Representational Redescription (RR) offers a solution to handling these processes of change without catastrophic failure of the system: additional resources may be recruited to represent the problem (typically at a higher level of analysis) while the original representation continues to function in service of behaviour (Karmiloff-Smith, 1992). The re-described representation eventually takes control of behaviour but may lead *en route* to intermittent errors.

Lexical-semantic development

In a lexical decision task, adults will respond faster to a target word, e.g., 'doctor', if it is preceded by a related word, e.g., 'nurse' (Meyer & Schvaneveldt, 1971). 'Nurse' is considered to prime 'doctor', providing evidence that the prime and target words spread activation between each other in the adult mental lexicon. Forward semantic priming tasks with toddlers involve the auditory presentation of pairs of related or unrelated words immediately prior to the presentation of a pair of pictures (Styles & Plunkett, 2009). The first word in the pair is the prime and the second word is the target. The picture pair includes the referent of the target word and an unrelated distractor object. Figure 8.1 depicts an example of a typical semantic priming trial, along with some common timing parameters used in these experiments. The variable monitored in the task is the toddler's propensity to fixate the target picture, measured in terms of the preference for the target picture over the distractor, or the latency to fixate the target picture. A priming effect is

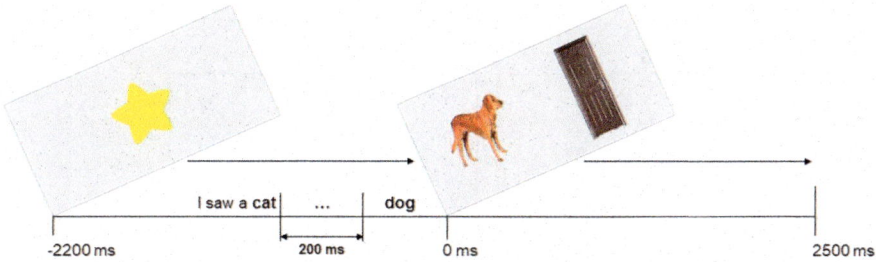

FIGURE 8.1   The forward semantic priming task (after Arias-Trejo & Plunkett, 2009)

observed when related word pairs produce shorter latencies to fixate the target picture or a greater preference for the target picture, as compared to unrelated word pairs. Priming effects have reported in a variety of studies of monolingual and bilingual toddlers for word pairs that are thematically related, taxonomically related or both (Arias-Trejo & Plunkett, 2009, 2013; Delle Luche et al., 2017; Singh, 2014; Styles & Plunkett, 2009). These priming effects are interpreted as an indicator of the toddler's sensitivity to the semantic relations between the word pairs, in much the same way as speeded lexical decision times are interpreted as indicating the presence of semantic relations between related word pairs in adults.

In one such toddler priming study, Arias-Trejo and Plunkett (2009) studied the impact of pairs of words that were both taxonomically and associatively related, e.g., 'cat' and 'dog', as compared to pairs of unrelated words, e.g., 'plate' and 'dog', on target looking. The experiment was performed with 18- and 21-month olds who were familiar with the visual and auditory stimuli used in the study. The 21-month olds demonstrated a clear priming effect: they looked longer and faster at the target object when presented with related word pairs than when presented with unrelated word pairs. No such priming effect was observed for the 18-month olds. Figure 8.2 shows the toddlers' preference for the target in the related and unrelated conditions in both age groups.

The authors interpreted these results as indicating that by 21 months of age, toddlers have started to construct a semantic network of word meanings, not apparent at 18 months of age. Follow-up work by Arias-Trejo and Plunkett (2013) demonstrated that by 24 months of age, toddlers demonstrated a sensitivity for word pairs that were only taxonomically related (e.g., 'plate' and 'bottle') or thematically related (e.g., 'plate' and 'cake'), indicating that both types of semantic relationship are involved in the construction of early semantic networks. Theoretical work based on graph-theoretic analyses of vocabulary growth, as measured by parental report (Hills, Maouene, Maouene, Sheya, & Smith, 2009), also provides converging evidence that toddlers have begun to develop a semantic network in which word meanings are interconnected. This body of research suggests that the emerging mental lexicon undergoes a process of reorganisation during the second year of life. Vocabulary growth is not merely the accrual of a large repertoire of words.

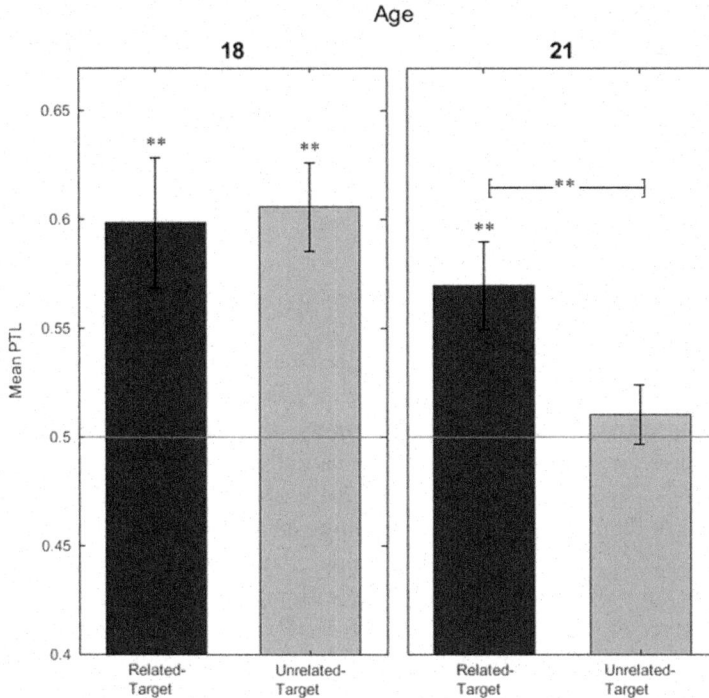

FIGURE 8.2 Proportion of target looking in related and unrelated conditions for 18- and 21-month-old toddlers. Adapted from Arias-Trejo and Plunkett (2009)

Here, I would like to focus attention on another comparison between the two age groups studied by Arias-Trejo and Plunkett (2009). A brief examination of Figure 8.2 reveals that 18-month olds are more successful in the task of referent identification than 21-month olds: across conditions, 18-month olds show a greater preference for the target picture. The primary locus of the difference is the degree of target preference in the unrelated condition: 18-month olds exhibit just as much target looking in the unrelated condition as in the related condition, whereas the 21-month olds exhibit no systematic target preference in the un-related condition (chance looking is 0.5). In fact, it is precisely the lack of a sys-tematic target preference in the unrelated condition that drives the priming effect in 21-month olds. Another way of describing this pattern of looking is that 18-month olds preferentially fixate the target picture just so long as it is named, while target looking in the 21-month olds is driven by the status of the prime word, i.e., whether it is related or unrelated. Note that other experimental con-ditions evaluated by Arias-Trejo and Plunkett (2009) demonstrated that the prime alone was not sufficient to drive target looking in the 21-month-old group: in the related condition, the target also had to be named. This pattern of results suggests that an unrelated prime *interferes* with the processing of the target word such that target recognition is impaired. No such impairment is observed with the 18-month olds.

One potential explanation of the superior behaviour of the 18-month olds is that they are slower to process words than the 21-month olds (Fernald, Swingley, & Pinto, 2001). Perhaps the onset of the target word interrupts processing of the prime word for the 18-month olds, preventing the prime word having any impact on the target word. Processing of the target word is not interrupted in this manner and so 18-month olds readily identify the target picture in the related and un-related conditions. For the 21-month olds, the prime word may be fully processed before the onset of the target word and so can achieve its full priming effect. It is unclear how much weight this account carries given that 18-month olds can extract sufficient information from the first 300 ms of a word to identify its re-ferent in a free-looking task (Swingley, Pinto, & Fernald, 1999).

An alternative explanation is that words are represented as unconnected *lexical islands* in the 18-month old's lexical-semantic system, precluding the possibility of semantic priming effects. On this account (proposed by Arias-Trejo & Plunkett, 2009), lexical entries in the 18-month-old lexicon are simply listings which do not encode any semantic relations. By 21 months of age, these lexical islands coalesce into a network of meanings. The construction of a semantic network by toddlers acts like a double-edged sword: the semantic network enables faster identification of a referent for related word pairs but it can also slow down identification when the prime is unrelated to the target (Styles & Plunkett, 2009), leading to inter-ference effects observed in the unrelated condition.

However, more recent evidence indicates that even 18-month olds are sensitive to semantic priming effects. Delle Luche, Durrant, Floccia, and Plunkett (2014) used the preferential looking procedure to determine whether 18-month olds can tell the difference between lists of words which are related in meaning and lists of words which are not related in meaning. This approach permits assessment of any priming effects in toddlers without the need to use pictures. If infants listen longer to one type of list than the other type of list, we can infer two things[1]: First, we can conclude that they must be computing the meanings of the words. Second, we can conclude that they are sensitive to the semantic relationships between the meanings of the words. The results show that at 18 months of age, infants look longer for the related lists, indicating that the related words are priming each other, evidence that is indicative of 18-month olds possessing a lexical-semantic network.

Note that unlike the forward semantic priming task (Arias-Trejo & Plunkett, 2009), it is not possible to conclude that the unrelated word lists produce shorter looking times because of interference effects. There is no baseline measure in the preferential listening task that is equivalent to chance looking at two pictures. Given the absence of interference effects between words in the forward priming task at 18 months, it would seem more parsimonious to suppose that the longer looking at the related lists in the preferential listening procedure is triggered by facilitative priming between words.

These findings undermine the Lexical Island Hypothesis described earlier which proposed that at 18 months old word meanings are represented in isolation from each other in the infant's lexical-semantic system. More recent evidence (Floccia, Delle Luche, Goslin, Hills, & Plunkett, 2016), using a similar technique, indicates that even 15-month olds are sensitive to meaning relations between words, particularly those that are associatively or thematically related. Overall, this body of research points to the existence of semantic networks as young 15 months of age. The earliest networks support facilitative priming between related words, whereas both facilitation and interference are apparent in the semantic networks that characterise the mental lexicons of infants from 21 months of age. We will return to a potential structure-building explanation of the emergence of the toddler's lexical-semantic system in a later section of the chapter.

Phono-lexical development

Phonological priming occurs in adults when a word influences the speed or accuracy of recognition of a similar-sounding word. For example, hearing the word 'car' influences the recognition of 'key'. Phonological priming can facilitate or interfere with recognition. Facilitation arises because of phonological overlap between two words: 'car' and 'key' have the same *onset* sounds. Facilitation is generally considered a *pre-lexical* effect. Interference is considered to be a *lexical level* effect: words stored in the mental lexicon and which sound similar compete with each other. Interference in the recognition of 'key' by 'car' is called a *cohort* effect (Marslen-Wilson, 1987). Cohort effects are widely reported in the adult word recognition literature and are particularly prominent for onsets that are shared by a large number of words (Johnson & Pugh, 1994; Marslen-Wilson & Tyler, 1980). Such interference effects are taken to be indicative of *phono-lexical* structure within the mental lexicon.

Cohort effects in the toddler's mental lexicon can be used to identify the presence and/or emergence of phono-lexical structure. Mani and Plunkett (2010) used a picture priming task to examine phonological priming in toddlers. They presented 18-month olds with an image of a name-known object (the prime image) followed by two images, one of which was labelled, i.e., the target image (see Figure 8.3). In half of the trials, the label for the prime image began with the same onset consonant as the label for the target image, e.g., bus—ball). In the other half of the trials, the label for the prime image was unrelated to the label for either the target or distracter image. Mani and Plunkett (2010) reported that 18-month olds looked longer at the target in related as compared to unrelated primed trials (see Figure 8.4). They argued that facilitation of target recognition resulted from phonological priming by the internally generated label for the prime image (bus) to the target word (ball) in related trials. The phonological priming effects observed were facilitatory, pointing to the operation of bottom-up pre-lexical priming. No cohort effects were found in the 18-month olds investigated in this study.

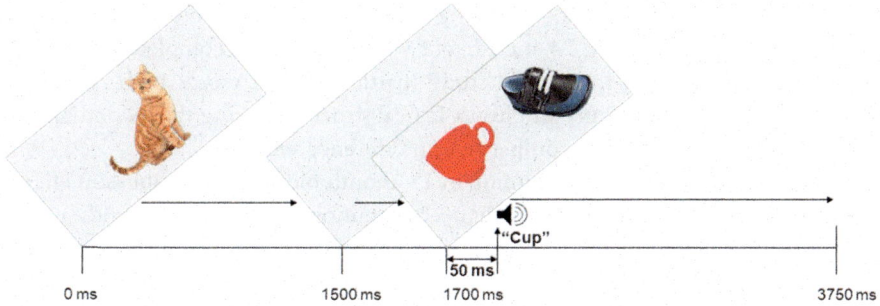

FIGURE 8.3 Timeline of a typical picture priming trial. Adapted from Mani and Plunkett (2010)

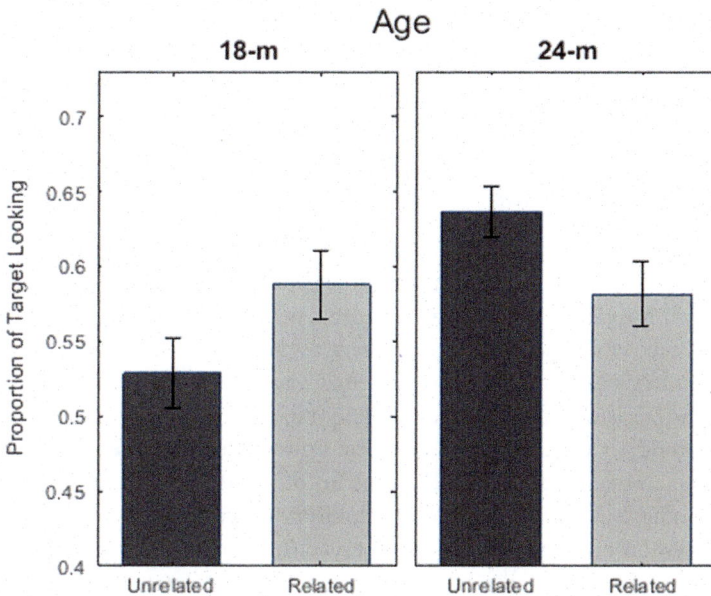

FIGURE 8.4 Mean proportion of time spent looking at the target as a function of trial type and age. Adapted from Mani and Plunkett (2010) and Mani and Plunkett (2011)

In a follow-up study, Mani and Plunkett (2011) tested 24-month-old toddlers using the same task and stimuli. They observed the opposite pattern of results: less target looking in the related condition than in the unrelated condition. Further examination of 24-month olds' performance on the task revealed that reduced looking in the related condition occurred when the prime label came from a *large* cohort of words, in particular, words that begin with /b/. The authors interpreted this result as indicating the operation of lexical-level cohort effects in the 24-month olds that were not apparent in the 18-month olds; large cohorts result in higher levels of competition between words than between words from small cohorts. Although these contrasting findings can be attributed in part to the relative sizes of the 18- and 24-month-old vocabularies, the complete absence of cohort effects in the 18-month-old study (particularly for words with /b/ as an onset) point to the emergence of phono-lexical structure during the second half of the second year, manifest as competitive interference effects.

Note that the phonological priming in 18-month olds yields a facilitation effect, e.g., words that begin with /b/ facilitate recognition of other words with a /b/. Negative phonological priming effects are only reported for 24-month olds, e.g., words that begin with /b/ now inhibit recognition of other words that begin with /b/. These results indicate that by 24 months of age, children's responses in these word recognition tasks approximates adult-like performance. This suggests that words begin to cluster together in the toddler lexicon based on their phonological properties, such that word recognition involves the activation and processing of phonologically related words. As was observed with lexical-semantic development, 24-month olds perform worse that 18-month olds, but here in a phonological priming task involved /b/ onset words, pointing to a reorganisation of the phono-lexical system during the second half of the second year.

Inhibition and reorganisation

The previous sections have documented a progressive reorganisation of the phono-lexical and lexical-semantic properties of the early mental lexicon. In each case, the progression is indexed by a transition from a period of relatively peaceful co-existence between words when there appears to be restricted evidence of structure building to competition/interference effects between words which impose structural constraints on lexical processes. The transition takes place during a time when many toddlers demonstrate a dramatic growth in their receptive and expressive language abilities, not least in the domain of vocabulary development. In this section, I argue that the emergence of inhibitory processes in the toddler's mental lexicon is responsible for the changes observed. In particular, I suggest that dramatic increases in the size of the toddler's vocabulary necessitate further reorganisation of the mental lexicon so that words, including both their sounds and their meanings, can be accurately and efficiently identified. The central process underlying this reorganisation is *inhibition*.

Phonological inhibition

Phono-lexical cohort effects in adults are commonly understood in terms of the influence of inhibitory connections between phonologically related words. Mutually exclusive lexical representations can be simultaneously activated in response to auditory input (e.g., cat vs. cap). Inhibitory connections can resolve this ambiguity rapidly and efficiently and reduce the number of word candidates highlighted for recognition. The TRACE model of word recognition (McClelland & Elman, 1986), and its Interactive Activation and Competition antecedent in the domain of visual word recognition (McClelland & Rumelhart, 1981), were amongst the first models to highlight the importance of inhibitory connections within the lexicon.

The emergence of phono-lexical cohort effects in toddlers can also be related to the emergence of inhibitory connections in the early mental lexicon. Eighteen-month olds show no evidence of cohort effects in a phonological priming task. In fact, Mani and Plunkett (2010) report only facilitation effects at this age which can be readily explained in terms of pre-lexical bottom-up processes. Therefore, we might assume that there are no inhibitory processes operating at this age. At 24 months, toddlers show evidence of lexical-level, cohort effects, pointing to the operation of inhibitory processes within the lexicon. The shift from phonological facilitation at 18 months and phono-lexical competition at 24 months might therefore be attributed to the emergence of lexical-level inhibitory processes.

This suggestion is clearly a *post hoc* explanation of the observed phenomena. What independent evidence is there for the emergence of phono-lexical inhibitory processes between 18 and 24 months of age? Swingley et al. (1999) report that 24-month olds are slower to identify the target referent of a spoken word in visual world task when both target and distractor pictures have labels with the same onset. These findings are consistent with lexical-level competition effects, but they may also be explained in terms of pre-lexical activation of the initial phonemes of the target and distractor labels which hinders target identification. It is noteworthy that Swingley et al. (1999) report that shared rhymes do *not* slow target recognition which one might otherwise expect if lexical-level effects were implicated.

Indirect evidence for the emergence of phono-lexical inhibitory processes comes from a computational study by Mayor and Plunkett (2014). These authors used the TRACE model of word recognition to model mispronunciation effects across a broad range of infant experimental studies and argued on the basis of TRACE simulations that vocabulary structure and content constrain infant word recognition in an experience-dependent fashion. These computational experiments systematically varied vocabulary size to reflect the increasing vocabulary size of infants over the age range considered. Recognition of an individual word in TRACE depended upon which other words it knows, how many words it knows and the frequency of exposure to the words themselves. However, successful replication of the experimental results with 18-month-old infants required that the inhibitory processes modelled in TRACE be switched off. In contrast, successful simulation of the findings of studies with 24-month olds required reinstatement of inhibitory processes in the model. The larger vocabulary of the "24-month old" model needed inhibition to eliminate inappropriate lexical candidates.

In TRACE, inhibitory connections between its representational components play an important role in structuring the processes of word recognition. A transition from an implementation of TRACE without inhibitory connections to one with inhibitory connections, as reported by Mayor and Plunkett (2014), can be thus be construed as a restructuring of phono-lexical space during the second half of the second year of life. Of course, as with any theoretical modelling endeavour, simulation of empirical data does not constitute proof of the theory embodied in the model, though it does add plausibility to the account offered here that inhibitory processes do not operate within the phono-lexical system at 18 months of age.

Semantic inhibition

As described earlier, lexical-semantic structure in adults is commonly inferred from semantic priming studies which demonstrate faster response times to a target word given a related prime, as compared with an unrelated prime (Meyer & Schvaneveldt, 1971). These are facilitative effects assumed to derive from spreading activation between related nodes in the adult lexical-semantic system. Inhibitory links have also been observed between and within semantic categories (e.g., Glaser and Düngelhoff (1984); Schriefers, Meyer, and Levelt (1990); Tipper (1985); Wheeldon and Monsell (1994)), most often when selective attention is required. For example, backward semantic inhibition (BSI) engages an inhibitory mechanism that can suppress attention to a previously attended item in a "backward" fashion (Fuentes, Vivas, & Humphreys, 1999a; Fuentes, Vivas, & Humphreys, 1999b; Weger and Inhoff (2006)): Adult responses in a modified lexical decision task were impaired when attending to words that belonged to a previously attended semantic category (e.g., Category A: Furniture →Category B: Animal → Category A: Furniture), compared to a previously unattended semantic category (e.g., Category C: Clothing → Category B: Animal → Category A: Furniture).

Previous studies with infants have demonstrated the existence of links in the lexical-semantic system: when presented with a prime label prior to the target label (as in Figure 8.1), 21-month olds' tendency to preferentially fixate the target picture in response to the target label was disrupted in an unrelated compared to a related condition. Hearing an unrelated prime interfered with subsequent processing of the target in a "forward" fashion (Arias-Trejo & Plunkett, 2009, 2013; Styles & Plunkett, 2009). No such disruption was observed in 18-month olds who successfully identified target referents in both related and unrelated conditions (Figure 8.2). A potential explanation of this pattern of results invokes the emergence of inhibitory connections in the lexical-semantic system after 18 months of age.

Several studies have explored semantic inhibition using an attention-switching paradigm with toddlers that mimic the logic of adult BSI studies (Chow, Aimola Davies, Fuentes, & Plunkett, 2016; Chow, Aimola Davies, Fuentes, & Plunkett, 2018). As seen in Figure 8.5, there were several types of trial: in *related* trials (ABA) a prime object was presented on screen and named immediately before the presentation of another object from a different category which was also named. In the third phase of the trial, two objects were presented side-by-side, one of which belonged to the same category as the prime object and the other was unrelated.

An example of an ABA trial would be Chair (Related-Prime) ➜ Chicken (Intervening-Word) ➜ Table (Target) vs. Flower (Distractor). In *unrelated* trials (CBA), the prime object was unrelated to all the remaining objects in the trial. Inhibition of the prime category was inferred when 24-month olds showed a reduced preference for the target picture in the ABA trials in comparison to the CBA trials. No inhibition was observed in baseline trials (not shown in Figure 8.5) when the semantic category B was replaced with a semantically neutral intervening stimulus (#_#: a checkerboard and sine-wave tone). In fact, A_A trials lead to a facilitatory effect in comparison to C_A trials. Importantly, when Chow et al. (2018) repeated the study with 18-month olds, they discovered that backward semantic inhibition was not apparent in the 18-month olds, unless they had relatively large vocabularies, in which case the pattern of inhibition was similar to 24-month olds.

These findings demonstrate that semantic inhibition emerges between the ages of 18 and 24 months, and suggest that an increased vocabulary may be an important driving force behind the emergence of inhibition, thus allowing efficient selective attention to currently relevant words and concepts. The finding parallels those for forward-semantic inhibition which is present at 21 but not 18 months of age (Arias-Trejo & Plunkett, 2009). It seems that a semantic inhibitory mechanism emerges to deal with a greater need for a more structured organisation of the mental lexicon, allowing for more efficient selection and deselection of items.

Mechanisms of inhibition

Chow et al. (2018) argue that vocabulary development is the driving force behind the emergence of semantic inhibitory effects, in much the same way that vocabulary development underlies the emergence of phono-lexical cohort effects late in the second year. In both cases, inhibitory processes increase the efficiency of target identification in an increasingly crowded lexical network. And in both cases, these inhibitory processes appear manifest after facilitatory processes have been

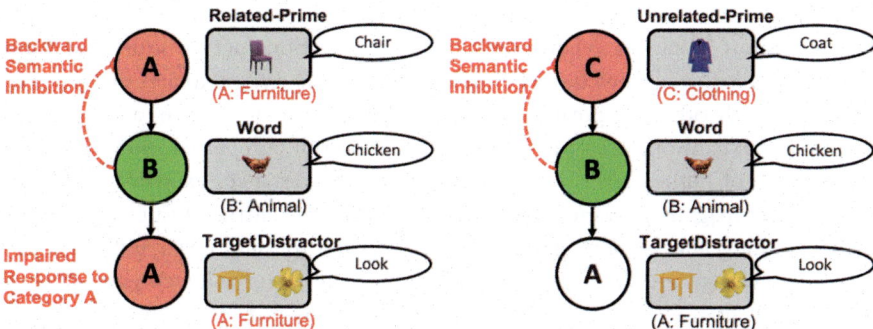

FIGURE 8.5 Backward semantic inhibition task in toddlers. ABA trials are related trials and CBA trials are unrelated trials. (Adapted from Chow *et al.* 2018)

identified in the phonological domain (Mani & Plunkett, 2010) and the semantic domain (Delle Luche et al., 2014). However, such findings beg the question as to the source of these inhibitory processes. Identifying vocabulary development as a correlate of the emergence of inhibitory processes does not identify the mechanism by which inhibition emerges, any more than observing that age is a correlate of the emergence of inhibitory processes.

Well over a century of research has appealed to inhibitory processes to explain psychological processes and illusions, Mach bands being a classical example in the visual domain (von Békésy, 1968). Lateral inhibition is a relatively well-understood and explored mechanism that has widespread application in the brain, so appealing to the operation of such processes is relatively uncontroversial. But how does inhibition 'get into' the phono-lexical-semantic system when, by all accounts to date, it does not appear to operate in the early lexical system? One solution is that inhibitory processes emerge through a process of maturation. In general, a posterior to anterior maturation of brain circuits is often assumed in developmental neuroscience. On this account, the anterior location of the phono-lexical-semantic system in the brain would predict a late emergence of inhibition. A first-order corollary of this assumption is that brain regions anterior to the phono-lexical-semantic system should exhibit manifestation of inhibitory processes later in development. However, Diamond's (1985) work on frontal lobe development in the context of the A–not–B error would appear to contradict this assumption. Pre-frontal cortical regions are implicated in the mechanisms of inhibitory control that enable infants to inhibit perseveratory responses in the A–not–B task before their first birthday, well before inhibitory processes are apparent in the lexical system.

A second possibility is that inhibitory processes are an inherent part of the developing lexical system but only become operational in the mental lexicon later in development. The finding that phono-lexical inhibition and lexical-semantic inhibition emerge around the same time and are related to increases in vocabulary size is consistent with this position. A *packing* metaphor whereby vocabulary items are stored in a limited-capacity lexical space may provide some explanatory leverage. Assume that inhibitory connections between lexical representations only act over short distances whereas excitatory connections act at short and long ranges. Words that are positioned close in lexical space have a greater opportunity to inhibit each other. A greater density of *packing* in lexical space, as a result of vocabulary expansion, then increases the likelihood of competitive inhibition. In the absence of higher levels of lexical density, long-distance, excitatory connections can propagate activation between related lexical entries, resulting in facilitative priming effects. Since lexical packing density (phonological and semantic) increases with age and vocabulary size, inhibitory effects in the lexicon are likely to emerge later than facilitatory effects. Of course, words that are 'proximal' to each other in lexical space may either inhibit or excite each other. Therefore, other factors such as representational similarity and/or featural overlap are needed to 'gate' the processes of excitation and inhibition. Generally, similar entities (shared features) should excite each other whilst dissimilar entities should inhibit each other, though an important role for inhibition is to eliminate highly similar competitors.

These considerations hardly provide a comprehensive account of the late emergence of inhibition in the mental lexicon, nor a specification of the nature of the mechanism(s) of inhibition itself. They are perhaps better interpreted as an appeal for further research to illuminate the nature of the processes underlying phono-lexical and lexical-semantic inhibition. One way to investigate the *packing* metaphor further would be to evaluate *local* patterns of excitation and inhibition between words. If inhibition only acts over short distances in lexical space, then it is possible that words that are very similar to each other, either from a phonological or semantic perspective, are more likely to manifest inhibitory connectivity than words that are dissimilar. On the other hand, if inhibitory connectivity does not respect phonological or semantic similarity, then a straightforward packing metaphor would seem inappropriate. Perhaps a critical mass of lexical items is required to trigger inhibition rather than any proximity metric. Clearly, further investigations are needed to explore these issues.

Conclusion

The theoretical thrust of this chapter has been to argue that the representational redescription of the early mental lexicon can be construed, first, as a process of establishing excitatory pathways between lexical items that are related to each other, both phonologically (Mani & Plunkett, 2010) and semantically (Delle Luche et al., 2014). Second, inhibitory processes emerge, again both phonologically (Mani & Plunkett, 2011) and semantically (Arias-Trejo & Plunkett, 2009; Chow et al., 2016), that limit unconstrained spread of excitation in the system and permit contextually appropriate target identification. The evidence suggests that the emergence of inhibitory connectivity is developmentally delayed and related to vocabulary growth (Chow et al., 2018). Excitatory connectivity is apparent much earlier in the mental lexicon, and probably as early as 15-months of age (Floccia *et al.* 2016).

The formation of excitatory and inhibitory connections in the toddler's mental lexicon is fundamental to its structural integrity and coherence. Without excitatory connectivity, lexical entries would function merely as islands in lexical space, akin to the listings in a dictionary with meaning definitions couched in terms of perceptual and conceptual features. The formation of excitatory connections highlights, and thereby restructures, the phonological and meaningful relationships between lexical entries. I suggest that the emergence of these connections heralds the onset of *phono-lexical-semantic* structure in the early mental lexicon. Without inhibitory connectivity in a rapidly expanding mental lexicon, spreading activation between closely related lexical items is in danger of destabilising the system, rendering word recognition and target identification difficult, if not impossible. The formation of inhibitory connections between words not only enables disambiguation of target representations in the early mental lexicon, but adds *depth* to the phono-lexical-semantic structure already triggered by the establishment of excitatory connections.

As we have seen, the establishment of excitatory and inhibitory connectivity in the early lexical can lead to reversals in performance in some tasks, e.g., accuracy and speed of target identification in visual world tasks. These reversals are closely related to success in building a vocabulary. Therefore, I contend that the prerequisite of fundamental *structural* change, in the form of progressive explicitation and restructuring of representational relationships *between words in the early lexicon*, is "success" rather than failure (adapted from Karmiloff-Smith, 1986, pg. 100).

Note

1 Just so long as other properties of the word lists such as word frequency, phonological and supra-segmental similarity, syllabic complexity etc., are appropriately controlled.

References

Arias-Trejo, N., & Plunkett, K. (2009). Lexical-semantic priming effects in infancy. *Philosophical Transactions of the Royal Society B, 364*, 3633–3647.

Arias-Trejo, N., & Plunkett, K. (2013). What's in a link: Associative and taxonomic priming effects in the infant lexicon. *Cognition, 128*, 214–227.

Berko, J. (1958). The child's learning of English morphology. *Word, 14*, 150–177.

Chow, J., Aimola Davies, A., & Plunkett, K. (2017). Spoken-word recognition in 2-year-olds: The tug of war between phonological and semantic activation. *Journal of Memory and Language, 93*, 104–134.

Chow, J., Aimola Davies, A. M., Fuentes, L. J., & Plunkett, K. (2016). Backward semantic inhibition in toddlers. *Psychological Science, 27*(10), 1312–1320.

Chow, J., Aimola Davies, A. M., Fuentes, L. J., & Plunkett, K. (2018). The vocabulary spurt predicts the emergence of backward semantic inhibition in 18-month-old toddlers. *Developmental Science, 22*(2), e12754.

Delle Luche, C., Durrant, S., Floccia, C., & Plunkett, K. (2014). Implicit meaning in 18-month-old toddlers. *Developmental Science, 6*, 948–955.

Delle Luche, C., Kwok, R., Durrant, S., Chow, J., Plunkett, K., & Floccia, C. (2017). The organisation of the bilingual lexicon: The impact of linguistic distance on semantic activation. In Paper presented at the *2017 meeting of the international association for the study of child language*. New Orleans.

Diamond, A. (1985). Development of the ability to use recall to guide action, as indicated by infants' performance on AB̄. *Child Development, 56*(4), 868–883.

Diamond, A., & Doar, B. (1989). The performance of human infants on a measure of frontal cortex function, the delayed response task. *Developmental Psychobiology: The Journal of the International Society for Developmental Psychobiology, 22*(3), 271–294.

Fernald, A., Swingley, D., & Pinto, J. P. (2001). When half a word is enough: Infants can recognize spoken words using partial phonetic information. *Child Development, 72*(4), 1003–1015.

Floccia, C., Delle Luche, C., Goslin, J., Hills, T., & Plunkett, K. (2016). The organization of the early lexicon: Evidence from auditory priming in 15- and 18-month olds. In Paper presented at the *2016 biennial international conference on infant studies*. New Orleans.

Fuentes, L. J., Vivas, A. B., & Humphreys, G. W. (1999a). Inhibitory mechanisms of attentional networks: Spatial and semantic inhibitory processing. *Journal of Experimental Psychology: Human Perception and Performance, 25*(4), 1114.

Fuentes, L. J., Vivas, A. B., & Humphreys, G. W. (1999b). Inhibitory tagging of stimulus properties in inhibition of return: Effects on semantic priming and flanker interference. *The Quarterly Journal of Experimental Psychology Section A, 52*(1), 149–164.

Glaser, W. R., & Düngelhoff, F.-J. (1984). The time course of picture-word interference. *Journal of Experimental Psychology: Human Perception and Performance, 10*(5), 640.

Hills, T. T., Maouene, M., Maouene, J., Sheya, A., & Smith, L. (2009). Categorical structure among shared features in networks of early-learned nouns. *Cognition, 112*(3), 381–396. Retrieved from http://dx.doi.org/10.1016/j.cognition.2009.06.002

Johnson, N., & Pugh, K. (1994). A cohort model of visual word recognition. *Cognitive Psychology, 26*(3), 240–346.

Karmiloff-Smith, A. (1986). From meta-processes to conscious access: Evidence from children's metalinguistic and repair data. *Cognition, 23*, 95–147.

Karmiloff-Smith, A. (1992). *Beyond modularity: A developmental perspective on cognitive science.* Cambridge, MA:MIT Press/Bradford Books.

Mani, N., & Plunkett, K. (2010). In the infant's mind's ear: Evidence for implicit naming in 18-month-olds. *Psychological Science, 21*(7), 908–913.

Mani, N., & Plunkett, K. (2011). Phonological priming and cohort effects in toddlers. *Cognition, 121*(2), 196–206.

Marcus, G., Pinker, S., Ullman, M., Hollander, J., Rosen, T., & Xu, F. (1992). Overregularisation in language acquisition. *Monographs of the Society for Research in Child Development, 57*(Serial No. 228).

Marslen-Wilson, W. D. (1987). Functional parallelism in spoken word-recognition. *Cognition, 25*, 71–102.

Marslen-Wilson, W. D., & Tyler, L. K. T. (1980). The temporal structure of spoken language understanding. *Cognition, 8*, 1–71.

Mayor, J., & Plunkett, K. (2011). A statistical estimate of infant and toddler vocabulary size from CDI analysis. *Developmental Science, 14*(4), 769–785.

Mayor, J., & Plunkett, K. (2014). Infant word recognition: Insights from TRACE simulations. *Journal of Memory and Language, 71*(1), 89–123.

McClelland, J., & Rumelhart, D. (1981). An interactive activation model of context effects in letter perception: Part 1. An account of the basic findings. *Psychological Review, 88*, 375–402.

McClelland, J. L., & Elman, J. L. (1986). The TRACE model of speech perception. *Cognitive Psychology, 18*, 1–86.

Meyer, D. E., & Schvaneveldt, R. W. (1971). Facilitation in recognizing pairs of words: Evidence of a dependence between retrieval operations. *Journal of Experimental Psychology, 90*, 227–235.

Schriefers, H., Meyer, A. S., & Levelt, W. J. (1990). Exploring the time course of lexical access in language production: Picture-word interference studies. *Journal of Memory and Language, 29*(1), 86–102.

Singh, L. (2014). One world, two languages: Cross-language semantic priming in bilingual toddlers. *Child Development, 85*, 755–766.

Styles, S., & Plunkett, K. (2009). How do infants build a semantic system? *Language and Cognition, 1*, 1–24.

Swingley, D., Pinto, J. P., & Fernald, A. (1999). Continuous processing in word recognition at 24 months. *Cognition, 71*, 73–108.

Tipper, S. P. (1985). The negative priming effect: Inhibitory priming by ignored objects. *The Quarterly Journal of Experimental Psychology, 37*(4), 571–590.

von Békésy, G. (1968). Mach-and Hering-type lateral inhibition in vision. *Vision Research, 8*(12), 1483–1499.

Weger, U. W., & Inhoff, A. W. (2006). Semantic inhibition of return is the exception rather than the rule. *Perception & Psychophysics, 68*(2), 244–253.

Wheeldon, L. R., & Monsell, S. (1994). Inhibition of spoken word production by priming a semantic competitor. *Journal of Memory and Language, 33*(3), 332–356.

9

REPRESENTATIONAL REDESCRIPTION: AN APPRECIATION OF ONE OF ANNETTE KARMILOFF-SMITH'S KEY CONTRIBUTIONS TO DEVELOPMENTAL SCIENCE

James L. McClelland

Since I first heard of the idea of representational redescription, I have been intrigued by it. Annette presented the idea in a talk she gave at UCSD when I was still on the faculty there, sometime around 1980. Annette's ideas on the subject are described in several places, perhaps most thoroughly in Karmiloff-Smith (1986), hereafter KS86. There, Annette laid out her view that our abilities progressed from an early stage in which the representations of knowledge are implicit through later stages where they are re-described, becoming accessible to explicit cognition and therefore to reflection and extension. Annette called this idea 'Representation Redescription (RR)'. It has been tantalising to me since I first heard of it. As we shall see, recent work in my laboratory is beginning to address some of the issues Annette hoped to address with the idea of representational redescription.

In her inimitably frank and direct manner, Annette made no bones about her reactions to my own work on cognitive development when she first heard about it, shortly after the publication of the PDP volumes (Rumelhart, McClelland, & the PDP research group, 1986) and her own 1986 paper. She was in the audience at Oxford in 1987 at a conference focused on the wave of new neural network modelling research that emerged in the 1980s, where I presented my connectionist simulations of developmental transitions in children's judgements of the roles of weight and distance in balance (McClelland, 1989). There I showed how a simple neural network, trained with examples of balance problems, could capture the developmental pattern that had been extensively documented by Siegler (1976). When confronted with balance problems where weight and distance vary on both sides of a balance scale, children first respond based only on the number of weights (choosing the side with more weights as the one that will 'go down'), before later reaching a stage where they consider distance from the balance point as well as weight in their judgments. She came up to me afterwards and let me know that I had addressed what she thought might be a part of the problem – the implicit knowledge aspect – but had failed to address what for her was the more important part, the redescription. Indeed, in her paper not long thereafter with Andy Clark (Clark & Karmiloff-Smith, 1993), hereafter CKS93, she and Andy explicitly discussed the role of connectionist models (including my balance scale model) in capturing the representational changes underlying the development of cognitive abilities. Discussing the representations learned in connectionist models, they wrote: 'Such representations remain implicit in the network's functioning. While this is the *endpoint* of learning in a connectionist network, in the human case it is the *starting* point for generating redescriptions of implicitly defined representations. In other words, current connectionist models account rather well for children's *initial* learning in a domain, but they do not yet adequately model the subsequent representational change posited by the RR model" (p. 488, the emphasis is in the original).

Why is representational redescription important and what is involved in it? This is discussed in KS86 pp 100–116. Four levels of representation are distinguished, although for simplicity, we can consider three, corresponding to the three phases of development Annette describes. In a first phase, implicit representations are formed, allowing, for example, a language learner to produce the correct article, marking it properly for gender, number and definiteness, when talking about an object or set of objects. During the second phase, implicit comparisons among procedures occur, and the child may go beyond what would be necessary, perhaps making errors or repairs thought to reflect the consequences of the comparison process. The child might say, to use Annette's example (pp. 113–114) 'un de mouchoir' to mean 'one handkerchief' to contrast the numerical function with non-specific reference ('un mouchoir', meaning 'a handkerchief'). During the third phase the additional marking would be dropped, so that the behaviour would look the same as the behaviour in phase one, but based on *'qualitatively different representations'* (p. 114, emphasis Annette's). The phase three, representations are thought to be potentially accessible to consciousness, and to be used in a planful way to allow the child to coordinate referential structure across a discourse, an ability lacking in children still in the earlier developmental phases. Two central assumptions are stressed in KS86: First, that this process is not at all age-related but recurs for each aspect of language or cognition in an experience-dependent fashion; and second, that it is not failure that drives the redescription process but success. Only after mastery at phase one can the child then reflect on the learned representations and progress to phase two. They do so not to correct mistakes but to build a new level of understanding.

While the details of the RR process as described in KS86 may be open to debate, it is clear that Annette was deeply insightful in calling for something beyond implicit knowledge to characterise children's (and adults') representations and cognitive operations. We do come to have a degree of strategic control over our thought processes, and exactly how we do so needs to be explained. The ideas have also been influential among neural network modellers who seek to understand the relationships between neural networks and conscious experience. The idea expressed in

CKS93 – that standard neural networks do not explicitly represent or manipulate their own knowledge – seems correct. The question, what might explicit knowledge be and how might it arise from an implicit foundation in a standard neural network has been explored extensively by Axel Cleeremans and his colleagues, building from his dissertation work with me on implicit learning (Cleeremans & McClelland, 1991).

Now, 30 years later, during the current resurgence of interest in neural networks, Andrew Lampinen, another colleague who completed his PhD in my laboratory, has been exploring how the knowledge in a neural network can be made accessible to manipulation (Lampinen & McClelland, 2020). A starting place for this work is the observation that a pattern of activation across a set of neuron-like processing units can be used to specify the strengths of the connections between neurons in a target network, relying on an intermediary network to do the translation. The pattern that specifies the connection strengths in the target network can be viewed as a more explicit representation of the implicit knowledge that might be in the connection weights of the target network, thus corresponding to a description of the knowledge in the connections. The next step is the observation that once we have a pattern of activation representing the knowledge in the connection weights, we can transform it using a neural network – the bread and butter of neural networks is transforming one pattern into another. One example Andrew has explored is learning to transform a strategy that wins at a game into a losing strategy (something we can do if we chose or if we are asked to). The neural network learns to transform the pattern that directs the target network to win a given game into a pattern that directs the network to lose it. After learning to do this for a subset of games, it can then transform the strategy for other games. Andrew has also explored using language to directly specify the pattern of activation that specifies the connection weights in the target network. A long-term possibility is that Andrew's architecture will allow us to build networks that can perform a wide range of tasks that could be specified through a linguistic representation.

Something still separates Andrew Lampinen's work from Annette's ideas about the representational redescription process. Annette saw this process as arising from within, as an active and constructive process engaged in by the learner. So far in Andrew's work, this active and constructive process is being carried out by Andrew; it does not arise from an active constructive process generated by the neural network itself. I believe human learners do have this ability, but we have yet to reach the point where it has been captured in our existing network architectures. We still have a long way to go, therefore, before fully realising Annette's vision. She has certainly given us a worthwhile target to aim for.

Bibliography

Clark, A., & Karmiloff-Smith, A. (1993). The cognizer's innards: A psychological and philosophical perspective on the development of thought. *Mind & Language 8*(4), 487–519.

Cleeremans, A., & McClelland, J. L. (1991). Learning the structure of event sequences. *Journal of Experimental Psychology: General, 120*, 235–253.

Karmiloff-Smith, A. (1986). From meta-processes to conscious access: Evidence from children's metalinguistic and repair data. *Cognition, 23*(2), 95–147.

Lampinen, A. K., & McClelland, J. L. (2020). Transforming task representations to perform novel tasks. *Proceedings of the National Academy of Sciences.* DOI:10.1073/pnas.2008852117

McClelland, J. L. (1989). Parallel distributed processing: Implications for cognition and development. In Morris, R. (Ed.)., *Parallel distributed processing: Implications for psychology and neurobiology* (pp. 8–45). New York: Oxford University Press.

Rumelhart, D. E., McClelland, J. L., & the PDP research group. (1986). *Parallel distributed processing: Explorations in the microstructure of cognition. Volumes I & II.* Cambridge, MA: MIT Press.

Siegler, R. S. (1976). Three aspects of cognitive development. *Cognitive Psychology, 8*(4), 481–520.

10

PROSPECTIVE AND LONGITUDINAL STUDIES OF THE EARLIEST ORIGINS OF LANGUAGE LEARNING IMPAIRMENTS: ANNETTE KARMILOFF-SMITH'S ONGOING LEGACY

April A. Benasich and Katherine Wolfert

Since the earliest times of Socrates, Plato and Aristotle, the puzzle of how we "become", how we move from conception to birth and on to a functioning, sentient being has fascinated philosophers and scientists. In Plato's Republic, Plato posited that knowledge is an active process through which we organize and classify our perceptions and that process then shapes the "mind". However, Plato and also Descartes definitely believed that certain things are inborn, that they occur naturally regardless of any environmental influences. Over time, pursuit of answers to this age-old question organized itself into a dichotomy – that of nature versus nurture; genes versus environment; pre-specified equations elaborated upon by experience (i.e., nativists such as Chomsky) versus a blank slate waiting to be written upon (i.e., John Locke's tabula rasa and behaviourists such as Skinner).

In the past, debates over the relative contributions of nature versus nurture often took a very rigid approach, with one side arguing that nature played the most important role and the other side vehemently suggesting that it was all nurture, all environment. Today, after years of research and often acrimonious and impassioned debate, most experts recognize that both factors play a critical role. Not only that, they also understand that nature and nurture interact in important ways throughout life. And yet, even today, if you examine closely contemporary debates in the biological sciences, in the social sciences, and in philosophy, you can still find strong echoes of that dichotomy. They are somewhat disguised with more sophisticated language and models, but are there nonetheless when you closely examine the purported mechanisms that are thought to influence behaviour.

When Annette Karmiloff-Smith (A K-S) entered this debate at the beginning of her career, she came with a particular point of view, based on her eclectic background and particularly her wide-ranging, idiosyncratic and curious, but very disciplined mind. She again asked the age-old question: How is it that human cognition emerges – and what needs to come together genetically and environmentally to permit language, memory and higher cognition to be established? How do these factors interact – and even more critically how might those interactive processes go awry to produce various developmental disabilities? Over the more than four decades of her career, Annette's elegant experimental work on the neurological and genetic bases of cognitive developmental disorders brought key cross-field insights that challenged long-held beliefs about how development unfolded.

Based on her work in the area of Williams syndrome, Down syndrome and other developmental research, A K-S and colleagues proposed a new perspective of development, strongly focused on non-linear dynamical models to capture the interaction of maturational factors, yes, under genetic control, but also taking account of the intensely complex, emergent and non-linear interactions with internal and external environmental factors (Elman et al., 1996). As A K-S notes in her 1998 paper:

> The neuroconstructivist approach to normal and atypical development fully recognizes innate biological constraints but, unlike the staunch nativist, considers them to be initially less detailed and less domain-specific as far as higher-level cognitive functions are concerned. Rather, *development itself* is seen as playing a crucial role in shaping phenotypical outcomes, with the protracted period of post- natal growth as essential in influencing the resulting domain specificity of the developing neocortex.
>
> *(p. 389-90, italics added)*

This "neuroconstructivist perspective" has begun to clarify a number of long-standing and intensely debated questions regarding the nature of developmental change and has had a major and sustained impact on the field.

The prevailing mindset at the time of her influential 1998 paper was that the brain is "modular", specifically that the brain is organized into specialized circuits or modules, which can be differentially damaged. Thus, disordered development was characterized by a damaged/broken module, often resulting in a "double dissociation", that is, a clear separation seen between damaged and intact sets of modules resulting in a deficit in only one or two processes or abilities. In contrast, A K-S tried to understand progressive developmental trajectories and the underlying cognitive processes that support them. In a number of studies she demonstrated that even behaviours that seem to fall "within the normal range" in children with developmental disorders are sustained by *different cognitive and brain processes* from the normal trajectory, and thus are NOT "intact" or typical.

The modular perspective suggested that specialized functions observed in older children and adults were innately specified and present from the earliest stages of development. A K-S tried to explain how a brain that "looked modular" could evolve from different mechanisms that were not innately specified (Karmiloff-Smith, 1992). Observing modularity through the lens of developmental change, A K-S suggested that modularization was a process that results in, rather than begins with, successively more developed and modularized knowledge representations over development. Mechanisms that started off as only "domain-relevant" were posited to become domain-specific through experience with processing varied inputs throughout development. Functions that appear highly modular later in development emerge dynamically as a function of learning and organization of knowledge.

Thus the current shift from quantifying developmental impairments solely at their end-state in school-aged children and adults, to an emphasis on studying such disorders prospectively and longitudinally to understand how alternative developmental pathways might lead to different phenotypical outcomes, can be laid directly at A K-S's door. In a more recent paper, A K-S noted that this change in perspective has important clinical implications, promoting the development of remediation strategies that target the developmental roots of deficits (Karmiloff-Smith, 2009). Such singular insights have had an enormous impact on other scientists worldwide and indeed have been seminal to our own developmental research.

Prospective longitudinal studies

Research in the Benasich lab strongly reflects the interdisciplinary cross-field approach that includes examination of developmental trajectories that A K-S espoused and championed to those she mentored and to her many collaborators (Benasich, Choudhury, Peters, & Ortiz-Mantilla, 2016). Her influence on me (AAB) and on my lab as a whole is evident at every turn. Given the technological explosion in neuroscience and the complexity of the research questions still to be answered in developmental cognitive neuroscience, my group focuses on both typical and atypical development using a prospective longitudinal approach. Our current research addresses the early neural processes necessary for normal cognitive and language development as well as the impact of disordered processing on infant neurocognitive status and later outcomes in high risk (e.g., family history of language-based learning impairments or autism; very low birth-weight preterms) or neurologically impaired infants. All of the prospective, longitudinal research is conducted on infants from 2 through 60 months and includes identifying biological substrates and potential early precursors and as well as using newly available computational neuroscience techniques to reach these goals.

We employ a multi-tiered, converging paradigms approach, including dense-array EEG (dEEG), source localization, oscillatory dynamics, auditory evoked response potentials, naturally sleeping MRI/fMRI and diffusion tensor imaging (DTI), non-invasive eye-gaze tracking (infra-red corneal reflection technology), as well as traditional infant behavioural paradigms, including operant head-turn, habituation and recognition memory tasks, and an array of standardized and research behavioural tasks. Further, we have developed state-of-the-art early assessment batteries (including both behavioural and electrophysiological measures), based on previous work in my lab, that allows evaluation of early cognitive and language development in nonverbal, motor-impaired children with early (or genetic) brain insult (e.g., Cantiani et al., 2016; Leevers, Roesler, Flax, & Benasich, 2005; Musacchia et al., 2018; Roesler et al., 2006). Many of our prospective, longitudinal studies use dEEG to address issues relating to the emergence of dynamic coordination, oscillatory signatures and synchrony in the developing brain and their role in organizing both lower-level sensory processing and higher-level function, such as language and cognition.

Emerging technology and analytic techniques are also currently being applied to allow us to collect information that provides a bridge between neural systems and behaviour across development. This has always been an important goal – and one that allows careful analysis of biological timing and cortical interactions as sensory information is processed and behaviour ensues. We have pioneered the analysis of developmental power-band and oscillation data in infants from 2 months through 4 years. Specifically, we are exploring the emergence and elaboration of dynamic coordination (oscillatory signatures) and synchrony in the developing brain and their role in organizing lower-level sensory processing as well as how such mechanisms impact higher-level functions, such as language and cognition.

In these studies, deliberate experimental design allows us to explore questions concerning change over time. A K-S reminds us that including infants in a study does not guarantee that the study is developmental in nature; truly developmental research examines change over time, whether in infancy or in older adulthood (Karmiloff-Smith, 2012). To that end, our studies incorporate both longitudinal and cross-sectional experimental groups in order to characterize the organization and course of early brain maturation. Together, the aforementioned techniques allow us to trace trajectories of typical development over the first years of life and identify divergent patterns early in development.

Origins of language learning impairments

By studying groups at higher risk of neurodevelopmental disorders, we aim to identify and develop treatments for individuals whose trajectories diverge during development. Individuals at high risk may be identified in infancy through genetic testing or reports of a history of the disorder within their immediate family. This approach captures a signature feature of A K-S's point of view as to how one might begin to observe the developmental process and identify individual differences in outcome.

A subset of the lab's work specifically examines the lower-level origins of language learning impairments, investigating ways in which rapid auditory processing deficits in infancy contribute to communication difficulties later in childhood and adulthood. Using converging neural and behavioural methodologies allows us to be more precise in describing the nature of deficits and how they evolve over time. For example, early work with infants at higher risk of language learning impairment (LLI) identified deviations in rapid auditory processing thresholds as early as 6–9 months-of-age (Benasich & Tallal, 2002; Choudhury,

Leppanen, Leevers, & Benasich, 2007). Performance on behavioural auditory processing tasks in early infancy predicted 35%–39% of the variance in expressive language measures at 16 months, providing better prediction than family history alone (Choudhury et al., 2007). Later studies using dEEG provided insight into the neural substrates of these basic auditory processing skills that involve processing and discriminating rapidly successive sounds in the tens-of-milliseconds range. Following infants longitudinally revealed atypical maturational trajectories of auditory event-related potential (ERP) peaks in children at higher risk of LLI, with N1 and P2 peaks emerging approximately 6 months after they were observed in children at lower risk (Choudhury & Benasich, 2011). These studies and others point to altered rapid auditory processing in infants at risk of LLI, and suggest that early interventions targeting these abilities could improve later language outcomes (see review: Benasich et al., 2016).

Another area of our research not only examines atypical maturational trajectories for children with a common neurodevelopmental disorder, Autism Spectrum Disorder (ASD), but also explores individual differences that emerge within a subgroup of children with ASD. Thus, we are studying 4- to 7-year-old children diagnosed with ASD who are minimally verbal (MV), defined as having an expressive vocabulary of five words or less (Cantiani et al., 2016; Roesler et al., 2013; Ortiz-Mantilla, Cantiani, Shafer, & Benasich, 2019). This is a challenging group to study, given the inability to adequately evaluate their cognitive and language skills using standardized assessments. Using a passive looking paradigm while recording EEG we found significant group differences between children with MV–ASD (4 to 7 years of age) and their typically developing age-matched controls on a non-verbal picture-word matching task. Early auditory and visual ERP components of the MV–ASD group were age-appropriate in morphology and amplitude, but slower than those of the typically developing control group. However, the later N400 component, elicited by semantically incongruent information, was absent in half of the children with ASD. Thus, minimally verbal children with ASD, as compared to typically developing age-matched peers, appear to process basic sensory inputs typically but a subset of this population diverges in their abilities to access higher-level linguistic semantic information (Cantiani et al., 2016).

Time/frequency analyses of oscillatory dynamics in the same population of children with MV–ASD revealed the underpinnings of these EEG findings (Ortiz-Mantilla et al., 2019). Oscillatory activity in the brain is widely observed at different levels and is thought to play a key role in processing neural information. Neuronal oscillations can be viewed as rhythmic changes in cortical excitability. Specifically, they represent periodic fluctuations in the activity of a single neuron or ensembles of neurons that reflect cyclic shifts in excitability of a neuronal population and thus control the population's excitatory response (Buzsaki, 2004; see Schroeder, Lakatos, Kajikawa, Partan, & Puce, 2008 for a review). Oscillations and the generation of synchronized neuronal activity play a crucial role in the activity-dependent self-organization of developing networks and also play an important role in the stabilization and pruning of connections and setting up of new networks (Liebe, Hoerzer, Logothetis, & Rainer, 2012; Luhmann et al., 2016).

Group differences were observed in two frequency bands important for visual and linguistic processing: theta (4–8 Hz) and gamma (>30 Hz). Among other roles, theta has been associated with memory-related processes, while gamma indexes attention and visual feature integration (Wang, 2010). Lower spectral power and phase coherence in theta and gamma bands were identified in occipital areas during both early and late visual processing of pictures of objects for the MV-ASD group as compared to typically developing controls. Children with MV-ASD showed quite limited processing at time frames typically associated with retrieval of cortical object representations and semantically related information. Moreover, the children in the MV-ASD group appeared to allocate fewer attentional resources to processing of the semantically related visual information. Importantly, however, we were able to categorize children in the MV-ASD group on a continuum – showing that a subset of MV-ASD children did a much better job of processing the early sensory information and also showed evidence of lagged but probable picture-word linkage (Cantiani et al., 2016). This type of research exemplifies insights that arise from A K-S's observation that rather than seeing a homogeneous modular deficit, each child will express deviation from a typical trajectory in somewhat different ways, depending on interactive processes across development.

Our most recent work examines the dynamics of early brain plasticity and the role of attention and sensory recruitment in the construction of cortical sensory maps in human infants. An early intervention developed in the lab provides infants with interactive experience processing spectrotemporally modulated non-speech sounds. This experience sharpens acoustic temporal processing abilities as well as attention in typically developing infants, with effects enduring months after intervention and generalizing to speech stimuli (Benasich, Choudhury, Realpe-Bonilla, & Roesler, 2014; Musacchia et al., 2017; Ortiz-Mantilla, Realpe-Bonilla, & Benasich, 2019). These techniques, particularly in translation, allow the possibility of remediating language disorders well before babies speak their first word and thus could have far-reaching implications. A patented interactive toy-like device – now in the prototype stage – has been designed to facilitate active technology transfer to home, daycare and clinic settings and hopefully provide "real-world" intervention solutions. This was an important goal for A K-S over her career. She recognized that the "end states" observed for Down and Williams syndromes neurocognitive phenotypes are emergent across developmental time, and thus "cognitive-level differences in older children and adults with neurodevelopmental disorders must be traced back to their more basic precursors in infancy and early childhood" (Edgin, Clark, Massand, & Karmiloff-Smith, 2015). Thus, once we identify particular deficits, then we must ask 'how can they be remediated?'

The process of identifying low-level, domain-relevant deficits early in development to explain a disorder that is seemingly domain-specific later in childhood, and developing early interventions to target these deficits, dovetails with the philosophy expressed in A K-S's 1998 paper. The research community studying developmental language disorders could be considered early adopters of this approach, but other neurodevelopmental disorders are also studied from this perspective. For example, research on autism spectrum disorder (ASD) increasingly incorporates prospective longitudinal studies on infants at high familial risk of the disorder (Jones, Gliga, Bedford, Charman, & Johnson, 2014; Szatmari et al., 2016). Interestingly, comparison between populations at risk of ASD and language disorders reveals overlap in the impairments observed early in infancy, a convergence that we are only beginning to understand (Riva et al., 2018). Part of the allure of neuroconstructivism is that it breaks down barriers between diagnoses imposed by modular approaches. Considering the progression of disorders over development opens up questions regarding the extent to which constellations of symptoms overlap, and when and how they diverge.

Computational modelling approaches

In A K-S's 1998 paper, she also emphasizes the importance of computational modelling: "One essential step toward a deeper understanding of developmental disorders is to model their various manifestations … Work of this nature is crucial in developing more constrained theories of developmental disorders". Neuroconstructivism is highly focused on using advanced computational approaches to characterize developmental trajectories. Thus, another significant influence on our lab has been willingness to conduct novel statistical analyses, including examination of dEEG phase/amplitude coupling and time/frequency signatures in infants across maturation, including low versus high gamma band EEG activity and theta/gamma coupling and their associations with language acquisition.

One initiative in our lab was accomplished via a collaboration with developmental physicists and the application of machine learning to identify selected features within infant spontaneous EEG in two infant groups who differed on risk as a function of familial history for language learning disorders (Zare, Rezvani, & Benasich, 2016). A classifier was trained on data from a subset of the participants, learning to identify features in the resting-state data that are characteristic of each group. Cross-validation confirmed that these features could be used to classify infants into high- and low-risk groups with 88% specificity, 91% precision and 79% accuracy. Application of an innovative combination of computational modelling approaches, to assess the ability of a novel, "automatic classification" algorithm to detect infants at highest risk for LLI, exemplifies the A K-S approach. Investigating multiple early paths to later impairment using converging methodologies permits earlier detection of developmental disorders that disrupt language acquisition. Hopefully, the next steps in this initiative will include trajectory analyses of prospective longitudinal EEG and MEG data from infant siblings of children with disorders such as LLI, autism spectrum disorder (ASD) and attention-deficit hyperactive disorder (ADHD) using spontaneous/resting EEG and MEG data employing similar machine learning procedures.

In other studies, we have explored the evolving dynamics of neural oscillatory patterns and the developmental neurobiology of cortical networks, linking these neural measures to subsequent linguistic and cognitive outcomes. By integrating insights gained across our own and other studies, we can further elucidate the role of oscillatory coordination in the developing brain and begin to define normative ranges of developmental circuits and their underlying neural substrates (e.g., Acar, Ortiz-Mantilla, Benasich, & Makeig, 2016, Heim, Choudhury, & Benasich, 2016; Jannesari et al., 2020; Musacchia, Ortiz-Mantilla, Realpe-Bonilla, Roesler, & Benasich, 2015; Musacchia *et al.*, 2017; Ortiz-Mantilla, Hämäläinen, Realpe-Bonilla, & Benasich, 2016).

Further, we have been using innovative approaches to studying naturally sleeping infants during non-rapid-eye-movement (NREM) sleep. In an NIMH funded study, we are examining the microstructure (i.e., spindles and slow waves) and microstates of sleep (Peters & Benasich 2020, under review). To our knowledge, we are the first and currently the only laboratory using a high-density electrophysiology setup to measure NREM sleep microstructure and microstates in infants over the first year of life, concurrent with behavioural measures. An accelerating number of studies have demonstrated that sleep plays a role in facilitating memory consolidation (e.g., Rasch & Born, 2013). However, little is currently known about the neurophysiology of infant sleep, so these studies will shed light on the typical maturation of sleep processes and their role in cognitive development. Recent research suggests that in 3- to 6-month-old infants sleep spindles may be a biomarker of information processing speed (Horváth, Hannon, Ujma, Gombos, & Plunkett, 2017; Simon et al., 2016) and in 6- to 8-month-old infants, sleep spindles and the length of daytime NREM2 sleep have been shown to be critically involved in the processing and transition of perceptual information into semantic information (Friedrich, Wilhelm, Mölle, Born, & Friederici, 2017).

Thus we are acquiring infant topographic brain activity during sleep using dense-array EEG (dEEG) concurrent with standardized developmental cognitive and behavioural measures. This high resolution, data-driven technique is used to pinpoint clusters of localized differences in spectral power on the infant scalp during sleep, which allows for robust and reliable correlational analyses with concurrent measures of behavioural development. We hypothesize that specific electrophysiological characteristics of sleep, including topography, microstructure (including morphology), network connectivity, and trajectory, will be predictive of concurrent and subsequent cognitive development.

Another study just completed examines more globally the dynamics and criticality of early brain development using avalanche theory (Beggs & Plenz, 2003), tracking the emergence of power-law behaviour in large-scale dynamics in infants across the first year of life. The theory of avalanche dynamics attempts to capture complex emergent properties that exist during "criticality", an "ideal" state of excitation-inhibition balance in which neural networks are poised to generate avalanches of spontaneous activity (Chialvo, 2010). During an avalanche, the firing of one neuron triggers action potentials in subsequent neurons, setting off a cascade of propagating neural activity (Beggs & Plenz, 2003). Large-scale cortical avalanches can be examined in the EEG signal as activation patterns across space and time that can be described by power laws. This approach has rarely been used to examine data from children but shows much promise for understanding deviations in early brain development. Indeed, in infants, scale-invariant neuronal activity has been observed in the EEG activity of preterm infants as early as 12 hours after birth (Iyer et al., 2015). Roberts and colleagues (Roberts, Iyer, Finnigan, Vanhatalo, & Breakspear, 2014) have used this signature to track recovery from burst suppression, periods of irregular neural activity, induced by hypoxia. Our analyses suggest that critical state dynamics (i.e., "criticality"), which theory and experiments have shown to be beneficial for numerous aspects of information processing in adults, including working memory (Seshadri, Klaus, Winkowski, Kanold, & Plenz, 2018), are maintained by the infant brain to process an increasingly complex environment during development (Jannesari et al., 2020, under revision, *Brain Structure and Function*).

The ongoing legacy of A K-S

Examination of both the scope and content of our research clearly reflects the influence of A K-S. Figure 10.1 summarizes the connections between the multi-tiered approach to developmental research that we've described, and the neuro-constructivist principles A K-S espoused. Reading her 1998 paper, 'Development itself is the key to understanding developmental disorders' is startling – it's 20 years old, and yet it is as current and important today as it was when first published. Perhaps the paper feels so relevant because its themes and A K-S's oft-repeated phrase, "taking development seriously", mirror a trend in contemporary neuroscience toward taking complex systems seriously. Dichotomous undercurrents still pervade discussions, but the field is moving away from simple explanations and embracing the subtleties of interactions between the brain's structure and function, genetics and changing environments. In these interactions, timing often emerges as a key player: regulating when and how these factors influence one another. Emphasis placed by neuroconstructivists on computational modelling has also become increasingly relevant, as emerging technologies allow us to examine the spatiotemporal dynamics of complex networks in greater detail. This trend toward celebrating complexity is particularly evident in developmental studies (D'Souza, D'Souza, & Karmiloff-Smith, 2017).

We've come a long way since A K-S's 1998 paper was published, but there remains much room for growth and innovation. In particular, the diagnosis and study of neurodevelopmental disorders to some extent still reflect the origins of their definitions in modular theories. As we discover similarities between the earliest origins of these disorders in infancy and comorbidity later in development, we may encounter a need to revise diagnostic criteria. This also captures the necessity of closely examining similarities in the biological origins of disorders that might be expressed differently in later life, thus looking like separate, distinct

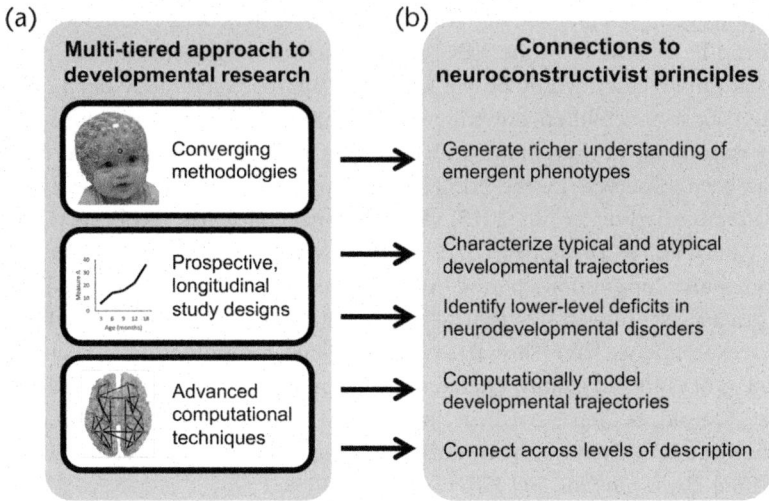

FIGURE 10.1   A summary of how the Benasich Lab's multi-tiered approach to developmental research maps onto neuroconstructivist principles. (A) We highlight several characteristics of our experimental approach, including the use of convergent methodologies such as EEG, fMRI, and behavioural methods, experimental design in the form of prospective longitudinal studies, and advanced computational techniques (e.g., Zare et al., 2016). (B) A K-S's influence on the lab and the field is evident in the connections between these techniques and neuroconstructivist principles. For example, the use of converging methodologies allows us to better understand emergent phenotypes. Prospective, longitudinal study designs enable characterization of developmental trajectories and identification of deficits in neurodevelopmental disorders present early in development. Advanced computational techniques allow for modelling of those trajectories and connection across multiple levels of description

disorders (Karmiloff-Smith et al., 2012; Scerif & Karmiloff-Smith, 2005; Benasich, Ribary, & Lupp, 2018). In a 2014 paper, A K-S and colleagues laid out important future directions and challenges that have yet to be fully implemented (Karmiloff-Smith, Casey, Massand, Tomalski, & Thomas, 2014). As was the case with "*Development Itself*" these guidelines seem prescient. Specifically, she continued to espouse an interdisciplinary, multimethod approach to neuroscience and then detailed five general issues that she and her co-authors felt must be at the heart of future research. They noted that (1) *converging methodologies*, such as electrophysiology and neurocomputational modelling, and (2) *longitudinal studies* are important tools for understanding individual differences in development. When considering theory, neuroscientists were reminded to (3) seek *connections across levels of description*, and (4) aim to understand the *emergence of phenotypes* through the interaction between genes and the environment. Finally, the authors suggested that (5) *interventions* should be designed with developmental timing and both risk and protective factors in mind (Karmiloff-Smith *et al.,* 2014).

As is evident in the research presented here from the Benasich lab and quite clear when considering the varying experimental approaches detailed in the accompanying chapters in this volume, Annette Karmiloff-Smith's point of view, deep insights and her challenge to all of us to think in a creative and fundamentally different way about how development unfolds has had a major and ongoing impact. Her ideas have fundamentally changed the way many of us approach development and developmental research – and that provides an ever-widening body of research that is now and will continue to influence both present and future scientists. That is Annette Karmiloff-Smith's ongoing legacy.

References

Acar, Z. A., Ortiz-Mantilla, S., Benasich, A., & Makeig, S. (2016). High-resolution EEG source imaging of one-year-old children. *Conference Proceedings: Annual International Conference of the IEEE Engineering in Medicine and Biology Society. IEEE Engineering in Medicine and Biology Society. Annual Conference, 2016*, 117–120. https://doi.org/10.1109/EMBC.2016.7590654

Beggs J. M., & Plenz D. (2003). Neuronal avalanches in neocortical circuits. *Journal of Neuroscience, 23*(35), 11167–11177.

Benasich, A. A., Choudhury, N. A., Realpe-Bonilla, T., & Roesler, C. P. (2014). Plasticity in developing brain: Active auditory exposure impacts prelinguistic acoustic mapping. *Journal of Neuroscience, 34*(40), 13349–13363. https://doi.org/10.1523/JNEUROSCI.0972-14.2014

Benasich, A. A., Choudhury, N., Peters, S., & Ortiz-Mantilla, S. (2016). Early screening and intervention of infants at high risk for developmental language disorders: The role of nonlinguistic rapid auditory processing. *Perspectives on Language and Literacy, 42*(1), 24–29.

Benasich, A. A., & Ribary, U. (Eds.), (2018). *Emergent Brain Dynamics: Prebirth to Adolescence*, Strüngmann Forum Reports, vol. 25, In J. Lupp, series editor. Cambridge, MA: MIT Press.

Benasich, A. A., & Tallal, P. (2002). Infant discrimination of rapid auditory cues predicts later language impairment. *Behavioural Brain Research, 136*(1), 31–49. https://doi.org/10.1016/S0166-4328(02)00098-0

Buzsaki, G. (2004). Large-scale recording of neuronal ensembles. *Nature Neuroscience, 7*, 446–451.

Cantiani, C., Choudhury, N., Yu, Y., Shafer, V., Schwartz, R., & Benasich, A. A. (2016). From sensory perception to lexical-semantic processing: An ERP study in non-verbal children with Autism. *PLoS One, 11*(8), e0161637. https://doi.org/10.1371/journal.pone.0161637

Chialvo, D. R. (2010). Emergent complex neural dynamics. *Nature Physics, 6*(10), 744–750. https://doi.org/10.1038/nphys1803

Choudhury, N., & Benasich, A. A. (2011). Maturation of auditory evoked potentials from 6 to 48 months: Prediction to 3 and 4 year language and cognitive abilities. *Clinical Neurophysiology, 122*(2), 320–338. https://doi.org/10.1016/j.clinph.2010.05.035

Choudhury, N., Leppanen, P. H. T., Leevers, H. J., & Benasich, A. A. (2007). Infant information processing and family history of specific language impairment: converging evidence for RAP deficits from two paradigms. *Developmental Science, 10*(2), 213–236. https://doi.org/10.1111/j.1467-7687.2007.00546.x

D'Souza, D., D'Souza, H., & Karmiloff-Smith, A. (2017). Precursors to language development in typically and atypically developing infants and toddlers: the importance of embracing complexity. *Journal of Child Language, 44*(3), 591–627. https://doi.org/10. 1017/S030500091700006X

Edgin J., Clark C. A., Massand E., & Karmiloff-Smith A. (2015). Building an adaptive brain across development: targets for neurorehabilitation must begin in infancy. *Frontiers in Behavioral Neuroscience, 9*(232). https://doi.org/10.3389/fnbeh.2015.00232. eCollection 2015.

Elman, J. L., Bates, E. A., Johnson, M. H., Karmiloff-Smith, A., Parisi, D., & Plunkett, K. (1996). *Rethinking Innateness: A Connectionist Perspective on Development.* MIT Press.

Friedrich, M., Wilhelm, I., Mölle, M., Born, J., & Friederici, A. D. (2017). The sleeping infant brain anticipates development. *Current Biology, 27*(15), 2374–2380.e3. https:// doi.org/10.1016/j.cub.2017.06.070

Heim, S., Choudhury, N., & Benasich, A. A. (2016). Electrocortical dynamics in children with a language-learning impairment before and after audiovisual training. *Brain Topography, 29*(3), 459–476. https://doi.org/10.1007/s10548-015-0466-y

Horváth, K., Hannon, B., Ujma, P. P., Gombos, F., & Plunkett, K. (2017). Memory in 3-month-old infants benefits from a short nap. *Developmental Science, 21*(3):e12587. https://doi.org/10.1111/desc.12587

Iyer K. K., Roberts J. A., Hellström-Westas L., Wikström S., Hansen Pupp I., Ley D., Vanhatalo S., & Breakspear M. (2015). Cortical burst dynamics predict clinical outcome early in extremely preterm infants. *Brain, 138*(8), 2206–2218.

Jannesari M., Saeedi A., Zare M., Ortiz-Mantilla S., Plenz D., Benasich A. A. (2020). Stability of neuronal avalanches and long-range temporal correlations during the first year of life in human infants. *Brain, Structure and Function, 225*(3), 1169–1183. https:// doi: 10.1007/s00429-020-02042-5. PMID: 32095900

Jones, E. J. H., Gliga, T., Bedford, R., Charman, T., & Johnson, M. H. (2014). Developmental pathways to autism: A review of prospective studies of infants at risk. *Neuroscience and Biobehavioral Reviews, 39*(100), 1–33. https://doi.org/10.1016/j.neubiorev.2013.12.001

Karmiloff-Smith, A. (1992). *Beyond Modularity: A Developmental Perspective on Cognitive Science.* Cambridge, MA: MIT Press

Karmiloff-Smith, A. (1998). Development itself is the key to understanding developmental disorders. *Trends in Cognitive Sciences, 2*(10), 389–398. https://doi.org/10.1016/S1364-6613(98)01230-3

Karmiloff-Smith, A. (2009). Nativism versus neuroconstructivism: Rethinking the study of developmental disorders. *Developmental Psychology, 45*(1), 56–63. https://doi.org/10. 1037/a0014506

Karmiloff-Smith A. (2012). Foreward: Development is not about studying children: The importance of longitudinal approaches. *American Journal on Intellectual and Developmental Disabilities, 117*(2), pp. 87–89.

Karmiloff-Smith A., Casey B. J., Massand E., Tomalski P., & Thomas M. S. (2014). Environmental and genetic influences on neurocognitive development: The importance of multiple methodologies and time-dependent intervention. *Clinical Psychological Science, 2*(5), 628–637. https://doi.org/10.1177/2167702614521188

Karmiloff-Smith A., D'Souza D., Dekker T. M., Van Herwegen J., Xu F., Rodic M., et al. (2012). Genetic and environmental vulnerabilities in children with neurodevelopmental disorders. *Proceedings of the National Academy of Sciences of the United States of America, 109*, 17261–17265. 10.1073/pnas.1121087109

Leevers, H. J., Roesler, C. P., Flax, J., & Benasich, A. A. (2005). The Carter Neurocognitive Assessment for children with severely compromised expressive language and motor skills. *Journal of Child Psychology and Psychiatry, 46*(3), 287–303. https://doi. org/10.1111/j.1469-7610.2004.00354.x

Liebe, S., Hoerzer, G. M., Logothetis, N. K., & Rainer, G. (2012). Theta coupling between V4 and prefrontal cortex predicts visual short-term memory performance. *Nature Neuroscience, 15*(3), p. 456.

Luhmann, H. J., Sinning, A., Yang, J. W., Reyes-Puerta, V., Stüttgen, M. C., Kirischuk, S., & Kilb, W. (2016). Spontaneous neuronal activity in developing neocortical networks: From single cells to large-scale interactions. *Frontiers in Neural Circuits, 10*(40). https://doi.org/10.3389/fncir.2016.00040

Musacchia, G., Ortiz-Mantilla, S., Choudhury, N., Realpe-Bonilla, T., Roesler, C., & Benasich, A. A. (2017). Active auditory experience in infancy promotes brain plasticity in Theta and Gamma oscillations. *Developmental Cognitive Neuroscience, 26*, 9–19. https://doi.org/10.1016/j.dcn.2017.04.004

Musacchia, G., Ortiz-Mantilla, S., Realpe-Bonilla, T., Roesler, C. P., & Benasich, A. A. (2015). Infant auditory processing and event-related brain oscillations. *JoVE (Journal of Visualized Experiments)*, (101), e52420–e52420. https://doi.org/10.3791/52420

Musacchia, G., Ortiz-Mantilla, S., Roesler, C. P., Rajendran, S., Morgan-Byrne, J., & Benasich, A. A. (2018). Effects of noise and age on the infant brainstem response to speech. *Clinical Neurophysiology: Official Journal of the International Federation of Clinical Neurophysiology, 129*(12), 2623–2634. https://doi.org/10.1016/j.clinph.2018.08.005

Ortiz-Mantilla, S., & Benasich, A. (2013). Neonatal electrophysiological predictors of cognitive and language development. *Developmental Medicine and Child Neurology, 55*. https://doi.org/10.1111/dmcn.12207

Ortiz-Mantilla S., Cantiani C., Shafer V. L., & Benasich A. A. (2019). Minimally-verbal children with autism show deficits in theta and gamma oscillations during processing of semantically-related visual information. *Scientific Reports, 9*(1), 5072. https://doi.org/10. 1038/s41598-019-41511-8

Ortiz-Mantilla, S., Hämäläinen, J. A., Realpe-Bonilla, T., & Benasich, A. A. (2016). Oscillatory Dynamics Underlying Perceptual Narrowing of Native Phoneme Mapping from 6 to 12 Months of Age. *Journal of Neuroscience, 36*(48), 12095–12105. https://doi. org/10.1523/JNEUROSCI.1162-16.2016

Ortiz-Mantilla, S., Realpe-Bonilla, T., & Benasich, A. A. (2019). Early interactive acoustic experience with non-speech generalizes to speech and confers a syllabic processing advantage at 9 months. *Cerebral Cortex, 29*(4), 1789–1801.

Peters, S., & Benasich, A. A. (2020). Sleep spindle topography in 6.5 month-old human infants is a sexually dimorphic biomarker of lateralized language development. Under review, *eLife*.

Rasch, B., & Born, J. (2013). About sleep's role in memory. *Physiological Reviews, 93*(2), 681–766. https://doi.org/10.1152/physrev.00032.2012

Riva, V., Cantiani, C., Mornati, G., Gallo, M., Villa, L., Mani, E., Saviozzi, I., Marino, C., & Molteni, M. (2018). Distinct ERP profiles for auditory processing in infants at-risk for autism and language impairment. *Scientific Reports, 8*(1). https://doi.org/10.1038/ s41598-017-19009-y

Roberts J. A., Iyer K. K., Finnigan S., Vanhatalo S., & Breakspear M. (2014). Scale-free bursting in human cortex following hypoxia at birth. *Journal of Neuroscience, 34*(19), 6557–6572.

Roesler, C. P., Flax, J., MacRoy-Higgins, M., Fermano, Z., Morgan-Byrne, J., & Benasich, A. A. (2013). Sensory desensitization training for successful net application and EEG/ERP acquisition in difficult to test children. *Communication Disorders Quarterly, 35*(1), 14–20.

Roesler, C. P., Paterson, S. J., Flax, J., Hahn, J. S., Kovar, C., Stashinko, E. E., & ... Benasich, A. A. (2006). Links between abnormal brain structure and cognition in holoprosencephaly. *Pediatric Neurology*, *35*(6), 387–394. https://doi.org/10.1016/j. pediatrneurol.2006.07.004

Scerif G., & Karmiloff-Smith A. (2005). The dawn of cognitive genetics? Crucial developmental caveats. *Trends in Cognitive Science*, *9*(3), 126–135.

Seshadri S., Klaus A., Winkowski D. E., Kanold P. O., Plenz D. (2018). Altered avalanche dynamics in a developmental NMDAR hypofunction model of cognitive impairment. *Translational Psychiatry*, *8*(1), 3. https://doi.org/10.1038/s41398-017-0060-z

Schroeder, C. E., Lakatos, P., Kajikawa, Y., Partan, S., & Puce, A. (2008). Neuronal oscillations and visual amplification of speech. *Trends in Cognitive Sciences*, *12*, 106–113.

Simon, K. N., Werchan, D., Goldstein, M. R., Sweeney, L., Bootzin, R. R., Nadel, L., & Gómez, R. L. (2016). Sleep confers a benefit for retention of statistical language learning in 6.5 month old infants. *Brain and Language*, *167*, 3–12. https://doi.org/10.1016/j. bandl.2016.05.002

Szatmari, P., Chawarska, K., Dawson, G., Georgiades, S., Landa, R., Lord, C., & ... Halladay, A. (2016). Prospective longitudinal studies of infant siblings of children with autism: Lessons learned and future directions. *Journal of the American Academy of Child and Adolescent Psychiatry*, *55*(3), 179–187. https://doi.org/10.1016/j.jaac.2015.12.014

Wang, X. J. (2010). Neurophysiological and computational principles of cortical rhythms in cognition. *Physiological Reviews*, *90*(3), 1195–1268. https://doi.org/10.1152/physrev. 00035.2008

Zare, M., Rezvani, Z., & Benasich, A. A. (2016). Automatic classification of 6-month-old infants at familial risk for language-based learning disorder using a support vector machine. *Clinical Neurophysiology*, *127*(7), 2695–2703. https://doi.org/10.1016/j.clinph. 2016.03.025

11

RETHINKING THE ATTENTION HOMUNCULUS THROUGH ATYPICAL DEVELOPMENT

Gaia Scerif

This chapter aims to guide my own reflections through three insights instigated by collaborations with Annette: a focus on atypical development, a call for clarity about theory and hypothesis making, and finally an interest in mechanistic understanding. After a brief journey explaining how I meandered into attention research, I cover our first collaboration: a foray into domain-general and specific processes via an investigation into an effect found across domains and commonly referred to as the semantic distance effect. I detail our findings on numerical and lexical comparisons in typically developing individuals and in individuals with Williams Syndrome (WS), a genetic condition characterised by striking weaknesses in numerical cognition and relative strengths in vocabulary because they illustrate how domain-general claims about comparison processes need to be situated in the context of domain-relevant cognitive demands. In the second section, I describe what this work has meant for me: taking development, theory and mechanisms into account to fully understand attention trajectories, outcomes and mechanisms. Again, I draw from collaborative work with Annette on the atypical development of attention and executive control in WS and fragile X syndrome (FXS), another genetic condition at very high risk for the emergence of complex attentional and broader cognitive profiles. I conclude with a brief summary of the lessons that this work has highlighted for me in the context of typical attention development: the attentional homunculus does not exist in isolation, but attentional processes need to be placed in the context of the domains with which they interact.

I am constantly fascinated by the extent to which, in the face of relatively similar visual and multisensory worlds, the focus of our attention shapes what we individually learn from these common environments, and is in turn shaped by our prior knowledge as well as interests. How did my journey in attention development begin? As a lesson to all of us who tutor and teach undergraduate students, beware of the directions that some of your reading lists may lead students towards! I began as an undergraduate student in biology with interests in animal behaviour, but the influence of inspirational mentors in psychology, my optional subject then, meant moving from marine mammals to non- and human primates. Instrumental in this migration was being taught by developmentalists along the way, from those contrasting young non-human primates and human children (Juan Carlos Gomez; Gómez, 2007), as well as social cognition in young children and infants (Kang Lee; e.g., Talwar & Lee, 2008). Juan Carlos Gomez in particular guided me to read for the first time the article that really convinced me I wanted to study developmental science (Karmiloff-Smith, 1998).

In the year 2000, I was therefore so excited to have landed at the Neurocognitive Development Unit in the Institute of Child Health at University College London, the team then led by Annette to focus on understanding atypical development. There, Annette listened quite patiently to my request to focus on attentional processes, at odds with most other work at the unit, as in those days, the majority of ongoing focused on language development. However, she first asked me to dip "my research toes" into a project investigating numerical and lexical comparisons in children, adolescents and adults with WS. That first experience will constitute the first section of this chapter because in that first summer of mentoring by Annette, as well as in discussions and work with Daniel Ansari and the other members of the Unit, three characteristics of their work already emerged very clearly, that would most definitely be echoed in the collaborations that followed. First, Annette's approach was truly developmental, making it very explicit that understanding developmental interactions across domains would be key to understanding atypical development (Karmiloff-Smith, 1998). Second, her focus was on the theoretical implications of all her empirical work: she continuously pushed us to consider and to explain to others what theoretical gains specific empirical findings would bring (following her theory-focused experimentation approach, Karmiloff-Smith, 1992). Finally, she reminded us that developmental psychologists should "not simply study children because we like them". By this, she explained, she meant that developmentalists should go beyond the description of children's behaviours, to attempt an understanding of mechanisms and what drives developmental change.

The second section of this chapter instead focuses on collaborations that developed after that initial summer. These collaborations led us to return to those three key characteristics of Annette's approach: developmental, theory-driven and mechanistic in focus. As our work together developed, longitudinal data collection meant really trying to study developmental change over time. But what was then, and is now, my theory of attention development? Annette once wrote "If you want to get ahead, get a theory" (Karmiloff-Smith & Inhelder, 1975). I took this to mean that empirical work should be grounded in theoretical positions that the work is set out to either support or revise entirely. Work with Annette urged me to return to this point over and over again: most theories of attention are based on adult models of attention, in which efficient attentional processes operate as a gateway or filter for incoming information. But how would this play out in a developmental context? To be honest, I did not know, when I first started working on attention with Annette and Jon Driver, my attention PhD mentor. However, a testament to Annette as a supervisor, not only did she point me to Jon as "the-must-have" attention supervisor, but she let me escape when I should have been writing my thesis, in 2003, and spend time in New York, to work, think and collect data with B.J. Casey, another giant in the developmental cognitive neuroscience of attentional control. That period most definitely sowed the seeds for the (much later) position statement on attention development, co-written with Dima Amso, my PhD office buddy while in New York (Amso & Scerif, 2015). In essence, Dima and I argued that adult models of attention highlight its role as a filter and bottleneck for incoming information, but that taking these processes into a developmental context suggests that, in turn and over time, attention itself is shaped by what is learnt. Finally, during the period at the Neurocognitive Development Unit, many a discussion led to considering what each of us meant by the word "mechanism"; not, overall, as the reduction of complex psychological phenomena to simpler physical phenomena (American Psychological Association, 2020) but rather as the investigation of multiple, possibly interacting, cognitive or neural factors underpinning the development of complex behaviours. I shall conclude with recent work in this direction, again inspired by Annette.

Probing the domain-generality of semantic distance effects via atypical cognition

Understanding conceptual organisation is a central aim of cognitive science. The bread and butter of this enterprise are the effects of manipulating semantic distance in speeded comparison tasks. In the context of numerical representations, the equivalent is the 'numerical distance effect' (Moyer & Landauer, 1967), 'NDE' henceforth, also known as the 'symbolic distance effect'. This is the finding that participants take longer to judge which of two simultaneously presented numbers is largest when these numerals are closer (e.g., 8 vs. 9) compared to further apart (e.g., 2 vs. 8). It has been interpreted as either indexing greater representational overlap on a 'mental number line' (Dehaene, Dehaene-Lambertz, & Cohen, 1998), where overlap increases as numerical distance decreases; or as an indicator of response-selection processes (Verguts, Fias, & Stevens, 2005) in which the similarity in representation-to-output mappings increases as numerical distance decreases (i.e., when comparing 1 vs. 2, mappings to 'larger on the left' or 'larger on the right' are more similar than when comparing 8 vs. 1). The NDE has been extensively used to study the neurocognitive basis of magnitude processing (e.g., Pinel, Dehaene, Riviere, & LeBihan, 2001; Verguts & Fias, 2004).

Similarly, research on the organisation of semantic categories has relied heavily on either lexical decision tasks or speeded matching of words and pictures of variable semantic overlap, resulting in the 'semantic/lexical distance effect', 'SDE' henceforth (Davis, Moss, Davies, & Tyler, 1999). It is therefore surprising that despite a burgeoning literature on distance effects in individual domains, cross-domain comparisons have focused primarily on contrasting stimuli across modalities (Bright, Moss, & Tyler, 2004; Piazza, Mechelli, Price, & Butterworth, 2006), discrete magnitudes (e.g., arrays of dots, Arabic numerals) and continuous extent (e.g., length, luminance, Holloway & Ansari, 2008; Cohen Kadosh et al., 2005). Fewer have investigated similarities and differences in semantic organisation across domains and stimulus types, as for example discrepancies in representational overlap for letters and numbers (Van Opstal, Gevers, De Moor, & Verguts, 2008).

Indeed, both numerical and lexical judgments require a comparison between variably overlapping semantic representations and a link to output connections during this comparison process, despite the surface differences across tasks typically employed to assess them. On the one hand, similarities in distance effects could be driven by the fact that comparisons for number and non-numerical material involve response selection processes (Gobel, Johansen-Berg, Behrens, & Rushworth, 2004). On the other hand, the magnitude comparison process may be influenced not solely by these domain-general response selection processes but also by number-specific factors constrained by the unique ordered representational features or representation-to-output mappings upon which magnitude comparisons operate (Dehaene et al., 1998; Verguts, et al., 2005). By contrast, lexico-semantic distance effects might emerge as a function of very different representational overlap, driven by functional and perceptual feature correlations across object concepts (Tyler & Moss, 2001).

In this context, Annette was central in suggesting that individuals with relative weaknesses in either numerical cognition or lexical semantics could help distinguish domain-specific and domain-general contributions to these comparison processes. In other words, if distance effects depended solely on domain-general attentional/response selection factors, one might expect equivalent effects of distance across domains. If, instead, domain-specific factors also play a role in influencing speeded decision-making, such individuals can be expected to show abilities that depend in part on the underlying competencies in each domain. Of note, populations such as individuals with developmental dyscalculia or with lexical impairments, although clinically of great interest, are identified on the bases of their difficulties and therefore present the circular problem that their difficulties are part and parcel of their diagnosis, rather than independent correlate of a specific aetiological factor. In other words, cross-domain comparisons in distance effects for dyslexia versus dyscalculia would be confounded because children are diagnosed with these conditions by virtue of domain-specific impairments in either language or number respectively. In contrast, Annette championed the view that conditions of identified genetic aetiology are a powerful tool to investigate developmental trajectories leading to dissociable cognitive strengths and weaknesses (Karmiloff-Smith, 1998) including attentional difficulties (Scerif & Baker, 2015).

Williams Syndrome is a rare genetic disorder occurring in 1 in 20,000 births, caused by a now well-understood hemizygous sub-microscopic deletion of some 24-28 genes on chromosome 7q11.23 (Donnai & Karmiloff-Smith, 2000).

Individuals with WS have a strikingly uneven cognitive profile with relative strengths in vocabulary and semantics but weaknesses in numerical cognition (e.g., Mervis & John, 2008; O'Hearn & Landau, 2007; Paterson, Brown, Gsodl, Johnson, & Karmiloff-Smith, 1999; Tyler et al., 1997). With respect to the processing of numerical magnitude in WS, Paterson, Girelli, Butterworth, and Karmiloff-Smith (2006) found that, when required to compare the magnitude of dot arrays of variable numerosity, error rates for comparisons between close rather than distant quantities were higher for the group with WS than for typically developing controls or for individuals with Down syndrome of equivalent verbal ability. These domain- and syndrome-specific difficulties are consistent with atypical cardinality understanding (Ansari et al., 2003), estimation (Ansari, Donlan, & Karmiloff-Smith, 2007), small versus large number discrimination (Van Herwegen, Ansari, Xu, & Karmiloff-Smith, 2008), and particularly severe delays in approximate numerical comparisons (Libertus, Feigenson, Halberda, & Landau, 2014).

In addition to the striking numerical cognition weaknesses, a substantial body of work on the developmental cognitive neuroscience of WS has highlighted atypical structure and function of frontoparietal circuits, including orbitofrontal and medial frontal cortices, as well as intraparietal sulcus (Meyer-Lindenberg, Mervis, & Berman, 2006), with the latter in particular implicated in the development of efficient adult numerical cognition (Ansari, 2008). Of note, while parietal involvement would predict specific weaknesses with NDE for people with WS, prefrontal abnormalities may instead lead to the prediction that weaknesses derive from a more general response selection impairment, and therefore apply to other tasks requiring comparisons (e.g., Gobel et al., 2004). The question of whether impaired numerical distance effects in WS are related to a general difficulty in making semantic comparisons or whether they are specific to the comparison of numerical magnitudes remained open.

Against this background, Annette, Daniel Ansari and I set out to explore the extent to which numerical distance effects reflect domain-specific processes or whether they provide a signature of semantic comparisons which is indistinguishable from the distance effects measured by the comparison of semantic categories in non-numerical domains. To achieve this, we used two tasks: one, a task requiring speeded comparisons of Arabic numerals and, second, a speeded word-picture comparison task designed to measure semantic processing for populations whose ability to do explicit lexical decisions is poor. We predicted that if numerical and lexico-semantic distance effects are driven by a single domain-general comparison process, individuals with WS would perform both tasks similarly. If instead domain-specific processes distinguish the decision processes involved in evaluating the distance of numerical symbols on the one hand, and lexical-semantic representations on the other, performance on the number comparison task and the lexical comparison task would break down differentially in WS.

Thirty-six participants took part in the study: 11 adolescents and adults with WS, individually matched to 22 typically developing children and adults on the bases of gender, chronological age ("CA controls") and verbal mental age, measured using the British Picture Vocabulary Scale ("MA controls", BPVS, Dunn, Dunn, Whetton, & Pintilie, 1982). For our number comparison task, participants were presented with pairs of single-digit Arabic numerals of varying numerosities, from 1 to 9. Pairs were either close together (representing a difference of 1, 2, or 3) or distant (a difference of 5, 6 or 7). A fixation cross preceded each pair of Arabic numerals and participants were asked to press the computer key located below the larger numeral. Both speed and accuracy of the comparison were emphasized and stimuli remained onscreen until a response was made. For our semantic comparison task, instead, word-picture pairs were selected to belong to one of four categories (animals, fruit/vegetables, vehicles and tools) and were matched across categories for lemma frequency, syllable length and letter length. After word onset, a picture appeared on screen and remained on display while participants decided whether the word and the picture matched. Word-picture pairs either matched (henceforth labelled "same" condition); mismatched and belonged to a different semantic category ("cross-category" condition, e.g., "cat" and picture of a bus); mismatched, but belonged to the same category and were either closely or distantly semantically related ("close" e.g., "cat" and picture of a dog or "distant", e.g. "cat" and picture of a cow conditions). Mean reaction times were calculated for the different conditions across the two tasks.

Figure 11.1 represents mean reaction times in the speeded word matching and magnitude comparison tasks. Inferential statistics revealed statistically significant effects of Distance and Group. Critically, these effects were moderated by a statistically significant three-way interaction of Domain Distance x Group. Individuals with WS found it harder to compare close numerals than close word-picture pairs, but did not differ when comparing distant pairs across domains,

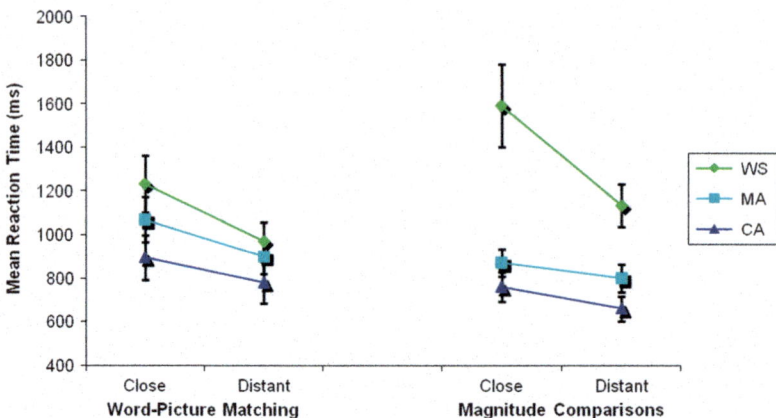

FIGURE 11.1 Mean reaction times (standard errors of the mean, S.E.M) in the speeded word-picture matching task and in magnitude comparison task for individuals with WS, MA and CA

whereas MA and CA controls responded to close and distant comparisons at similar speeds across domains. In other words, participants with WS were slower overall; they showed bigger modulation by the distance of the comparison (close vs. distant), but this effect was dependent on the domain – it was exaggerated for numerical comparisons compared to lexical comparisons. This evidence therefore was not compatible with domain-general difficulties alone having an impact on comparisons across domains. This was of our two initial dichotomous hypotheses, but WS difficulties realised in particular for numerical comparisons.

Relationships in the strength of semantic representations across domains also had the potential to be informative. Correlational analyses between semantic distance effects for the two tasks were carried out for the three groups of participants. Although MA and CA controls produced comparable distance effects across the lexical comparison task and the magnitude comparison task, variability in scores across these measures did not correlate significantly for either group, providing further evidence that the two types of semantic representations are independent in typically developing individuals. Interestingly, for individuals with WS, the larger the distance effect for the lexical comparison task, the larger the distance effect for Arabic numerals, suggesting an atypical relationship in performance across tasks. Figure 11.2

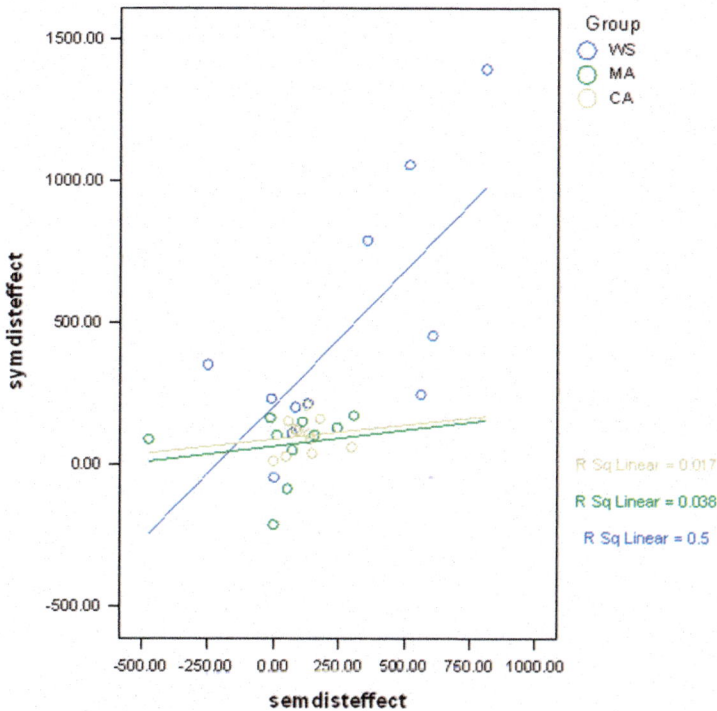

FIGURE 11.2   Correlation between the distance effect for the lexical-semantic comparisons (on the x-axis) against the distance effect for the numerical comparison task (y-axis) for the three groups

illustrates this relationship across groups. The correlation between distance effects across domains for the WS group remained significant after partialling out variability in vocabulary score, chronological age and visuo-spatial construction abilities, which may have contributed to this shared variability. This atypical relationship is interesting because it nuanced our findings with respect to our initial hypotheses. Participants with WS found difficult numerical comparisons harder than semantic comparison, suggesting some kind of domain-specific dissociation, but their individual levels of performance across the two tasks were related, suggesting a possible additional common factor helping them succeed (or putting them at risk) on both tasks.

In summary, then, taking a cross-domain approach, we investigated distance effects across tasks requiring the comparison of differences in either numerical magnitude or the semantic similarity of lexical items in typically developing adolescents and adults, as well as in individuals with Williams Syndrome. If domain-general comparison processes were truly at the core of distance effects across domains, these effects would operate similarly within groups. If, by contrast, these were to break down differentially, we would need to consider more carefully the constraints imposed by each comparison process across domains. Individuals with WS were differentially more affected by the requirement to compare numerical magnitudes than typically developing individuals and more than would be expected given their vocabulary knowledge, whereas they did not differ from controls for the comparison of lexical items. These findings are therefore consistent with typical lexical semantics (e.g., Tyler et al., 1997) and weaknesses across aspects of numerical cognition in WS (e.g., Ansari et al., 2007; Ansari et al., 2003; Paterson et al., 1999; Paterson *et al.* 2006; Van Herwegen et al., 2008).

In turn, these differences have implications for an understanding of numerical and semantic distance effects. It has been argued that distance effects reflect task difficulty and domain-general response selection (Gobel et al., 2004) and that, as such, distance effects do not index specific characteristics of a given domain (whether language or number). Alternative accounts, however, stipulate that the numerical distance effect emerges from number-specific factors, be they differing degrees of representational overlap along the mental number line (Dehaene et al., 1998), or the number-specific monotonic increases in similarity for representation-to-output mapping as distance increases (Verguts et al., 2005). The present data argue against a domain-general view, as distance effects can be affected by group membership differentially and therefore index differences in the way in which particular classes of stimuli are compared. These differences may be driven by domain-specific characteristics associated with the ordered organisation of numerical symbols and/or of their representation-to-output mappings.

Further support for a differentiation across semantic representations comes from the fact that distance effects do not correlate across typically developing participants. However, despite the much larger group-level impairment for numerical comparisons than for lexical comparisons in WS, these two indices did instead correlate for individuals with WS, a relationship that could not be accounted for by individual differences in chronological age, verbal or nonverbal skills. This atypical relationship is intriguing: it suggests that despite the striking differences across domains in WS there seem to remain some residual common constraints to distance effects. For example, individuals with WS who are most severely affected by close comparisons may be recruiting additional domain-general decision-making or memory resources that slow them down across both types of comparisons differentially more than other individuals with WS. A further non-mutually exclusive alternative may be that over developmental time these two comparisons processes differentiate over typical development, but for individuals with WS, additional challenges recruit individual skills in decision-making or memory into the process. It is possible that, in younger typically developing children, we may have seen similar correlations in comparison processes. These atypical correlational findings therefore deserve further investigation. The differences across domains point to behavioural profiles conditioned by the domain in which the comparison is taking place. However, in the WS group alone, the correlations of distance effects across domains seem to suggest a domain-general constraint. In turn, this suggests that the initially dichotomous hypotheses we started with (either domain-general only or domain-specific only) are not easily reconciled with the data. The data seem to have characteristics of both domain generality and domain specificity.

As a whole, the finding that numerical, but not lexical, semantic distance effects differ in individuals with WS and controls points to a clear differentiation across comparison processes. To an attention scientist, these findings are incompatible with an attentional control homunculus that is either fully efficient or not: one must rather think of cognitive processes involved in the comparisons as dependent on the domain-relevant demands that are characteristic of numerical representations and lexical representations, respectively. In the next section, a brief overview of further collaborations with Annette shows how those three initial characteristics of Annette's approach continued to influence my approach to attention.

Beyond the attention homunculus: Theory of development, outcomes and mechanisms

Very cogently, Annette made the point that developmental disorders have often been described in terms of domain-specific strengths and weaknesses, but that this view implicitly (or more explicitly) ignores the point that developmental disorders are developmental in nature (Karmiloff-Smith, 1998; Thomas & Karmiloff-Smith, 2002). In contrast, uneven profiles may well be the developmental outcome of gradually changing interactions with the environment, with attentional processes as the gateways to interactions with the environment (Karmiloff-Smith et al., 2012), rather than seemingly independent islets of impaired and intact ability. For example, in the context of visual exploration of quantities in the environment, eye-movement patterns that focus on detail, rather than on global characteristics of the stimuli surrounding them, may predispose young children with WS, compared to young children with DS, to extract less efficiently the properties that characterize number, and in turn, put them at risk of not being able to ground numerical symbols into well-established representations of quantity (Karmiloff-Smith et al., 2012). Interestingly, the experience with numerical and lexical distance described in section 1 suggested that particular constraints of the to-be-learnt domain must also be at play, not simple domain-general attentional biases. If attentional differences could operate to shape development of domain-specific skills, should we consider attention itself as a homunculus that exists in isolation and independently of the skills whose development it drives? Work in collaboration with Annette, after that first summer research assistantship, really questioned this view.

After that initial work with individuals with WS, I moved to working with a different model system, fragile X syndrome (FXS). This was because it really appealed to my early training as a biologist: it is caused by the silencing of a single X-linked gene, labelled FMR-1, so it is genetically relatively "simple". It is also the most common cause of genetically inherited learning difficulties, suggesting a tantalizing link between FMR-1 and the cognitive profile associated with FXS. FMR-1 codes for a pretty important but generally expressed protein, FMRP, that acts as a regulator of activity-dependent changes in neural connectivity. Our work at the Unit, as well as that of many other researchers across the world (e.g., Scerif et al., 2004; Sullivan et al., 2006), clearly demonstrated severe attention difficulties in FXS from infancy into childhood and later adulthood, above those that may be expected, given developmental delay in this population. Other cognitive domains (social cognition) are also affected very severely (e.g., Roberts, Hatton, Long, Anello, & Colombo, 2012), whereas others (such as aspects of visual processing and of receptive language) less severely so (see Doherty & Scerif, 2017 for a review). Why would the silencing of this pretty general-purpose gene have more profound effects on some cognitive processes compared to others? The intriguingly uneven profile, coupled with the general-purpose role of FMRP, made Annette and I speculate that key would be to understand why initially general changes in neurocomputational properties may affect the development of cognitive functions that require the experience-dependent shaping of short and long-range connections over prolonged periods of time to be more severely affected than others (Scerif & Karmiloff-Smith, 2005). But of course, the gap between this speculative hypothesis, theoretical models of attention and empirical evidence was open wide.

In addition, the gap between the average attentional skills of the children I saw, and the extent to which the boys differed from each other in attentional and non-attentional ways staggered me. After a home visit a long way away, or a testing session in the lab at the Unit, Annette often asked for a debriefing chat on qualitative and observational insights that the data alone could not outline. One of the most fascinating (and least studied) observation from testing sessions was the individual differences across boys with the condition. After going back to the Unit, and summarising the "average level of performance" in the young boys with FXS I had met, I kept on going back to their similarities and differences. What precisely predicted these attentional and non-attentional differences, especially given that FXS is a seemingly genetically simple condition, monogenic in its aetiology? Longitudinal work was most definitely needed to understand these differences and their association with cognitive outcomes.

In addition, these observations pointed to my need to study attention-in-context, not attention out-of-context: Could at least some of the differences in profiles depend on attention allocation to visual as opposed to auditory stimuli? We agreed that part of the puzzle would be solved by investigating longitudinal trajectories of attention and their relations to behavioural and cognitive outcomes for young children with FXS and so we set out to design our first longitudinal study. Much of this work is now published (Cornish, Cole, Longhi, Karmiloff-Smith & Scerif, 2012; Cornish, Cole, Longhi, Karmiloff-Smith, & Scerif, 2013; Scerif, Longhi, Cole, Karmiloff-Smith, & Cornish, 2012; Doherty et al., in press), but the exercise in and of itself highlighted a number of stark challenges and emerging solutions when studying atypical attention longitudinally. Among the challenges, some were methodological: how can we study where attention difficulties stem from, especially in young children who may not have complex expressive or motoric skills? By necessity, we needed to modify standardised assessment tools to study and track not only attentional skills but also lower-level perceptual skills across modalities. In addition, when studying potential outcomes of attentional differences, should we study symptoms as behaviours reported by parents and teachers? And if so, in which particular domain of symptomatology? For example, we found that visual and audiovisual attention predicted reported ADHD symptoms a year later (Scerif et al., 2012); in contrast, individual differences in auditory processes were stronger predictors of reported autism symptoms (Cornish et al., 2012).

Among the solutions, again methodological advances presented themselves: user friendlier touch-screen technology (Scerif et al., 2004) and later eye-trackers (Guy et al., in press) changed the ways in which we studied attention. This was very much in the footsteps of how Annette used these technologies to study cognitive processes underlying atypical behaviours, as in the example concerning eye-movements and extraction of magnitude information in very young children with WS or DS (Karmiloff-Smith et al., 2012) and later examples by Annette's team (D'Souza, D'Souza, Johnson, & Karmiloff-Smith, 2015). For example, although many researchers, including us, had reported atypical social anxiety symptoms in FXS, recently we used more dynamic and temporally sensitive cognitive and eye-tracking paradigms to show that, when viewing complex naturalistic scenes that include people, for children and adolescents with FXS, initial gaze aversion reduces with increased familiarity, and social landmarks within scenes are used when searching and remembering scenes. In turn, these new methods suggest that overall performance when dealing with complex scenes is underpinned by alternative means of attending, learning and remembering for children with FXS.

Concluding thoughts on moving beyond the attention homunculus: Where next?

I think that Annette would continue to push me to take a developmental, theory-driven and mechanistic approach. Annette would very much want us to continue to test developmental trajectories explicitly, as indeed Annette, Kim Cornish, Ann Steele and I attempted together in the case of children with DS and WS, for whom we investigated typical and atypical predictors of literacy development (Steele, Scerif, Cornish, & Karmiloff-Smith, 2013). We are also investigating more explicitly the longitudinal relationships between attentional skills and emerging other cognitive skills. This is something that more recently my team and I have engaged in, reviving a collaboration with Daniel Ansari, in typically developing pre-schoolers (Coolen et al., submitted).

I also think that, while longitudinal data gained us many insights when we worked together, Annette would continue to push us to pursue the theoretical relevance of this work. Dima Amso and I argued that adult models of attention highlight its role as a filter and bottleneck for incoming information. However, taking the operation of these processes into a developmental context suggests that, in turn and over time, attention itself is shaped by what is learnt. Indeed, we have since accumulated empirical evidence supporting this theoretical suggestion. For example, children's attentional orienting is guided by their memory representations (Nussenbaum, Scerif, & Nobre, 2019) and distracting information presented while learning influences attentional orienting both at encoding and recall, for adults (Doherty, Patai, Duta, Nobre, & Scerif, 2017) and children (Doherty, Fraser, Nobre, & Scerif, 2019a), leaving neural traces of the encoding process on markers of preparing one's attention orienting (Doherty et al., 2019b).

Finally, Annette would also expect us to try and understand the cognitive and neural mechanisms underpinning the temporal correlations that ultimately are modelled in longitudinal studies. Indeed, none of our data on individual differences are mechanistic, but they suggest an interface between domain-general cognitive comparisons and domain-relevant demands that may apply to lexical and ordered symbolic representations. I have continued to return to questions about how these representations may be ordered and how attentional control factors may constrain their development, and in turn be influenced by them: we have employed experimental paradigms that require participants to learn about new symbols, to then deploy comparisons on these newly learnt representations (Merkley & Scerif, 2015). This process has led us to understanding how these comparisons are learnt most effectively: in young adults and relatively old children being taught individual mappings/comparisons is sufficient, whereas for youngest children being provided ordinal information is helpful (Merkley & Scerif, under review). This is speculative, but it may be that, taken together, a combination of mechanisms lead to representations that are both ordered and flexibly linked, but that in individuals with Williams syndrome these linkages are more focused on the ordinal aspects, not on their deeper understanding, as has been suggested for their understanding of the counting sequence (Ansari et al., 2003).

In conclusion, Annette influenced all these research questions and more. For me as an attention scientist, collaborative work with her reminded me constantly that one should construe of attention development as dynamic, rather than the maturation of an independent control "homunculus". Attention influences learning and memory in specific domains, but it is in turn influenced by them. Ultimately, though, something I have not touched upon at all but would like to close on is what Annette did as more than a scientist: she was a role model, a supportive mentor, friend and ... a great deal of fun. She is sorely missed, but she continues to inspire us.

Acknowledgements

This chapter reflects on Annette Karmiloff-Smith's influence on me as a researcher over the years. Indeed, a whole section of what follows was co-written with her, as the outcome of our first experience of working together, when I was a fresh summer research assistant working with her. I have returned to that initial experience many times as a reminder of how domain-specific demands interact with domain-general processes. I also thank Daniel Ansari, who was a key collaborator on that project, for all his input on questions pertaining to symbolic processes specifically and the development of numerical cognition since then. Lorraine Tyler and Sarah Paterson provided respectively the language and numerical stimuli used to probe the domain-generality of distance effects reported early in this chapter, and Michael Thomas commented on numerous versions of the findings reported in the first section. As always I am indebted to all participants, their families and schools, without whom any of the ideas presented here could not have been developed. The Williams Syndrome Foundation and the Fragile X Society were instrumental in providing contact with individuals, families and schools and in supporting this research programme as a whole.

References

American Psychological Association (2020). *The APA Dictionary of Psychology*. Entry on "mechanistic theory".

Amso, D., & Scerif, G. (2015). The attentive brain: Insights from developmental cognitive neuroscience. *Nature Reviews Neuroscience, 16*(10), 606–619.

Ansari, D. (2008). Effects of development and enculturation on number representation in the brain. *Nature Reviews Neuroscience, 9*(4), 278–291. doi: 10.1038/nrn2334.

Ansari, D., Donlan, C., & Karmiloff-Smith, A. (2007). Typical and atypical development of visual estimation abilities. *Cortex, 43*(6), 758–768.

Ansari, D., Donlan, C., Thomas, M. S. C., Ewing, S. A., Peen, T., & Karmiloff-Smith, A. (2003). What makes counting count? Verbal and visuo-spatial contributions to typical and atypical number development. *Journal of Experimental Child Psychology, 85*(1), 50–62.

Bright, P., Moss, H., & Tyler, L. K. (2004). Unitary vs multiple semantics: PET studies of word and picture processing. *Brain and Language, 89*(3), 417–432.

Bunn, E. M., Tyler, L. K., & Moss, H. E. (1998). Category-specific semantic deficits: The role of familiarity and property type reexamined. *Neuropsychology, 12*(3), 367–379.

Cohen Kadosh, R. C., Henik, A., Rubinsten, O., Mohr, H., Dori, H., van de Ven, V., et al. (2005). Are numbers special? The comparison systems of the human brain investigated by fMRI. *Neuropsychologia, 43*(9), 1238–1248.

Cornish, K., Cole, V., Longhi, E., Karmiloff-Smith, A., & Scerif, G. (2012). Does attention constrain developmental trajectories in fragile X syndrome? A 3-year prospective longitudinal study. *American Journal on Intellectual and Developmental Disabilities, 117*, 103–120.

Cornish, K., Cole, V., Longhi, E., Karmiloff-Smith, A., & Scerif, G. (2013). Mapping developmental trajectories of attention and working memory in fragile X syndrome: developmental freeze or developmental change? *Development and Psychopathology, 25*, 365–376.

Cornish, K., Cole, V., Longhi, E., Karmiloff-Smith, A., & Scerif, G. (2013). Do behavioural inattention and hyperactivity exacerbate cognitive difficulties associated with autistic symptoms? Longitudinal profiles in fragile X syndrome. *International Journal of Developmental Disabilities, 59*, 80–94.

Coolen, I., Merkley, R., Dove, E., Dowker, A., Mills, A., Murphy, V., von Spreckelsen. M., Ansari, A., & Scerif, G. (submitted). Domain-general and domain-specific influences on emerging numerical cognition: Contrasting uni-and bidirectional prediction models.

Davis, M. H., Moss, H. E., Davies, P. D., & Tyler, L. K. (1999). Spot the difference: Investigations of conceptual structure for living things and artifacts using speeded word-picture matching. *Brain and Language, 69*(3), 411–414.

Dehaene, S., Dehaene-Lambertz, G., & Cohen, L. (1998). Abstract representations of numbers in the animal and human brain. *Trends in Neurosciences, 21*(8), 355–361.

Doherty, B. R., & Scerif, G. (2017). Genetic syndromes and developmental risk for autism spectrum and attention deficit hyperactivity disorders: Insights from fragile X syndrome. *Child Development Perspectives, 11*, 161–166.

Doherty B. R., Patai E. Z., Duta M., Nobre A. C., Scerif G. (2017). The functional consequences of social distraction: Attention and memory for complex scenes. *Cognition, 158*, 215–223.

Doherty B. R., Fraser A., Nobre A. C., Scerif G. (2019a). The functional consequences of social attention on memory precision and on memory-guided orienting in development. *Developmental Cognitive Neuroscience, 36*, 100625. doi: 10.1016/j.dcn.2019.100625.

Doherty B. R., van Ede F., Fraser A., Patai E. Z., Nobre A. C., Scerif G. (2019b). The functional consequences of social attention for memory-guided attention orienting and anticipatory neural dynamics. *Journal of Cognitive Neuroscience, 31*(5), 686–698. doi: 10. 1162/jocn_a_01379

Donnai, D., & Karmiloff-Smith, A. (2000). Williams syndrome: From genotype through to the cognitive phenotype. *American Journal of Medical Genetics, 97*(2), 164–171.

D'Souza D., D'Souza H., Johnson M. H., Karmiloff-Smith A. (2015). Concurrent relations between face scanning and language: A cross-syndrome infant study. *PLoS One, 10*(10), e0139319. doi: 10.1371/journal.pone.0139319.

Dunn, L. M., Dunn, L. M., Whetton, C., & Pintilie, D. (1982). *British Picture Vocabulary Scale*. Windsor: NFER-Nelson.

Gobel, S. M., Johansen-Berg, H., Behrens, T., & Rushworth, M. F. S. (2004). Response-selection-related parietal activation during number comparison. *Journal of Cognitive Neuroscience, 16*(9), 1536–1551.

Gómez J. C. (2007). Pointing behaviors in apes and human infants: a balanced interpretation. *Child Development*, *78*(3), 729–734.

Holloway, I. D., & Ansari, D. (2008). Domain-specific and domain-general changes in children's development of number comparison. *Developmental Science*, *11*(5), 644–649.

Holloway, I. D., & Ansari, D. (2009). Mapping numerical magnitudes onto symbols: The numerical distance effect and individual differences in children's mathematics achievement. *Journal of Experimental Child Psychology*, *103*(1), 17–29.

Karmiloff-Smith, A. (1992). *Beyond modularity: A developmental perspective on cognitive science*. Cambridge: MIT Press/Bradford Books.

Karmiloff-Smith A. (1998). Development itself is the key to understanding developmental disorders. *Trends in Cognitive Sciences*, 1998 Oct 1, *2*(10), 389–398.

Karmiloff-Smith, A., & Inhelder, B. (1975). If you want to get ahead, get a theory. *Cognition*, *3*, 195–212.

Karmiloff-Smith A., D'Souza D., Dekker T. M., Van Herwegen J., Xu F., Rodic M., Ansari D. (2012). Genetic and environmental vulnerabilities in children with neurodevelopmental disorders. *Proceedings of the National Academy of Sciences of the United States of America*, 2012 Oct 16, *109*(Suppl 2), 17261–17265. doi: 10.1073/pnas.1121087109.

Libertus M. E., Feigenson L., Halberda J., Landau B. (2014). Understanding the mapping between numerical approximation and number words: evidence from Williams syndrome and typical development. *Developmental Science*, 2014 Nov, *17*(6), 905–919. doi: 10.1111/desc.12154.

Merkley, R., & Scerif, G. (2015). Continuous visual properties of number influence the formation of novel symbolic representations. *Quarterly Journal of Experimental Psychology*, *68*(9), 1860–1870.

Merkley, R., Shimi, A., & Scerif, G. (2016). Electrophysiological markers of newly acquired symbolic numerical representations: The role of magnitude and ordinal information. *ZDM Mathematics Education*, *48*, 279. doi: 10.1007/s11858-015-0751-y.

Mervis, C. B., & John, A. E. (2008). Vocabulary abilities of children with Williams syndrome: Strengths, weaknesses, and relation to visuospatial construction ability. *Journal of Speech, Language, and Hearing Research*, *51*(4), 967–982.

Meyer-Lindenberg, A., Mervis, C. B., & Berman, K. F. (2006). Neural mechanisms in Williams syndrome: a unique window to genetic influences on cognition and behaviour. *Nature Reviews. Neuroscience*, *7*(5), 380-393. doi: 10.1038/nrn1906.

Moyer, R. S., & Landauer, T. K. (1967). Time required for judgements of numerical inequality. *Nature*, *215*(5109), 1519-1520.

Nussenbaum K., Scerif G., & Nobre A. C. (2019). Differential effects of salient visual events on memory-guided attention in adults and children. *Child Development*, *90*(4), 1369–1388. doi: 10.1111/cdev.13149.

O'Hearn, K. & Landau, B. (2007). Mathematical skill in individuals with Williams syndrome: evidence from a standardized mathematics battery. *Brain and Cognition*, *64*(3), 238–246. doi: 10.1016/j.bandc.2007.03.005.

Paterson, S. J., Brown, J. H., Gsodl, M. K., Johnson, M. H., & Karmiloff-Smith, A. (1999). Cognitive modularity and genetic disorders. *Science*, *286*(5448), 2355–2358.

Paterson, S. J., Girelli, L., Butterworth, B., & Karmiloff-Smith, A. (2006). Are numerical impairments syndrome specific? Evidence from Williams syndrome and Down's syndrome. *Journal of Child Psychology and Psychiatry*, *47*(2), 190–204.

Piazza, M., Mechelli, A., Price, C. J., & Butterworth, B. (2006). Exact and approximate judgements of visual and auditory numerosity: An fMRI study. *Brain Research*, *1106*, 177–188.

Pilgrim, L. K., Moss, H. E., & Tyler, L. K. (2005). Semantic processing of living and nonliving concepts across the cerebral hemispheres. *Brain and Language, 94*(1), 86–93.

Pinel, P., Dehaene, S., Riviere, D., & LeBihan, D. (2001). Modulation of parietal activation by semantic distance in a number comparison task. *NeuroImage, 14*(5), 1013–1026.

Roberts, J. E., Hatton, D. D., Long, A. C. J., Anello, V., & Colombo, J. (2012). Visual attention and autistic behavior in infants with fragile X syndrome. *Journal of Autism and Developmental Disorders, 42*(6), 937–946. doi: 10.1007/s10803-011-1316-8.

Scerif G., Karmiloff-Smith A. (2005). The dawn of cognitive genetics? Crucial developmental caveats. *Trends in Cognitive Sciences,* 2005 Mar, *9*(3), 126–135.

Scerif G., Baker K. (2015). Annual research review: Rare genotypes and childhood psychopathology--uncovering diverse developmental mechanisms of ADHD risk. *Journal of Child Psychology and Psychiatry and Allied Disciplines,* 2015 Mar, *56*(3), 251–273. doi: 10.1111/jcpp.12374.

Scerif, G., Cornish, K., Wilding, J., Driver, J., & Karmiloff-Smith, A. (2004). Visual search in typically developing toddlers and toddlers with Fragile X or Williams syndrome. *Developmental Science, 7*(1), 116–130. doi: 10.1111/j.1467-7687.2004.00327.x

Scerif, G., Longhi, E., Cole, V., Karmiloff-Smith, A., & Cornish, K. (2012). Attention across modalities as a longitudinal predictor of early outcomes: The case of fragile X syndrome. *Journal of Child Psychology and Psychiatry, 53,* 641–650.

Steele, A., Scerif, G., Cornish, K. M., & Karmiloff-Smith, A. (2013). Learning to read in Williams syndrome and Down syndrome: Syndrome-specific precursors and developmental trajectories. *Journal of Child Psychology and Psychiatry, 54*(7), 754–762.

Sullivan, K., Hatton, D., Hammer, J., Sideris, J., Hooper, S., Ornstein, P., & Bailey, D. B. (2006). ADHD symptoms in children with FXS. *American Journal of Medical Genetics, Part A, 140*(21), 2275–2288. doi: 10.1002/ajmg.a.31388.

Talwar V., & Lee, K. (2008). Social and cognitive correlates of children's lying behavior. *Child Development,* 2008 Jul-Aug, *79*(4), 866–881.

Thomas, M., & Karmiloff-Smith, A. (2002). Are developmental disorders like cases of adult brain damage? Implications from connectionist modelling. *The Behavioral and Brain Sciences. 25*(6), 727–787. doi: 10.1017/s0140525x02000134.

Tyler, L. K., KarmiloffSmith, A., Voice, J. K., Stevens, T., Grant, J., Udwin, O., et al. (1997). Do individuals with Williams syndrome have bizarre semantics? Evidence for lexical organization using an on-line task. *Cortex, 33*(3), 515–527.

Tyler, L. K., & Moss, H. E. (2001). Towards a distributed account of conceptual knowledge. *Trends in Cognitive Sciences, 5*(6), 244–252.

Van Herwegen, J., Ansari, D., Xu, F., & Karmiloff-Smith, A. (2008). Small and large number processing in infants and toddlers with Williams syndrome. *Developmental Science, 11*(5), 637–643.

Van Opstal, F., Gevers, W., De Moor, W., & Verguts, T. (2008). Dissecting the symbolic distance effect: Comparison and priming effects in numerical and nonnumerical orders. *Psychonomic Bulletin & Review, 15*(2), 419–425.

Verguts, T., & Fias, W. (2004). Representation of number in animals and humans: A neural model. *Journal of Cognitive Neuroscience, 16*(9), 1493–1504.

Verguts, T., Fias, W., & Stevens, M. (2005). A model of exact small-number representation. *Psychonomic Bulletin & Review, 12*(1), 66–80.

12

WHAT HAS CHANGED IN 18 YEARS? REFLECTIONS ON ANSARI & KARMILOFF-SMITH (2002)

Daniel Ansari

Introduction

In every life there are certain people who represent real turning points. Meeting Annette Karmiloff-Smith and having the opportunity to work under her supervision was one of those significant turning points in my life for which I will eternally grateful. Annette was my doctoral supervisor. I worked under her supervision from 1999 to 2003 and continued collaborating with Annette long after that. She pushed me out of my comfort limits, made me think harder than anybody had done before and inspired me to have the confidence to try being a scientist. To this day, I take much inspiration from things that Annette said to me. What I remember most about Annette is the incredible commitment to mentorship of trainees at all levels that Annette demonstrated day in and day out. Annette put her trainees first. She supported us, gave us detailed, constructive feedback on our writing and presentations, told us to get our act together when it was necessary to do so and worked hard to build a cohesive and collaborative team of researchers ranging from undergraduates to post-doctoral fellows. Furthermore, Annette used her high international reputation in the field to promote us and generously open doors for her trainees. In what follows I will describe an example of this, which kickstarted my career and probably helped me land my first post-PhD academic job.

I had been a member of Annette's Neurocognitive Development Unit at the Institute of Child Health in London for just over a year and was very much still trying to find my feet. She had been invited by *Trends in Cognitive Sciences* to contribute a review paper and wanted to offer me the opportunity to take the lead on writing a review on what we knew at the time about the typical and atypical development of numerical and mathematical skills (the focus of my PhD thesis). While I was, of course, thrilled to have this incredible opportunity, I have to confess that I was also very daunted by this opportunity, not least because Annette published a truly spellbinding review entitled: "Development itself is the key to understanding developmental disorders" in the same journal just a few years prior (Karmiloff-Smith, 1998). This was a paper which I had read during my undergraduate degree and was already having a huge influence on researchers working on developmental disorders in multiple fields (according to Google Scholar this paper has been cited over 1,400 times). The idea of writing a similar paper was very intimidating. In the end, the process of writing the review paper turned out to be a tremendous learning experience. Annette provided me with detailed, constructive comments on countless drafts, and I feel writing that paper helped me become a better scientific writer (of course a process that is still ongoing). She also contributed a tremendous amount to the theoretical underpinnings of the paper.

In the paper, Annette and I discussed what was then known about the typical and atypical development of numerical cognition. Broadly speaking, consistent with the neuroconstructive perspective articulated by Annette and others (Elman et al., 1996; Mareschal et al., 2007), we argued that in order to truly understand the origins of developmental disorders, a developmental perspective, rather than one coming from adult neuropsychology, was necessary.

Our review paper entitled: "Atypical trajectories of number development: a neuroconstructivist perspective" was published in 2002 (Ansari & Karmiloff-Smith, 2002). Eighteen years later and in the context of this *Festschrift*, I would like to return to the outstanding questions (a section included in all *Trends in Cognitive Sciences* review papers) that Annette and I raised at the end of the article and reflect on how far we have (or not) come in the 18 years since. The aim here is not to review, in detail, the background that motivated each of the questions. Nor is it to comprehensively reflect on how much of the question has been answered, but to provide a reflection on what has been achieved, how new research informs the questions we posed, what still remains to be answered and what approaches may be most fruitful in doing so.

Outstanding questions from Ansari & Karmiloff-Smith (2002):

Do analog and symbolic systems of number develop independently or become progressively integrated or separated over developmental time?

The extent to which our symbolic numerical abilities (e.g., use of count words and Arabic numerals) are grounded in an analog system for the non-symbolic representation of numerical quantity (e.g., sets of items) remains a question of significant debate. When Annette and I posed this question in 2002 the dominant hypothesis was that humans are born with (1) a system for the exact representation of small sets and (2) an analog system for the representation of large sets (nowadays often referred to as the 'Approximate Number System') and that these systems scaffold the acquisition of symbolic numerical abilities (such as learning the meaning of number words). Since 2002, a growing body of evidence has challenged this dominant hypothesis (for reviews see Carey & Barner, 2019; Leibovich & Ansari, 2016; Núñez, 2017). It has shown that, contrary to the prediction that children learn the meaning of numerical symbols by connecting them to pre-existing representations, the development of symbolic number appears to be independent of the analog magnitude system. Indeed, the development of symbolic numerical abilities influences the analog magnitude system rather than the other way around (e.g., Lyons, Bugden, Zheng, De Jesus, & Ansari, 2018; Matejko & Ansari, 2016; Mussolin, Nys, Content, & Leybaert, 2014). In other words, when children learn numerical symbols it changes their processing of analog magnitudes rather than analog magnitude developmentally scaffolding the development of symbolic numerical abilities.

Beyond correlational approaches, training studies, which provide a more direct test of the potential causal influence of analog, non-symbolic numerical processing on symbolic number learning, have not yielded convincing evidence that training the analog system in young children leads to improvement in symbolic number processing (for reviews see: Inglis, Batchelor, Gilmore, & Watson, 2017; Szűcs & Myers, 2017).

These and other data suggest that the answer to the question we posed in 2002 is that the analog and symbolic systems of number develop independently. However, it is perhaps too early to provide such a firm answer to our 2002 question. There is still much work to be done. Especially, we need more longitudinal data starting in infancy and tracking developmental changes in the analog and symbolic systems from an earlier age to truly understand whether these systems develop independently and/or become progressively integrated. Moreover, it is possible that there may be different groups of children who rely on symbolic and analog representations to different extents at different points in their developmental trajectory.

Annette saw data on infants' early cognitive abilities as the key to understanding development because they provide us with information on the starting state of any cognitive system, including number. As she pointed out repeatedly, the study of development allows us to provide accounts of how children move from the starting state to the end-state and, importantly, that information about the starting state of a system does not necessarily tell us about the end-state without taking development into account (Paterson, Brown, Gsödl, Johnson, & Karmiloff-Smith, 1999). However, as we pursue further answers to the above question Annette and I posed in 2002, we need to carefully evaluate the existing data we have from infants on the origins of analog numerical magnitude processing. Recently, Rachael Smyth and I conducted a meta-analysis of the existing data in infants' numerical abilities (using a tool called p-curve which assesses the evidential value of a particular set of studies that address a common hypothesis). What we found is that the evidential value for the hypothesis that infants can discriminate between analog magnitudes is weak and that published studies suffered from low statistical power (Smyth & Ansari, 2020). Therefore, in my view, before we pursue further studies of the relationship between the starting and end-state of numerical development and how different systems relate to one another, developmentally, we ought to carefully assess, with adequate statistical power, what infants know about analog magnitudes.

What developmental changes explain why young children do not have a full grasp of the meaning of counting, despite the surprising numerical abilities of preverbal infants?

The second outstanding question that Annette and I raised in our 2002 paper is very closely related to the first. At the time all the data pointed to, as we put it, 'surprising numerical abilities' among pre-verbal infants. Fei Xu and Elizabeth Spelke had just published the first report of data that suggested that 6-month-old infants could discriminate between large sets of dots (Xu & Spelke, 2000). In particular, they had demonstrated that 6-month-old infants could discriminate between 6 and 18 but not between 8 and 12 dots. These were the first data to suggest that young infants had an analog system. At the same time, data on how children learn the meaning of number words, and in particular the relationship between number words and quantities (e.g., that the number word 'five' refers to sets of five items), suggested that children's understanding of number words develops gradually and it is not until around 3 and a half years of age that children understand that the function of counting is to enumerate the total number of items in a set (also referred to as cardinality). Thus, there existed a significant gap between what was being discovered about infant numerical abilities and children's understanding of the meaning of counting.

Eighteen years on, we still have not come to a full understanding of the developmental changes that allow children to 'bridge the gap' between early sensitivity to number and an understanding of the meaning of counting. And, given my reflection on the first question we posed back in 2002 (see above), we need to question whether this is even the right question to ask. Not only have researchers, since 2002, amassed data to suggest that there may not necessarily be a strong connection between the analog representation of numerical magnitude and the development of symbolic number skills (including the understanding of the meaning of counting), but, moreover, it is unclear to what extent we may have overestimated the numerical abilities of young infants.

What factors explain the co-morbidity of dyslexia and dyscalculia? Are common brain/cognitive systems affected, are both particularly vulnerable to atypical development or is number even more at risk than reading?

When we asked this question in 2002, dyscalculia and dyslexia were understood primarily as separate specific learning disorders (though Annette was always very critical of the term 'specific' disorders, much like she did not like the term 'domain-specific'; see Karmiloff-Smith, 2015). According to the 'specific' characterization, a child with developmental dyscalculia presents with significantly below average numerical and mathematical skills but otherwise normal cognitive competencies, including intelligence and reading abilities. Similarly, a child with developmental dyslexia was thought to have below-average reading abilities, but normal intelligence and math ability. While there is no doubt that there are some children who present with such pure profiles (Landerl, Bevan, & Butterworth, 2004), the evidence that comorbidity of reading and math difficulties is common among children is overwhelming. Importantly, since 2002, it has been repeatedly shown that similar genes affect reading and math development (Kovas & Plomin, 2007; Daucourt, Erbeli, Little, Haughbrook, & Hart, 2020) and that common genetic vulnerability may explain, in part, the substantial comorbidity of dyslexia and dyscalculia. Furthermore, recent brain imaging studies that have used cutting-edge multivariate analytic methods have shown that the patterns of brain activity that are correlated with doing numerical tasks are highly similar in children with dyscalculia and dyslexia and, furthermore, that the brain activation patterns of children with 'pure dyscalculia' are indistinguishable from those with comorbid dyscalculia and dyslexia (Peters, Bulthé, Daniels, Op de Beeck, & De Smedt, 2018; Moreau, Wilson, McKay, Nihill, & Waldie, 2018). Data such as these have lead researchers to argue that in order to make progress in understanding developmental disorders of reading and math it is necessary to focus on common underlying factors rather than view them as completely separate disorders (e.g., Peters & Ansari, 2019)

In children with dyscalculia, are their other abilities really normal or can we detect subtle representational impairments?

As discussed in answer to the previous questions, it has become abundantly apparent that children with dyscalculia frequently present with comorbidities with other learning disorders such as dyslexia and Attention Deficit Hyperactivity Disorder (ADHD). In addition to comorbidities with other learning difficulties, researchers have shown that children with dyscalculia often present with severe impairments of visuo-spatial cognition, especially visuo-spatial working memory (Fias, Menon, & Szucs, 2013; Szűcs, 2016; Szucs, Devine, Soltesz, Nobes, & Gabriel, 2013). Furthermore, children with developmental dyscalculia have been shown to have impairments in logical reasoning (Morsanyi, Devine, Nobes, & Szűcs, 2013). Eighteen years since Annette and I posed this question the answer to it, the evidence now clearly points to the second alternative: children with dyscalculia do not just have impairments in numerical and mathematical processing but in other domains too. These impairments might provide the key to understanding the developmental origins of developmental dyscalculia as well as point to candidate mechanisms that could explain the comorbidity of dyscalculia with other learning difficulties.

Which aspects of numeracy deficits in genetic disorders are due to syndrome-specific causes and which to syndrome-general impairments?

My reading of the literature suggests that 18 years on we do not have a good answer to this question. We do know that syndrome-specific causes of numeracy deficits at the genetic level are unlikely. This is because our understanding of the role of genetic variants in learning has changed substantially. When Annette and I wrote our review in 2002, the 'candidate gene' approach was still highly influential. In other words, researchers were associating one (or a small number of) genetic variants with a cognitive phenotype. Annette had always been critical of this approach (for a review see Scerif & Karmiloff-Smith, 2005). Since then it has become clear that variability in cognitive phenotypes is explained by variants across many, many genes (also referred to as polygenic effects), with each variant explaining a tiny fraction of the overall variance in, say, numerical abilities; and, importantly, that these genetic variants do not have domain-specific effects because the same genes explain variability in, for example, both reading and mathematics (Kovas & Plomin, 2007; Hagenaars et al., 2016). It is unlikely, therefore, that studying numerical abilities in children with genetic disorders will lead us to isolate specific genes for specific math abilities. However, it is possible that the developmental mechanisms that link genotype to phenotype could lead to differences in the mathematical difficulties experienced by children with different developmental disorders. To date, I do not think that we have a clear answer to this question. A recent systematic review (Van Herwegen & Simms, 2020) of the numerical and mathematical difficulties experienced by children with Williams Syndrome (WS, the genetic disorder that was the focus of much of Annette's work from the mid-nineties onwards) suggests that, as of now, we do not even have enough research to fully characterize the mathematical difficulties experienced by children with WS alone. Given that our knowledge of mathematical learning difficulties within genetic disorders is still so limited, it seems likely that we are still a long way off from understanding how exactly mathematical difficulties differ between disorders.

Concluding thoughts

In the above, I have provided some reflections on a set of outstanding questions pertaining to the typical and atypical development of numerical and mathematical cognition that Annette and I raised some 18 years ago. This process has illustrated to me how slow progress is made in science and how many of the questions only have, at best, partial answers. Our quest for more knowledge to inform these questions clearly continues and many aspects of the questions we outlined remain unanswered. At the same time, I think it is a fair reflection of the literature to say that the analog magnitude system does not play as greater role in shaping the development of numerical and mathematical abilities as previously assumed. In addition, this process has reminded me of the sharpness and clarity of Annette's thinking and writing. I and countless others benefitted in immeasurable ways from her mentorship and guidance and owe her tremendous gratitude. Moreover, it has reminded me of how forward looking Annette was in her thinking and questioning. She was never shy to challenge received wisdom, such as the notion of 'specific learning disorders' or 'specific genes' for 'specific cognitive functions'. Annette was always ahead of the times. I feel so fortunate to have been given an opportunity to work with her on a theoretical review so early on in my career.

References

Ansari, D., & Karmiloff-Smith, A. (2002). Atypical trajectories of number development: A neuroconstructivist perspective. *Trends in Cognitive Sciences*, 6(12). https://doi.org/10.1016/S1364-6613(02)02040-5

Carey, S., & Barner, D. (2019). Ontogenetic origins of human integer representations. *Trends in Cognitive Sciences*. https://doi.org/10.1016/j.tics.2019.07.004

Daucourt, M. C., Erbeli, F., Little, C. W., Haughbrook, R., & Hart, S. A. (2020). A meta-analytical review of the genetic and environmental correlations between reading and attention-deficit/hyperactivity disorder symptoms and reading and math. *Scientific Studies of Reading*, 24(1), 23–56. https://doi.org/10.1080/10888438.2019.1631827

Elman, J. L., Bates, E. A., Johnson, M. H., Karmiloff-Smith, A., Parisi, D., & Plunkett, K. (1996). Rethinking innateness: A connectionist perspective on development. Cambridge, MA: MIT Press.

Fias, W., Menon, V., & Szucs, D. (2013). Multiple components of developmental dyscalculia. *Trends in Neuroscience and Education*, 2(2), 43–47.

Hagenaars, S. P., Harris, S. E., Davies, G., Hill, W. D., Liewald, D. C. M., Ritchie, S. J., ... Deary, I. J. (2016). Shared genetic aetiology between cognitive functions and physical and mental health in UK Biobank (N=112 151) and 24 GWAS consortia. *Molecular Psychiatry*. https://doi.org/10.1038/mp.2015.225

Inglis, M., Batchelor, S., Gilmore, C., & Watson, D. G. (2017). Is the ANS linked to mathematics performance? *Behavioral and Brain Sciences*, 40, e174. https://doi.org/10.1017/S0140525X16002120

Karmiloff-Smith, A. (1998). Development itself is the key to understanding developmental disorders. *Trends in Cognitive Sciences*. https://doi.org/10.1016/S1364-6613(98)01230-3

Karmiloff-Smith, A. (2015). An alternative to domain-general or domain-specific frameworks for theorizing about human evolution and ontogenesis. *AIMS Neuroscience*, 2(2), 91–104. https://doi.org/10.3934/Neuroscience.2015.2.91

Kovas, Y., & Plomin, R. (2007). Learning abilities and disabilities: Generalist genes, specialist environments. *Current Directions in Psychological Science*, 16(5), 284–288. https://doi.org/10.1111/j.1467-8721.2007.00521.x

Landerl, K., Bevan, A., & Butterworth, B. (2004). Developmental dyscalculia and basic numerical capacities: A study of 8-9-year-old students. *Cognition*. https://doi.org/10.1016/j.cognition.2003.11.004

Leibovich, T., & Ansari, D. (2016). The symbol-grounding problem in numerical cognition: A review of theory, evidence, and outstanding questions. *Canadian Journal of Experimental Psychology*, 70(1). https://doi.org/10.1037/cep0000070

Lyons, I. M., Bugden, S., Zheng, S., De Jesus, S., & Ansari, D. (2018). Symbolic number skills predict growth in nonsymbolic number skills in kindergarteners. *Developmental Psychology*. https://doi.org/10.1037/dev0000445

Mareschal, D., Johnson, M., Sirios, S., Spratling, M., Thomas, M. S. C., & Westermann, G. (2007). *Neuroconstructivism: How the brain constructs cognition.* Oxford: Oxford University Press.

Matejko, A. A., & Ansari, D. (2016). Trajectories of symbolic and nonsymbolic Magnitude processing in the first year of formal schooling. *PLoS One, 11*(3), e0149863. https://doi. org/10.1371/journal.pone.0149863

Moreau, D., Wilson, A. J., McKay, N. S., Nihill, K., & Waldie, K. E. (2018). No evidence for systematic white matter correlates of dyslexia and dyscalculia. *NeuroImage. Clinical, 18,* 356–366. https://doi.org/10.1016/j.nicl.2018.02.004

Morsanyi, K., Devine, A., Nobes, A., & Szűcs, D. (2013). The link between logic, mathematics and imagination: evidence from children with developmental dyscalculia and mathematically gifted children. *Developmental science, 16*(4), 542–553. https://doi. org/10.1111/desc.12048

Mussolin, C., Nys, J., Content, A., & Leybaert, J. (2014). Symbolic Number Abilities Predict Later Approximate Number System Acuity in Preschool Children. *PLoS One, 9*(3), e91839. https://doi.org/10.1371/journal.pone.0091839

Núñez, R. E. (2017). Is there really an evolved capacity for number? *Trends in Cognitive Sciences.* https://doi.org/10.1016/j.tics.2017.03.005

Paterson, S. J., Brown, J. H., Gsödl, M. K., Johnson, M. H., & Karmiloff-Smith, A. (1999). Cognitive modularity and genetic disorders. *Science.* https://doi.org/10.1126/science. 286.5448.2355

Peters, L., & Ansari, D. (2019). Are specific learning disorders truly specific, and are they disorders? *Trends in Neuroscience and Education, 17,* 100115. https://doi.org/10.1016/J. TINE.2019.100115

Peters, L., Bulthé, J., Daniels, N., Op de Beeck, H., & De Smedt, B. (2018). Dyscalculia and dyslexia: Different behavioral, yet similar brain activity profiles during arithmetic. *NeuroImage: Clinical.* https://doi.org/10.1016/j.nicl.2018.03.003

Scerif, G., & Karmiloff-Smith, A. (2005). The dawn of cognitive genetics? Crucial developmental caveats. *Trends in Cognitive Sciences, 9*(3), 126–135. https://doi.org/10. 1016/j.tics.2005.01.008

Smyth, R. E., & Ansari, D. (2020). Do infants have a sense of numerosity? A p-curve analysis of infant numerosity discrimination studies. *Developmental Science.* https://doi. org/10.1111/desc.12897

Szűcs, D. (2016). Subtypes and comorbidity in mathematical learning disabilities (pp. 277–304). https://doi.org/10.1016/bs.pbr.2016.04.027

Szucs, D., Devine, A., Soltesz, F., Nobes, A., & Gabriel, F. (2013). Developmental dyscalculia is related to visuo-spatial memory and inhibition impairment. *Cortex, 49*(10), 2674–2688. https://doi.org/10.1016/j.cortex.2013.06.007

Szűcs, Dénes, & Myers, T. (2017). A critical analysis of design, facts, bias and inference in the approximate number system training literature: A systematic review. *Trends in Neuroscience and Education.* https://doi.org/10.1016/j.tine.2016.11.002

Van Herwegen, J., & Simms, V. (2020). Mathematical development in Williams syndrome: A systematic review. *Research in Developmental Disabilities, 100,* 103609. https://doi.org/ 10.1016/j.ridd.2020.103609

Xu, F., & Spelke, E. S. (2000). Large number discrimination in 6-month-old infants. *Cognition.* https://doi.org/10.1016/S0010-0277(99)00066-9

13

QUO VADIS MODULARITY IN THE 2020s?

Michael S.C. Thomas and Daniel Brady

Both of us had the privilege of being postdoctoral fellows for Annette: one of us (MT) when Annette set up her own research unit in 1998 after leaving MRC Cognitive Development Unit, the other (DB) in Annette's final project, an investigation of infant development in Down syndrome (DS) and its link with lifelong elevated risk for Alzheimer's disease in the syndrome (Karmiloff-Smith et al., 2016). In this chapter, we consider the idea of *modularity* because it figured in both our collaborations with Annette (henceforth, AKS) – but between the two collaborations, the usage of the concept had changed.

For MT, his first publication with AKS was a commentary that challenged the use of the Fodorian sense of modularity (Fodor, 1983) to explain deficits in developmental disorders, as it featured in four papers then appearing in the journal *Learning and Individual Differences*. We entitled our commentary, 'Quo vadis modularity in the 1990s?' (or, for non-Latin speakers, 'modularity, where are you going?'; Thomas & Karmiloff-Smith, 1999). For DB, modularity featured in his use of Graph Theoretic methods to analyse time series correlations between the electroencephalogram (EEG) data recorded from electrodes spread across the scalps of infants with DS as they watched a video. Graph Theory was deployed to understand more about the functional connectivity of these infants' brains - which regions appeared to be working with other regions - and the relation of functional connectivity both to the infants' chronological age and to their cognitive ability.

For this Festschrift volume, as we enter the third decade of the 21st century, we can ask this question once again: *quo vadis, modularity?* In this chapter, we will see that over the decades and with the advance of neuroscience, the notion of modularity has been recast: from an *a priori* design principle which it would make good sense for potential cognitive systems to employ, to a data-driven concept based on how the brain is actually working.

Modularity as an advantageous *a priori* design principle

In 1998, when MT joined AKS's new research group, the Neurocognitive Development Unit, based at London's Institute of Child Health, AKS set him three goals. The first was to advance research on neurodevelopmental disorders, pursing ideas in her seminal paper *The Key to Developmental Disorders Is Development Itself* (Karmiloff-Smith, 1998). This involved investigating the genetic syndrome Williams syndrome (WS), characterised by overall learning disability, but relative strengths in language and face recognition, and weaknesses in visuospatial cognition. But, AKS asked, was language truly a strength? Was language developing typically, in-dependently of the rest of cognition? How come it was not at chronological age level? How did language fit with the overall pattern of learning disability? This line of work was reasonably successful: a set of studies suggested that language skills in WS were broadly in line with overall mental age, but with subtle patterns of un-evenness – that is, atypicality – such as in semantic compared to phonological skills (Thomas et al., 2001; Laing et al., 2005; Ansari et al., 2003; Thomas & Karmiloff-Smith, 2005; Thomas, Forrester, & Richardson, 2006; Thomas, 2006). A similar pattern was found in detailed investigations of face processing, although the beha-vioural strength here was more marked, and there was greater evidence for atypical underlying cognitive processes (Annaz, Karmiloff-Smith, Johnson, & Thomas, 2009; Grice et al., 2001; Karmiloff-Smith et al., 2004).

The second goal AKS set for MT was to begin a programme of computational modelling. Since there existed at the time reasonable computational models of ty-pical cognitive development, based on artificial neural networks, was it possible to build computational models of developmental disorders, such as of language de-velopment in WS? This work was reasonably successful as well. MT built a model of the acquisition of morphosyntax in WS, where different types of atypical constraint (arising from the structure of semantic representations, phonology representations, or computational properties learning to mediate the relationship between language representations) could shape developmental trajectories in ways that did or did not reflect those observed in individuals with WS (Thomas & Karmiloff-Smith, 2003a). The model made more concrete the theoretical notion that deficits in develop-mental disorders were the result of atypical constraints acting on the developmental process by specifying that process in neurocomputational terms.

The third goal proved more challenging. AKS wanted MT to build a com-putational model of *representational redescription* (RR). RR was AKS's account of how children could develop cognitive flexibility. It was based on the proposal that there was a general phase of cognitive development lying beyond behavioural mastery, where the knowledge underlying mastery of a given ability becomes both more accessible to conscious awareness (or more 'explicit') and also more open to flexible application (Karmiloff-Smith, 1992). For MT, this work did not go so well. He ran into a stumbling block when attempting to think about RR in computational terms. Was RR, on the one hand, a process *intrinsic* to a cognitive system that was acquiring some behaviour, such that even when the system had achieved behavioural mastery, it somehow went on to reorganise its knowledge into a form that could be 'offered up' to other cognitive systems, such as those involved in cognitive control, verbal report, or metacognition? What would drive this reorganisation, if mastery was already achieved? Or was RR, on the other hand, a process concerned with the enabling of latent, pre-existing *connectivity between* different cognitive systems, so that once the internal representations of the target system began to exhibit interesting internal structures or covariations relevant to other systems (cognitive control, verbal report, etc.), these latent connections would begin to strengthen? Could RR involve both intrinsic and connectivity elements? The dilemma was hard to resolve, but theoretical clarification was necessary prior to proceeding with implementation in a computational model.

Meanwhile, the attention of both AKS and MT was focused on atypical development, and RR moved to the backburner. In her final years, AKS was perhaps on the verge of returning to RR: she noted the possibility that brain imaging might test the RR hypothesis through studying emerging hierarchical network structure in 'resting state' functional brain imaging data (Karmiloff-Smith, 2010); and viewed Dehaene's proposal of a global workspace for flexible use of knowledge (Dehaene et al., 2014) as a possible brain basis for the concept of RR (Karmiloff-Smith, 2015). It is fantastic to see McClelland and his team (this volume) taking RR ideas forward in a neurocomputational framework, as well as Hughes and Edgin (this volume), considering a link between RR and sleep, and Plunkett (this volume) consider RR in relation to vocabulary development. No competing theory has emerged in the meantime to explain the developmental emergence of explicit knowledge and cognitive flexibility in children.

However, behind all three of these goals lay a pervasive idea that was key to AKS's thinking at the time – that of *modularity*. There were, in the 1990s, explanations of uneven cognitive profiles in developmental disorders that made reference to innate modules. For AKS, this was the wrong way to go. For her, modules were an outcome of development, not a precursor to it (Karmiloff-Smith, 1992, 1994, 1998, 2009, 2015). Multiple convergent methods within cognitive neuroscience, including computational modelling and functional brain imaging, would be necessary to specify the nature of developmental processes. And even when modularity had emerged in a given domain, somehow development could move beyond it, so that children could begin to show insight into their own behaviour.

Let us take a moment to recapitulate the original, Fodorian sense of what a module was (Fodor, 1983), around which this debate revolved. The theory was based on a cognitivist approach that sought to cleave *cognition* from *perception*, and consider each as a decomposable into a set of specialised sub-processes. Fodor constructed his arguments by appealing to what would constitute a sensible way to design a cognitive system, grounded by (limited) empirical reference to intuitive phenomena such as visual illusions, and supported by thought experiments. The key idea was *encapsulation of information*.[1] Specialised and dedicated processing systems would contain their own dedicated information; these systems would be impervious to information from outside in the rest of the cognitive system. Intuitively, it is evident that we continue to see visual illusions even when we know they are illusions. For Fodor, this is because low-level vision is a module – fast, automatic, self-contained, independently delivering its results to the rest of cognition. Indeed, it would be a poor design for a perceptual system to rely on background knowledge: if you see a tiger leaping at you while out shopping, you need to duck, not think about whether it's likely you'll encounter a tiger in the shopping mall. Fodor distinguished perceptual modules from the central cognitive system. The central system is involved in thought and reasoning, where in his view background knowledge is crucial to its operation. Should you go to the shops today? A vast range of factors might be relevant (including the chance of tigers). Fodor suggested that somewhere in between perception and the central system, there might be *cognitive modules* for self-contained cognitive domains such as language, or perhaps syntax. Along with properties of fast processing, automatic operation and encapsulated information, these cognitive modules would likely be innately specified, hardwired (that is, neurally specific), and not assembled from other cognitive components. Modularity amounted to a *cluster* concept, a set of properties that Fodor thought would likely hang together within the architecture of the cognitive system.

The idea of cognitive modules fitted with what was observed in cases of specific loss of abilities following acquired focal brain damage in adults – as if a bit of the cognitive system were implemented by a bit of the brain, and damage to that bit of the brain could produce a specific deficit in the corresponding ability. But in 1990s, some researchers were trying to apply this adult-based framework to uneven cognitive profiles in developmental disorders, such as autism, developmental language disorder (then called Specific Language Impairment), dyslexia, and prosopagnosia. For AKS, however, Fodor's framework left out a key aspect: development. The increasing understanding of functional brain development, as articulated in *Rethinking Innateness* (Elman et al., 1996; see Johnson, this volume) suggested that although modularity may be characteristic of the adult state, it was likely a product of developmental processes from an initially less differentiated and specialised system. Therefore, the static adult framework was inappropriate to characterise developmental disorders. AKS and MT spent some time making this argument more concrete by comparing the different profiles of deficits in computational models of developmental deficits versus models of acquired deficits (Thomas, & Karmiloff-Smith, 2002; Thomas, & Richardson, 2006; Thomas *et al.* 2006).

In one sense, Fodor's notion of modularity is hard to argue against – and that sense is *falsification*. The six proposed properties of a module were not necessary and sufficient conditions, just properties likely to cluster together, based on what would be a sensible way to design a cognitive system (and some supporting intuitive phenomena). The absence of one or more feature from the cluster for some target behaviour would not be enough to reject the idea. (For example, visual word recognition is learned, not innate, but shows other characteristics of modular function – it is fast, automatic, can be specifically lost after acquired brain damage in alexia – but its lack of innateness would not serve to falsify the cluster concept of modularity; maybe some modules can be learned, Fodor would say.) So the idea of modularity was never rejected.

The main flaw with the original concept of modularity, and indeed of the cognitivist approach of the 1980s and early 1990s more generally, was that it dealt in ways that cognitive systems *could* work – hence the close contemporary affiliation of cognitive theory with artificial intelligence research. However, it did not necessarily deal with the way the cognitive system *does work*, which is constrained by its actual biology and the attendant evolutionary history that shaped the biology. The actual way the cognitive system works has become more apparent with progress in cognitive neuroscience over the intervening years. The brain operates via content-specific systems and modulatory systems (both focal and diffuse) which coordinate their operation; it operates via distributed networks of interacting systems which are configured dynamically according to task and context; hierarchical systems learn to extract ever higher-order invariances from sensory input or to generate motor output; cortical systems are continuously interacting with limbic systems whose structures reflect the deeper evolutionary goals of the species through the emotions. The increasing influence of cognitive neuroscience led some commentators to revise how the notion of modularity should usefully be deployed. Already by the end of the century, Coltheart (1999) was arguing that modularity should be ascribed to a cognitive system solely on the basis of the system being *domain specific*. This did not even preserve the notion of information encapsulation that was the motivating factor in Fodor's original conception.

Modularity as a data-driven description of brain structure and function

Let us pause and let some time pass. AKS has pursued her neuroconstructivist agenda, using cross-syndrome comparisons in an attempt to triangulate the developmental constraints that operate on cognition, in domains such as language and face recognition, and has continued to integrate wider sets of developmental cognitive neuroscience methods. MT has left to broaden his approach to thinking about different types of cognitive variation beyond disorders, such as intelligence and giftedness, and how to link cognitive neuroscience findings to education. Less is heard of modularity, and less still of RR. We come to Annette's final project.

The London Down Syndrome (LonDownS) Consortium was launched in 2013, funded by a Wellcome Trust Strategic grant. Its aim was to investigate the link between Alzheimer's disease and DS. Adults with DS are at greater risk of developing Alzheimer's disease (AD) compared to the general population, and the onset of AD tends to occur at an earlier age (Head, Powell, Gold, & Schmitt, 2012). One reason is that one of the key genes implicated in AD, the amyloid beta precursor protein (*APP*) gene, lies on chromosome 21; there are three copies of chromosome 21 in DS, potentially disrupting the gene dosage for all the genes on this chromosome. The consortium took a lifespan perspective, both seeking to understand the degenerative disease of AD and to predict as early as in infancy which individuals with DS are most at risk of the disease in adulthood. AKS worked with the infant and child cohorts, investigating whether risk factors for AD would be reflected in early developmental trajectories (the answer: sometimes, sometimes not; D'Souza et al., 2020; Thomas et al., 2020 submitted). She also addressed the difficulties of developing cognitive tests for infants and children with DS and how to match these with the tests being developed for the mouse models of the syndrome (see D'Souza et al., Chapter 16).

The LonDownS project was a large affair, five years in length including longitudinal follow-ups of children and adults, and a multidisciplinary team set to investigate dementia in adults, mouse models of DS, genetics of AD risk, in vitro modelling of DS neurons cultured from induced pleuripotent stem cells, and of course, AKS's study of infancy in DS, with the largest UK cohort that she could manage to put together (over 100, in the end, with around half of those being followed up longitudinally) (see Karmiloff-Smith et al., 2016). AKS embraced the complexity of development, instead of attempting to simplify, seeking to measure her cohort across multiple behaviours, with multiple methods (behavioural, eye-gaze monitoring, brain imaging, genetics), including measures of parent-child interaction and sleep. DB, the second author of this chapter, joined her team to focus on the analysis of brain imaging data for the infants with DS and typically developing controls.

It is here that the notion of modularity resurfaces, in a different guise. A new technique has emerged to analyse both structural and functional brain imaging data, one that focuses on networks, construed within the mathematical framework of *Graph Theory*. We will go into a little more detail in the following paragraphs to convey the technical innovations of Graph Theory and the potential link of imaging data to cognition. Graph Theory takes data summarising the correlations between many variables and translates these into graphs, where each variable is a node in the network, and the patterns of correlations are reflected in the links joining the nodes. For brain imaging data, the nodes may be parcellated brain regions (structural analyses), signals from optical or electrical sensors spread over the scalp, or blood oxygenation levels from specific brain regions (functional analyses) (see Sporns, 2014). In structural analyses, the key question is whether two regions are connected, while for functional analyses, the key question is whether the activity of pairs of nodes is correlated over time, and therefore presumably involved in similar processes.

Once a network has been generated from a set of correlational data, a range of metrics can be generated describing the graph. Three of the most widely used measures are whether a network is characterised by *hubs*, the *average path length* between any two nodes, and the level of *modularity* (Vertes & Bullmore, 2015). Hubs are nodes that have a number of links that greatly exceeds the average, suggesting they are key locations for communication between less connected areas. The average path length is a measure of efficiency, indicating how many links have to be traversed to move from any node to any other node. Short average path length indicates a network architecture that facilitates fast communication. Modularity in graph theory is a measure that assesses the division of a network into groups (also called *clusters* or *communities*): networks with high modularity have dense connections between the nodes within modules but sparse connections between nodes in different modules (Newman, 2006). In functional terms, the activity of a set of nodes (call it, group A) may be highly correlated among each other, but uncorrelated to another group (B), while the nodes in B in turn show high correlations among each other. Figure 13.1 shows examples of networks exhibiting low modularity and high modularity, respectively. In brain terms, therefore, a module now comes to mean sets of brain areas which appear to be strongly physically connected among themselves but not to other regions (structural modularity), or sets of regions whose activity seems to be highly correlated to each other but not to other regions (functional modularity). Modularity in this sense indeed appears to be a property of biological systems (Newman, 2006; Sporns, 2014).

Analyses of brain imaging data employing graph theoretic techniques have found that graph metrics appear to correspond to interesting cognitive dimensions. When Crossley, Mechelli, and Vertes (2013) examined co-activated regions across a large set of functional imaging studies, they found a coactivation network that was modular, with occipital, central, and default-mode modules predominantly coactivated by specific cognitive domains including, respectively, perception, action, and emotion. Networks appear to alter across development and are sensitive to differences in atypical development (Morgan, White, Bullmore, & Vertes, 2018; Vertes & Bullmore, 2015; Whitaker et al., 2016). When Hilger, Ekman, Fiebach, and Basten (2017) correlated measures of general intelligence to networks

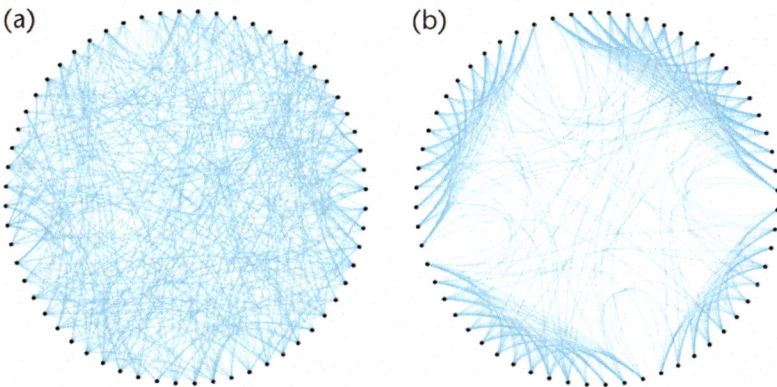

FIGURE 13.1 Illustrative network analysis results. Both networks have the same number of nodes and vertices, but network (a) has low modularity while network (b) has high modularity

constructed from resting-state functional magnetic resonance imaging data in a sample of over 300 individuals, they found brain networks characterised by substantial modularity. Intelligence was not associated with global features of modularity, such as the number or size of modules. However, intelligence was associated with node-specific measures of within- and between-module connectivity, particularly in frontal and parietal brain regions that have previously been linked to intelligence.

AKS, DB and their team acquired resting-state EEG data from the cohort of infants and toddlers with DS as they watched a short 2-minute video. The prospect was that this kind of sensitive measure might detect differences in underlying brain processes reflecting later risk/resilience to AD even before they are manifest in behaviour, on the basis that in typical development, genes conferring additional risk for AD have been observed to influence white matter growth as early as 2–25 months of age (Dean, Jerskey, & Chen, 2014). The analyses of these data are still on-going, and we present here some of the provisional results to illustrate the method, and also some of its limitations.

In EEG analyses, the small voltages changes produced at the scalp by brain activity are measured in the millisecond range. This activity can be broken down into the amount of energy produced (power) and phase in different frequency bands, typically delta (0.1–3 Hz), theta (4–7 Hz), alpha (8–15 Hz), beta (16–31 Hz), and, gamma (32–100 Hz). Comparing the phase of the activity between nodes (in the case of EEG this is usually sensors) can be used as a way of quantifying the degree of connectedness between the nodes, and undertaking this for all combinations of nodes allows creation of a network or graph. Specifically, the degree of correlation between nodes was determined by the weighted Phase lag index (https://www.nbtwiki.net/doku.php?id=tutorial:phase_lag_index), so the graph measures were not built from the overall energy at a particular frequency band (the power) but instead the similarity between phases of EEG activity at the nodes.

The goal here was to explore whether individual differences in the properties of these networks correlated to measures of cognition. Figure 13.2 shows scatter plots linking each network's cluster coefficient in each frequency band to the children's receptive language ability, as measured by the Mullen test (Mullen, 1995). It is evident here that no relationship was present for any frequency band, despite extensive variation in language ability. Figure 13.3 focuses on just one band, beta, and examines the change in five different network metrics across chronological age.[2] Cluster coefficient, global and local efficiency each reduced with age, while mean shortest path length increased, and modularity showed no change. Figure 13.4 contrasts the correlation of modularity and mean shortest path length with overall mental age on the Mullen: there was no relationship with modularity, but an increase in shortest path length with increasing mental age. These results replicate the developmental sensitivity of network metrics and the absence of a correlation of overall modularity with intelligence observed by Hilger et al. (2017). The increase in shortest path length both with age and with mental age is

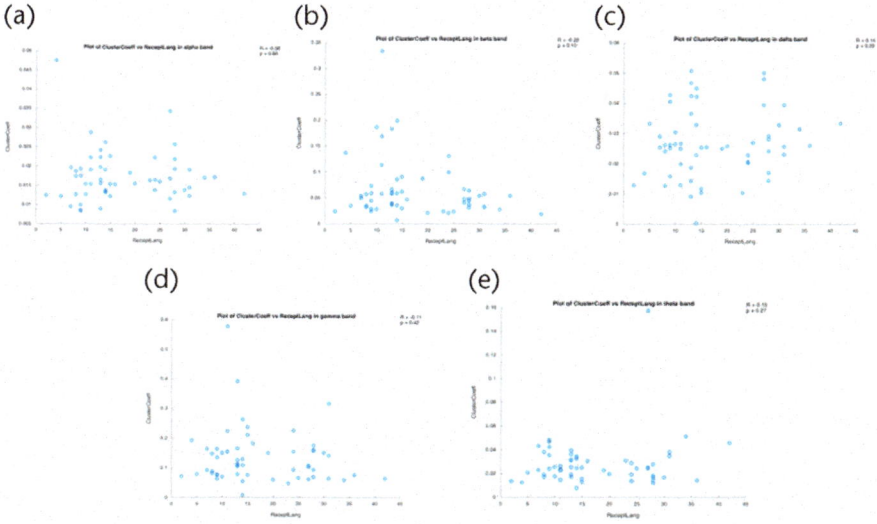

FIGURE 13.2 Example network analysis results from EEG data as infants and toddlers with DS watched a short video. Graphs were generated on 40 seconds of data for around 40 children for 5 frequency bands (delta, theta, alpha, beta, and gamma). Graphs were proportionally thresholded so 20% of strongest connections remain. Data show the relationship between the cluster coefficient (the degree to which nodes in a graph tend to cluster together) and receptive language ability at different frequency band: (a) alpha (b) beta (c) delta (d) gamma, and (e) theta

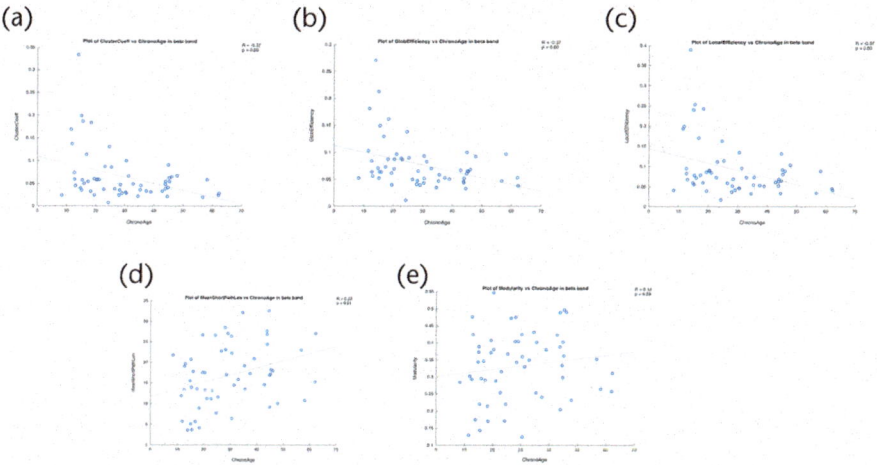

FIGURE 13.3 Example network analysis results from EEG data as infants and toddlers with DS watched a short video. Data show the relationship between different graph metrics and chronological age in the beta frequency band. (a) Cluster coefficient (b) global efficiency (c) local efficiency (d) mean shortest path length, and (e) modularity

(a)

Plot of Modularity vs MentalAge in beta band

R = 0.10
p = 0.44

(b)

Plot of MeanShortPathLen vs MentalAge in beta band

R = 0.33
p = 0.01

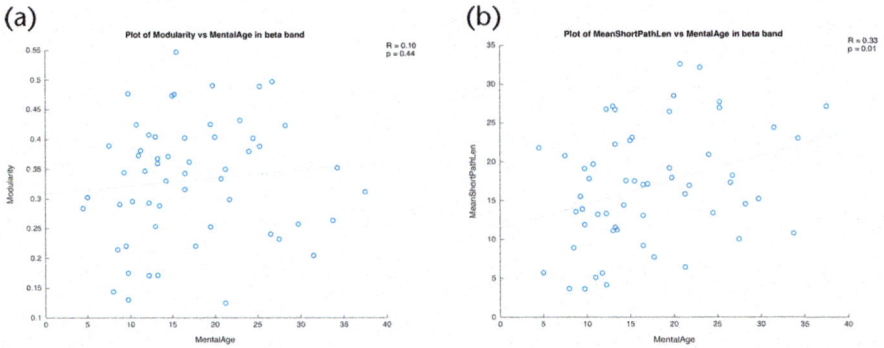

FIGURE 13.4 Relationship between (a) the modularity metric and mental age in the beta frequency band; and (b) the mean shortest path length and mental age in the beta band

suggestive of worsening efficiency in the network, which is consistent with the emerging developmental delay observed in DS across early childhood (D'Souza et al., in preparation).

However, these data remain provisional because they need to be compared to the patterns observed in typical development (currently underway) and because of their sensitivity to decisions made in constructing the graphs. For instance, EEG data are noisy: decisions need to be made in the quality of the signal that is included into the analyses; in unweighted graphs, links are either present or absent: decisions need to be made on strength of the correlation between any two nodes required before a link is generated. Moreover, as with any neuroscience data generating a large number of comparisons, there is the challenge of how to correct for elevated risk of Type 1 statistical errors (false alarms). Many of the other comparisons in other frequency bands were null. In these analyses, we have 5 different measures of cognition from the Mullen test plus chronological age, 5 different graph measures, and 5 different frequency bands (150 comparisons). Without an *a priori* hypothesis, the appropriate significance level (alpha) would be .0003 rather than the standard of .05, which represents a stiff test for noisy data. The sense is that such analyses might give fleeting glimpses into emerging architectures, but would stand in need of replication in other samples before being viewed as robust.

Final thoughts

We see here, then, the notion of modularity recast. It is now a data-driven concept based on how the brain is actually working, rather than an *a priori* design principle, one that would seem to make sense for potential cognitive systems to employ. While functional brain imaging presumes some degree of domain-specificity of local neural processing, overall function emerges from the interaction of many regions. Moreover, these regions may not necessarily align with the modules or component processes articulated in cognitive theories (Price & Friston, 2005). Since modularity is now data driven, it is an empirical issue whether it is a property that alters across development or varies with intellectual ability.

In our view, the routine use of brain imaging data to complement cognitive theories of development makes it unlikely we will return to the 1980s Fodorian notion of modularity, as research proceeds into the 2020s. For the mammalian brain, evolution does not appear to have followed the principle of componential design that might have yielded the benefits Fodor had in mind. Developmental studies suggest that specialisation is in part an emergent property influenced either by experience or internally generated neural activity, so called *interactive specialisation* (Johnson, this volume). The on-line interactivity of different brain regions characterised by graph theory does not seem consistent with the key idea of information encapsulation championed by Fodor.

However, this reality does not negate Fodor's key insight – we still need to characterise how, in brain terms, fast perception is different to slow thinking, why optical illusions are resistant to knowledge, or why the arcane knowledge underlying the processing of syntax appears unconscious and specific to language. To take one contemporary puzzle: when teenagers spend hundreds of hours playing action video games, they change their brains and enhance their visual perception skills, particularly for peripheral attention. Yet, despite this massive effort, and the observed behavioural change, brain imaging suggests that no enhancement has occurred to 'bottom-up', automatically triggered, exogenous attention, only to voluntarily exerted top-down attention (Altarelli, Green, & Bavelier, 2020). Why should that be? What is different about exogenous attention? Altarelli et al. report that it remains largely unknown why initial orienting under the control of exogenous attention is not altered by intensive action video game play, despite it being advantageous to improve this skill for game performance. They argue one possible explanation is that orienting develops very early during development and is in a large part mediated by sub-cortical structures, such as the nucleus basalis or the pulvinar; these structures may be less plastic later in life than cortical brain regions that support top-down attentional mechanisms. This account seems close to the kind of dimension of cognition that Fodor had in mind: a fast, automatic, early developing, modular perceptual function, to be contrasted with a slower, more strategic and informationally rich system exerting top-down control. Similarly, we are yet to construct convincing explanations for the phenomena that led AKS to propose representational redescription in the early part of her research career – that in many cognitive domains (but not all), expertise can be accompanied by explicit knowledge and greater flexibility in a phase of development extending beyond behavioural mastery of the task itself.

These challenges, posed by theoretical adversaries such as AKS and Fodor, remain for the future. Yet it is our belief that the multi-disciplinary approach advocated by AKS, with development at its core, is best positioned to yield the answers we seek.

Acknowledgements

We would like to acknowledge all of Annette's team on the LonDownS project who contributed to this research including George Ball, Joanne Ball, Amelia Channon, Charlotte Dennison, Dean D'Souza, Hana D'Souza, Maria Eriksson, Jennifer Glennon, Kate Hughes, Amanda Lathan, Esha Massand, Beejal Mehta, Lidija Nikolova, Jessica Schulz, Laura Simmons, Margaux Trombert, and Kate Whitaker. The scripts for collection of the resting state EEG data were kindly shared by the BASIS project.

The LonDownS project is supported by a Wellcome Trust Strategic Award [098330/Z/12/Z], the Baily Thomas Charitable Fund, and the Jerome LeJeune Foundation. The London Down Syndrome (LonDownS) Consortium principal investigators are Andre Strydom (chief investigator), Department of Forensic and Neurodevelopmental Sciences, Institute of Psychiatry, Psychology and Neuroscience, King's College London, London, UK, and Division of Psychiatry, University College London, London, UK; Elizabeth Fisher, Department of Neurodegenerative Disease, Institute of Neurology, University College London, London, UK; Dean Nizetic, Blizard Institute, Barts and the London School of Medicine, Queen Mary University of London, London, UK, and Lee Kong Chian School of Medicine, Nanyang Technological University, Singapore, Singapore; John Hardy, Reta Lila Weston Institute, Institute of Neurology, University College London, London, UK, and UK Dementia Research Institute at UCL, London, UK; Victor Tybulewicz, Francis Crick Institute, London, UK, and Department of Medicine, Imperial College London, London, UK; Annette Karmiloff-Smith, Birkbeck University of London, London, UK (deceased); Michael Thomas, Birkbeck University of London, London, UK; and Denis Mareschal, Birkbeck University of London, London, UK.

Notes

1 Fodor (1985): 'A **module** is (inter alia) an informationally encapsulated computational system – an inference-making mechanism whose access to background information is constrained by general features of cognitive architecture, hence relatively rigidly and relatively permanently constrained' (p. 3).
2 Graph Theory network metric definitions: *Cluster coefficient*: A measure of the degree to which nodes in a graph tend to cluster together. To what extent are the nodes to which a given node is connected, connected to each other? *Global efficiency*: The efficiency of a network measures how efficiently it exchanges information. Global efficiency quantifies the exchange of information across the whole network where information is concurrently exchanged. *Local efficiency*: The local efficiency quantifies a network's resistance to failure on a small scale. The local efficiency of a node reflects how well information is exchanged by its neighbours when it is removed. *Mean shortest path length*: The average number of steps along the shortest paths for all possible pairs of network nodes, which indexes the information flow across a network. *Modularity*: **Modularity** measures the strength of division of a network into groups. Networks with high modularity have dense connections between the nodes within modules, but sparse connections between nodes in different modules.

References

Altarelli, Green C. S., & Bavelier, D. (2020). Action video games: from effects on cognition and the brain to potential educational applications. In M. S. C. Thomas, D. Mareschal & I. Dumontheil (Eds.), *Educational Neuroscience: Development Across the Lifespan*. London: Psychology Press.

Annaz, D., Karmiloff-Smith, A., Johnson, M. H., & Thomas, M. S. C. (2009). A cross-syndrome study of the development of holistic face recognition in children with autism, Down syndrome and Williams syndrome. *Journal of Experimental Child Psychology*, *102*, 456–486.

Ansari, D., Donlan, C., Thomas, M. S. C., Ewing, S., Peen, T., & Karmiloff-Smith, A. (2003). How children with Williams syndrome understand why counting counts. *Journal of Experimental Child Psychology*, *85*, 50–62.

Coltheart, M. (1999). Modularity and cognition. *Trends in Cognitive Sciences*, *3(3)*, 115–120.

Crossley, N. A., Mechelli, A., Vertes, P. E., et al. (2013). Cognitive relevance of the community structure of the human brain functional coactivation network. *Proceedings of the National Academy of Sciences USA*, *110*(28), 11583–11588.

Dehaene, S., Charles, L., King, J. R., & Marti, S. (2014). Toward a computational theory of conscious processing. *CurrOp Neurobio*, *25*, 76–84.

D'Souza, H. D., Mason, L., Mok, K. Y. , Hardy, J., Karmiloff-Smith, A.Mareschal, D. the LonDownS Consortium Strydom, A., & Thomas, M. S. C. (2020), Differential associations of apolipoprotein E ε4 genotype with attentional abilities across the life span of individuals with down syndrome, *JAMA Network Open*, *3*(9), e2018221. doi: 10.1001/jamanetworkopen.2020.18221

D'Souza, H., Karmiloff-Smith, A., Mareschal, D., & Thomas, M. S. C. (2019). *The Down syndrome profile emerges gradually over the first years of life*. Manuscript submitted for publication.

Dean, D. C. 3rd, Jerskey, B. A., Chen, K., et al. (2014). Brain differences in infants at differential genetic risk for late-onset Alzheimer disease: A cross-sectional imaging study. *JAMA Neurology*, *71*(1), 11–22.

Elman, J., Karmiloff-Smith, A., Bates, E., Johnson, M., Parisi, D., & Plunkett, K. (1996). *Rethinking Innateness: A Connectionist Perspective on Development*. Cambridge, MA: MIT Press.

Fodor, J. A. (1983). *The Modularity of Mind*. Cambridge, MA: MIT Press.

Fodor, J. A. (1985). Precis of the modularity of mind. *Behavioral and Brain Sciences*, *8*, 1–42.

Grice, S. J., Spratling, M. W., Karmiloff-Smith, A., Halit, H., Csibra, G., de Haan, M., et al. (2001). Disordered visual processing and oscillatory brain activity in autism and Williams syndrome. *Neuroreport*, *12*, 2697–2700.

Head, E., Powell, D., Gold, B. T., & Schmitt, F. A. (2012). Alzheimer's Disease in Down Syndrome. *European Journal of Neurodegenerative Disease*, *1*(3), 353–364.

Hilger, K., Ekman, M., Fiebach, C. J., & Basten, U. (2017). Intelligence is associated with the modular structure of intrinsic brain networks. *Scientific Reports*, *7*, 16088. http://doi.org/10.1038/s41598-017-15795-7

Karmiloff-Smith, A. (1992). *Beyond Modularity: A Developmental Perspective on Cognitive Science*. Cambridge, MA: MIT Press/Bradford Books.

Karmiloff-Smith, A. (1998). Development itself is the key to understanding developmental disorders. *Trends in Cognitive Sciences*, *2*(10), 389–398. http://doi.org/10.1016/S1364-6613(98)01230-3

Karmiloff-Smith, A. (2010). Neuroimaging of the Developing Brain: Taking "Developing" Seriously. *Human Brain Mapping*, *31*, 934–941.

Karmiloff-Smith, A. (2013). Challenging the use of adult neuropsychological models for explaining neurodevelopmental disorders: Developed versus developing brains. *The Quarterly Journal of Experimental Psychology*, *66*, 1–14.

Karmiloff-Smith, A. (2015). An alternative to domain-general or domain-specific frameworks for theorizing about human evolution and ontogenesis. *AIMS Neuroscience*, *2*(2), 91–104.

Karmiloff-Smith, A., (2009). Nativism versus neuroconstructivism: Rethinking the study of developmental disorders. *Developmental Psychology*, *45*(1), 56–63.

Karmiloff-Smith, A., Al-Janabi, T., D'Souza, H., Groet, J., Massand, E., Mok, K., ... Strydom, A. (2016). The importance of understanding individual differences in Down syndrome. *F1000Research*, *5*, 389. http://doi.org/10.12688/f1000research.7506.1

Karmiloff-Smith, A., Thomas, M. S. C., Annaz, D., Humphreys, K., Ewing, S., Grice, S., Brace, N., Van Duuren, M., Pike, G., & Campbell, R. (2004). Exploring the Williams syndrome face processing debate: The importance of building developmental trajectories. *Journal of Child Psychology and Psychiatry and Allied Disciplines*, *45*(7), 1258–1274.

Laing, E., Grant, J., Thomas, M. S. C., Parmigiani, C., Ewing, S., & Karmiloff-Smith A. (2005). Love is... an abstract word: The influence of lexical semantics on verbal short-term memory in Williams syndrome. *Cortex*, *41*(2), 169–179.

Morgan, S. E., White, S. R., Bullmore, E. T., & Vertes, P. E. (2018). A network neuroscience approach to typical and atypical brain development. *Biological Psychiatry: Cognitive Neuroscience and Neuroimaging*, *3*(9), 754–766.

Newman, M. E. J. (2006). Modularity and community structure in networks. *Proceedings of the National Academy of Sciences of the United States of America*, *103*(23), 8577–8696.

Mullen, E. M. (1995). *Mullen Scales of Early Learning*. Circle Pines, MN: American Guidance Service Inc.

Price, C. J., & Friston, K. J. (2005). Functional ontologies for cognition: The systematic definition of structure and function. *Cognitive Neuropsychology*, *22*(3), 262–275. http://doi.org/10.1080/02643290442000095

Sporns, O. (2014). Contributions and challenges for network models in cognitive neuroscience. *Nature Neuroscience*, *17*, 652–660. http://doi.org/10.1038/nn.3690

Thomas, M. S. C. (2006). Williams syndrome: Fractionations all the way down? *Cortex*, *42*, 1053–1057.

Thomas, M. S. C. & Karmiloff-Smith, A. (1999). Quo vadis modularity in the 1990s? *Learning and Individual Differences*, *10*(3), 245–250.

Thomas, M. S. C. & Karmiloff-Smith, A. (2002). Are developmental disorders like cases of adult brain damage? Implications from connectionist modelling. *Behavioral and Brain Sciences*, *25*(6), 727–788.

Thomas, M. S. C. & Karmiloff-Smith, A. (2003a). Modelling language acquisition in atypical phenotypes. *Psychological Review*, *110*(4), 647–682.

Thomas, M. S. C. & Karmiloff-Smith, A. (2003b). Connectionist models of development, developmental disorders and individual differences. In R. J. Sternberg, J. Lautrey, & T. Lubart (Eds.), *Models of Intelligence: International Perspectives*, (pp. 133–150). American Psychological Association.

Thomas, M. S. C., & Johnson, M. H. (2018). Karmiloff-Smith from Piaget to neuroconstructivism. In: A. Karmiloff-Smith, M. S. C. Thomas & M. H. Johnson, M. H. (Eds.), *Thinking Developmentally from Constructivism to Neuroconstructivism: Selected Works of Annette Karmiloff-Smith*. London: Routledge.

Thomas, M. S. C., & Karmiloff-Smith, A. (2005). Can developmental disorders reveal the component parts of the human language faculty? *Language Learning and Development*, *1*(1), 65–92.

Thomas, M. S. C., & Richardson, F. (2006). Atypical representational change: Conditions for the emergence of atypical modularity. In Y. Munakata, & M. Johnson (Eds.), *Attention & Performance XXI: Processes of Change in Brain and Cognitive Development* (pp. 315–347). Oxford: Oxford University Press.

Thomas, M. S. C., Dockrell, J. E., Messer, D., Parmigiani, C., Ansari, D., & Karmiloff-Smith, A. (2006). Speeded naming, frequency and the development of the lexicon in Williams syndrome. *Language and Cognitive Processes, 21*(6), 721–759.

Thomas, M. S. C., Forrester, N. A., & Richardson, F. M. (2006). What is modularity good for? In Proceedings of The 28th Annual Conference of the Cognitive Science Society (pp. 2240–2245), July 26-29, Vancouver, BC, Canada.

Thomas, M. S. C., Grant, J., Barham, Z., Gsodl, M., Laing, E., Lakusta, L., Tyler, L. K., Grice, S., Paterson, S. & Karmiloff-Smith, A. (2001). Past tense formation in Williams syndrome. *Language and Cognitive Processes, 16* (2/3), 143–176.

Thomas, M. S. C., Ojinaga Alfageme, O., D'Souza, H., Patkee, P. A., Rutherford, M. A., Mok, K. Y., Hardy, J., the LonDownS Consortium, & Karmiloff-Smith, A. (2020). A multi-level developmental approach to exploring individual differences in Down syndrome: genes, brain, behaviour, and environment, *Research in Developmental Disabilities, 104*, 103638. doi: 10.1016/j.ridd.2020.103638. Epub 2020 Jul 10. PMID: 32653761; PMCID: PMC7438975.

Thomas, M. S. C., Purser, H. R. M., & Richardson, F. M. (2013). Modularity and developmental disorders. In: P. D. Zelazo (Ed), *Oxford Handbook of Developmental Psychology*. Oxford: Oxford University Press.

Vertes, P. E., & Bullmore, E. T. (2015). Annual Research Review: Growth connectomics–the organization and reorganization of brain networks during normal and abnormal development. *Journal of Child Psychology and Psychiatry, 56*(3), 299–320.

Whitaker, K. J., Vertes, P. E., et al. (2016). Adolescence is associated with genomically patterned consolidation of the hubs of the human brain connectome. *Proceedings of the National Academy of Sciences USA, 113*(32), 9105–9110. DOI: 10.1073/pnas.1601745113

14

WHAT CAN NEURODEVELOPMENTAL DISORDERS TELL US ABOUT DEVELOPMENTAL PATHWAYS? REALISING NEUROCONSTRUCTIVIST PRINCIPLES NOW AND IN THE FUTURE

Emily K. Farran

Annette as a mentor, a collaborator and a friend

I first met Annette in October 1998 when I was 3 weeks into my PhD, at a national convention held by the Williams Syndrome Foundation, UK. Having spent those first 3 weeks reading a lot of her papers, I was very much in awe of her. I had several questions to ask her, and while I was daunted to be introduced to Annette, I was delighted that she gave her time so generously to someone at such an early stage in their career. I was equally awe-struck when she spent the evening dancing like the best of us at the Williams syndrome disco. On that one day, I witnessed Annette's generosity of character and her enthusiasm for fun, which I was to see many times over the subsequent years.

Beyond my PhD I began to collaborate with Annette. She helped me to get that coveted first grant, and agreed to give a keynote speech at the first conference that I organised. My most cherished memories are of the year that we spent co-editing a book together (Farran & Karmiloff-Smith, 2012). The book began life as a book on Williams syndrome (WS), with Annette as one of the contributors. Annette (rightly!) pointed out that I'd simply gathered my friends together to write chapters, and that it needed a theoretical focus. I asked Annette to come on board as a co-editor (she later told me that she had broken her vow to never write an edited book). Annette's suggestions were invaluable. We used Williams syndrome as a case study to demonstrate neuroconstructivist principles, a theoretical focus which gave the book a broader purpose. The intense collaboration was absolutely enthralling. I witnessed Annette's brilliance in her ideas, and the sheer speed at which she could write fabulously insightful pieces of work. The book was 'born' in the same month as my first daughter. Annette asked me to weigh the book, and she put together a slide for her talks announcing the birth of both book and baby, along with their birthweights and names. She was not only an immensely intelligent woman, but she had a brilliant sense of humour! Annette may no longer be with us, but she still guides us in many ways. While those of us who were lucky to have known Annette strive to mentor like her, and can only hope to achieve Annette's sense of style, this book is about her impact on the academic world. With this in mind, in the sections below, I use quotes from Annette's papers to illustrate neuroconstructivist principles predominantly from within my domain of expertise, visuospatial cognition.

The neuroconstructivist approach explains development as a dynamic process which occurs within the context of genetic, neural, biological and environmental constraints on the developing brain. In this way, domain specificity emerges over developmental time through repeated pairings of domain–relevant brain areas (i.e., areas best suited to processing certain kinds of input) to specific sensory input, via interactions with other multi-level processes (genetic, environmental). Annette first applied neuroconstructivist principles to neurodevelopmental disorders in her seminal paper, "Development itself is the key to understanding developmental disorders" (Karmiloff-Smith, 1998). Among other principles, she explained how genetic differences are most likely to impact low-level cognitive outcomes, rather than specific domains. Syndrome-specific cognitive phenotypes, therefore, result from the cascading effects of this genetic impact (among other inputs) across different domains as development occurs. This importance placed on the developmental process is a central tenet to the principles listed below.

Visuospatial cognition

Visuospatial cognition refers to the ability to perceive and interact with our visual world, for example, being able to individuate objects, to perceive the location and shape of objects and to understand the relationships between them. It provides us with skills such as the ability to read a map, to search for an object in a visual array (e.g., finding your coat on a coat rack), and to draw and interpret diagrams and pictures. Visuospatial ability also applies to large scale knowledge. That is, we use visuospatial ability to navigate to work and back, to reorient when we get lost and to know when to get off the bus. Here I will provide examples from my own data and others, of neuroconstructivism within the domain of visuospatial cognition, each example inspired by quotes from Annette's papers. To pay homage to much of the work that Annette and I collaborated on, I will predominantly use examples from individuals with WS.

Static versus dynamic approaches to neurodevelopmental disorders

> "...the roots of development are often critical for understanding the dynamic trajectory"
>
> (Karmiloff-Smith, 2009).

Early studies of neurodevelopmental disorders discussed the brain of individuals with neurodevelopmental disorders as having parts intact and parts impaired. This assumes that brain-behaviour associations are pre-determined and does not take into account input from genetic, developmental or environmental factors, that is, it ignores neuroplasticity. Annette (e.g., Karmiloff-Smith, 2009) highlighted that the brain is not a static unchanging structure, but that it is self-organising and developmentally responsive to interactions across multiple levels. For example, children exposed to more parental spatial language, develop stronger spatial language and better spatial abilities (Pruden, Levine, & Huttenlocher, 2011).

Annette was keen to emphasise that precursors or antecedents to later development provide a window to understanding interacting developmental processes. In her research she demonstrated that the deficit in number processing in WS relates to atypical early attentional processes (Karmiloff-Smith et al., 2012) and that in WS, in contrast to typical development (Bates, Benigni, Bretherton, Camaioni, & Volterra, 1979), joint attention is not a precursor to the emergence of first words (Laing et al., 2002).

A current example of the dynamic approach to understanding development is the strong interest in determining early markers of later impairment which is being investigated via prospective longitudinal studies of infants at risk of a behaviourally defined disorder such as Autism (see Jones, Gliga, Bedford, Charman, & Johnson, 2014). These studies have discovered, for example, that early deficits in attention disengagement and an atypical neural response to shifts in eye gaze are associated with a later diagnosis of Autism (Elsabbagh et al., 2013; 2013; also see Bedford et al., 2017). We are also currently experiencing an explosion of research into early motor behaviour as a predictor of later development. A longitudinal association between early motor abilities and later expressive language has been documented in both typical and atypical development (Iverson 2010; Leonard & Hill, 2015), while atypical motor development at 6 months is another early predictor of later Autism status (LeBarton & Landa, 2019; also see Bedford, Pickles, & Lord, 2016). This cascading influence of early motor development extends beyond the language domain. Clearfield (2004), for example, report that, in typical development, the onset of crawling and walking marks step changes in the development of spatial cognition (see Newcombe, 2019 for further examples).

The above "roots of development" enable a better understanding of typical development, as well as the developmental trajectory that leads to the emergence of a specific phenotype. With reference to the relationship between motor and language skills, for example, it is considered that this reflects a developmental pathway in which a growth in motor skill increases children's ability to explore objects, which in turn, supports their ability to learn about the properties of objects and to engage others in play and thus language experiences (LeBarton & Landa, 2019). Subtle differences in this early development can have a cascading impact on later development, hence why poor motor development predicts disruption in later expressive language development. Relatedly, interventions that are targeted at these early markers, as opposed to the domain of impairment, have the potential to have downstream impact on phenotypic outcomes (e.g., Jones, Dawson, Kelly, Estes, & Webb, 2017). Furthermore, for behaviourally defined disorders such as autism, these early markers provide an opportunity to detect risk of later diagnosis (Bedford et al., 2017).

In my lab, we are currently investigating whether the spatial deficit in WS has its origins, at least in part, in the impaired development of motor abilities in this group. This stems from knowledge of the association between motor and spatial development discussed above (Clearfield, 2004). Our initial investigations, which relied on retrospective reports of the emergence of motor milestones in WS, demonstrated that motor milestones were substantially delayed in WS, and that large-scale spatial abilities in adulthood were impaired in WS. These two deficits, however, did not show an association (Farran et al., 2018). Motor milestone measures are relatively course, and a long retrospective element can decrease the reliability of these measures; the next steps in this programme of research is to follow infants with WS longitudinally to prospectively document motor milestone achievement and motor quality, and to determine their impact on the develop-ment of spatial cognition. Annette often said that there wasn't enough research into infants with WS. She was a collaborator on the study above, and she would have been delighted to know that we're developing this research in this way. In view of the association between motor ability and expressive language, we also plan to determine the broader impact of early motor deficits in WS, such as the predictive relationship with later expressive language. Given that joint attention (which, although predominantly social, has a motor element) is not predictive of the emergence of language in WS (Laing et al., 2002), it is possible that the re-lationship between motor and language domains is not fixed, but that language can develop via alternative developmental pathways. The extent to which alternative pathways are fully compensatory, however, has implications for intervention design.

The future of the dynamic approach to neurodevelopmental disorders.

Individual differences in cross-sectional data do not always reflect longitudinal developmental processes (Purser, Thomas, Farran, Karmiloff-Smith, & Van Herwegen, 2019; although see Jarrold, Baddeley, Hewes, & Phillips, 2001), thus longitudinal data is important if we are to truly understand the dynamics of de-velopment. Over the next decade, longitudinal studies will reveal the importance of early sensory, attentional and motor development (as well as other factors) on later development. There are now many prospective longitudinal studies of children at risk of diagnosis of behaviourally defined disorders (examples from the Centre for Brain and Cognitive Development [CBCD], Birkbeck, where Annette worked are: The British Autism Study of Infant Siblings [BASIS]; and Studying Autism and ADHD Risks [STAARS]). These programmes enable us to under-stand the infant start-state and, vitally, to document the developmental processes which lead to the phenotypic characteristics of that disorder. However, there are few longitudinal studies of genetically diagnosed disorders. Paradoxically, in contrast to "at-risk" groups, for genetic disorders a confirmatory diagnosis can be made before the behavioural phenotype fully emerges. Until recently, the rarity of genetically defined disorders has provided a stumbling block. The LonDownS consortium (https://www.ucl.ac.uk/london-down-syndrome-consortium/) is one of the first truly multi-level, developmental investigations into Down Syndrome (DS). Annette led the infant strand of this research collaboration, and while she was not able to see the fruition of her work, this longitudinal cohort of infants and toddlers with DS, in concert with the other strands of the LonDownS consortium (an adult strand, genetics, cellular modelling and mouse modelling), really paves the way for understanding multi-level dynamic development in DS. It truly is a project which "takes development seriously" (Karmiloff-Smith, 1998).

While large longitudinal data is expensive to generate, at a minimum as researchers, we should work collaboratively to agree common protocols of background or baseline tasks that will feature in all testing batteries across multiple labs. This would enable pooling of data across labs in a cost-effective manner, and over time will generate longitudinal datasets. Indeed, in response to the lack of longitudinal research with individuals with WS, a number of researchers in the UK have formed the Williams Syndrome Development (WISDom) group. We have pooled together our data (including Annette's) from individuals with WS, across decades of research, with plans to collate and collect further data. Preliminary analyses have demonstrated that cross-sectional development of verbal and non-verbal ability does not necessarily mirror that observed longitudinally (Purser *et al.* 2019). This emphasises the impact of individual differences when working with cross-sectional data. Further research using this dataset will enable us to better understand the *longitudinal* cross-domain dynamics of development in WS. This will help to determine which cascading developmental trajectories observed in the typical population are fixed, and which aspects of development can emerge via alternative developmental pathways, akin to the example of joint attention and language above.

Cross-domain associations

"Do ... systems start out prespecified in the infant brain, already dissociated into separate subsystems, or do these subsystems emerge over time through early cross-domain interactions as different brain circuits progressively specialized for different ... functions?"

(Karmiloff-Smith et al., 2012)

The brain is not modular at the infant start-state. Optimum development involves extensive interaction across domains. Many of the developmental cascades discussed in the section above are demonstrations of cross-domain interactions (e.g., cross-domain interactions between motor and language domains). For neurodevelopmental disorder populations, cross-domain interaction can also be influenced by their phenotypic cognitive profile. A relative strength in their cognitive profile can be used to bootstrap an area of weakness, while a relative weakness can negatively impact the development of subdomains of an area of strength.

Individuals with WS demonstrate an uneven cognitive profile within which visuospatial cognition is poor relative to verbal abilities. In WS, while visuospatial cognition has scarcely been investigated in infancy (Farran, Brown, Cole, Houston-Price, & Karmiloff-Smith, 2008), there is clear evidence of cross-domain developmental interactions between the visuospatial domain and other domains from data from child and adult participants. For example, for inhibition tasks, number tasks and working memory, stronger performance is observed in WS when a verbal strategy can be applied, than when the task is reliant on visuospatial processing (Atkinson et al., 2003; Ansari et al., 2003; Jarrold, Baddeley, & Hewes, 1998).

A well-documented cross-domain interaction in typical development is the ability to represent visual images verbally, which Bruner (1966) describes as a shift from an iconic to a symbolic representation which occurs at about 6 years. This is supported by the working memory literature; from a similar age, children begin to verbally rehearse visual information (Baddeley, Gathercole, & Papagno, 1998). In my research we have shown that individuals with WS can employ a verbal coding strategy during simple spatial navigation tasks (Farran, Blades, Boucher, & Tranter, 2010; Farran et al. 2012b), in a manner akin to the verbal rehearsal of visuospatial information documented above. While this does not reflect an atypical strategy, it demonstrates that spatial performance can be augmented in WS via compensation from verbal ability, a relative strength within the WS cognitive profile.

Within the domain of number, Ansari et al. (2003) measured the cardinality principle in children with WS. This principle refers to knowledge that the total number of objects in a set is the same as the last number in the counting sequence of those objects, and emerges at about 3 years in typically developing (TD) children (Wynn, 1990). Ansari et al. (2003) compared the performance of children with WS (aged 6–11 years) with that of TD children (aged 2–5 years) on two measures of cardinality. The two groups performed at the same level for both tasks, and showed similar progression with task difficulty (i.e., numbers 1 to 3 were easier than numbers 4 to 6). However, correlational analyses demonstrated an association between verbal ability and task performance for the WS group, but between visuospatial ability and task performance for the TD group. Thus, although the performance of the two groups initially looked the same, by considering the contributions of verbal and visuospatial abilities it is possible to conjecture that the development of cardinality does not proceed along a typical developmental trajectory in WS. That is, in WS, poor visuospatial ability is compensated for by their comparatively stronger verbal abilities. Ansari et al. (2003) also point out that verbal abilities are not able to completely compensate; performance was below their verbal mental age. This suggests that visuospatial deficits (and possibly other constraints such as attention; Karmiloff-Smith et al., 2012) remain a constraint on performance because of the visuospatial demands of the task.

The developmental trajectory approach can also reveal cross-domain interactions (Thomas et al., 2009). This approach involves building task-specific cross-sectional developmental trajectories via regression analyses of the predictive power of potential pre-requisite mechanisms (e.g., spatial ability, verbal ability, attention, memory) against an outcome measure of interest. If development has proceeded in an atypical manner for the atypical group, developmental trajectory analysis might reveal the presence of a compensatory mechanism, or a limiting factor on development. We have used this approach to explore the contributing mechanisms to large-scale spatial navigation in DS and WS (e.g., Farran et al., 2012a; Farran et al., 2015; Purser et al., 2014). Annette was always very enthusiastic about my navigation research. I think this was for two reasons. First, there is cross-species relevance – one of Annette's infamous studies, which often appeared on her slides, was her ball pool Morris water maze (a task that has been extensively used to study navigation in rodents), which she used with toddlers with WS and partial deletion patients. Second, due to the real-world application of navigation research. Annette took care to spend time with the families of the participants that she worked with, and provided advice and help to countless families (as many of us have continued to do so). Research with a direct application to the families, therefore, had strong appeal.

Using developmental trajectory analysis in our navigation research, we determined that the ability to learn a route from A to B (route knowledge; Siegel, & White, 1975) in a virtual environment was related to non-verbal ability (as measured by the Ravens Coloured Progressive Matrices, Raven, 1993) for individuals with WS, individuals with DS and TD children, but that group differences were revealed with respect to input from attentional mechanisms and long-term memory (Purser et al., 2014). For the DS and WS groups, poorer inhibition was associated with making more route learning errors, while for the TD group, sustained attention was a better predictor of route learning ability. Furthermore, long-term memory was a strong contributor to route learning in the DS and TD groups, but not the WS group. Thus, while large-scale navigation is broadly considered as a spatial task, it is impacted by different attentional mechanisms in atypical development (in this case DS and WS), compared to typical development. This might relate to strengths and weaknesses for each group; for both DS and WS group, inhibition is weak, particularly so for WS (Carney, Brown, & Henry, 2013). Thus, we deemed poor inhibition to be a limiting factor to the development of large-scale navigation in these groups. In contrast, for the TD group, the inhibition demands (e.g., inhibiting turning down an incorrect path, and inhibiting attending to non-unique landmarks such as streetlights) of the task did not challenge their ability to learn the route. These analyses, therefore, revealed important mechanistic cross-domain differences across syndromes in how a task is completed.

The future of cross-domain associations

Cross-domain interactions occur as part of typical and atypical developmental trajectories. Equally, task completion is influenced by an individual's strengths and weaknesses. Where a phenotypic cognitive profile is uneven, as in many groups with neurodevelopmental disorders, we have observed that tasks are completed in an atypical manner, by compensating for areas of weakness using areas of relative strengths. This is further evidence of the self-structuring nature of development. However, a current weakness in the majority of neurodevelopmental disorder research is the small sample sizes. Thus, where the statistical techniques that are used with data from typical developing children now take a more mechanistic or process-based approach (i.e., they determine what the underlying processes are that drive performance in an outcome measure), this is not so easy in neurodevelopmental research. Because of the difficulty achieving sufficient power, much of the research with neurodevelopmental disorder groups relies on analyses of group means and group differences. Multiple regression and mediation analyses, as examples, enable one to determine the relative contribution of a range of potential contributing mechanisms to performance and thus would pinpoint cross-domain interactions at a fine-grained level. One way to accomplish strong group numbers is to run multi-centre studies. The EU-AIMS (https://www.eu-aims.eu/) and AIMS 2 TRIALS (https://www.aims-2-trials.eu/) studies of autism are examples of this within neurodevelopmental disorder research.

Associations can be as informative as dissociations

> *"…researchers need to recall that similar behavioral outcomes may stem from very different cognitive/brain causes…"*
>
> *(Karmiloff-Smith, 2009)*

The cross-domain interactions described above provide insight into the use of compensatory strategies in neurodevelopmental disorders. I advocate an analytical approach to understanding task performance. An understanding of why and how a particular level of performance is achieved can uncover dissociations, where performance might otherwise have been described as typical or simply delayed. Techniques such as eye-tracking, error analysis, and determining input from underlying mechanistic processes are examples of such an analytical approach. Jo Camp, in her PhD (Camp, 2014), employed the Tower of London task (TOL: Shallice, 1982) as a measure of planning ability with individuals with WS, individuals with Down syndrome and TD children. In this task, the participant is presented with three different coloured beads on three pegs of different heights. They are shown a goal state and asked to rearrange the beads one at a time to reach the goal configuration of beads. Camp (2014) demonstrated that despite similar overall TOL scores (a composite score comprised of: errors; rule violations; and number of moves made) across the groups, this was driven by different behaviour.

The WS group produced more absolute errors than the TD group (this included perceptual errors and reaching the maximum number of 20 moves without reaching the solution). The increased number of perceptual errors is indicative of the limitations in WS in visuospatial perception. The DS group, in contrast, lost points by violating the rules of the task. Equally, while the TD and WS group demonstrated a relationship between TOL performance and planning ability, this was only evident for the easier trials in the DS group, suggesting a lack of engagement with the task once it became difficult. This suggests that poor performance on this task, for the WS group, was a reflection of both poor planning and poor visuospatial cognition; their phenotypic visuospatial deficit limited performance on this task for this group in a manner that was not evident for the DS and TD groups.

Face processing is another area in which an analytical approach can reveal atypical processing despite seemingly typical performance. Early studies of face processing in WS employed standardised face processing tasks such as the Benton Facial Perception Test (BFPT: Benton, Hamsher, Varney, & Spreen, 1983) to reveal performance in the normal range for people with WS (e.g., Bellugi, Sabo, & Vaid, 1988). However, these tasks lack the sensitivity required to detect subtle atypicalities in processing. In the typical population, a holistic/configural strategy is employed to process faces, e.g., processing the whole face or the spatial distance between the eyes and nose (Young, Hellawell, & Hay, 1987). Where studies have taken a more analytical approach, evidence suggests that individuals with WS do not process faces in a typical manner, but rely on featural information, i.e., the individual components of a face (eyes, nose, mouth) to recognize faces (Deruelle, Mancini, Livet, Casse-Perrot, & deSchonen, 1999). Annette and her team investigated configural processing capabilities in WS in depth (Karmiloff-Smith et al., 2004). Across a series of studies, they demonstrated that individuals with WS are less sensitive to configural information than TD children and adults, and show an atypical developmental trajectory for configural information. Instead, individuals with WS are reliant on a featural face processing strategy (where a task allows) to process faces.

We have recently shown that the featural processing strategy employed to process face identity also extends to face expression processing in WS (Ewing, Farran, Karmiloff-Smith, & Smith, 2017). This collaboration was borne out of one of Michael Thomas' and Annette's DNL lab meetings. When I came to London, I received a fantastic welcome from Michael and Annette, and absolutely relished the weekly DNL meeting. Marie Smith presented the innovative "bubbles" technique (Gosselin & Schyns, 2001) at one of these meetings; we went on to use this technique in WS. Using this technique, on each trial, a face is visible through randomly presented circular apertures or "bubbles". The participant uses the information visible through the bubbles to make their judgement. In this study, participants were asked to judge facial expression from a partially concealed face. Hence, we were able to determine the specific information from each face that was used to make the facial expression judgement. Our results showed that the WS group required the same number and size of bubbles as 11-year-old TD children but that they rely on a less integrated set of visual information to categorise facial expressions than either TD adults or TD children, that is, a relatively more featural processing style. This ability to achieve a similar level of performance to controls, via an alternative processing route is consistent with the results mentioned above in the context of face identity processing. Face processing, therefore, represents an elegant example of associations being as informative as dissociations. One further example within this domain, for which I have fond memories of Annette, relates to the impaired social interaction observed in both WS and Autism. This is nicely described by Debbie Riby in our book (Riby, 2012). When editing the book, I wrote 'editor summaries' at the end of each chapter, and an absolute favourite sentence of Annette's was the idea that "…whilst *decreased* attention to faces in Autism can be detrimental to social interaction, in WS *increased* attention to faces can have the same effect" (Farran & Karmiloff-Smith, 2012).

The Raven's Coloured Progressive Matrices (RCPM; Raven, 1993) and Raven's Standard Progressive Matrices (RSPM; Raven, 1976) are commonly used measure of non-verbal reasoning, which are often used with individuals with neurodevelopmental disorders. This is based on the assumption that these tasks measure the same processes and constructs as in the typical population. This assumption has been tested in WS, DS and Autism (Facon, & Nuchadee, 2010; Soulieres et al., 2009; Van Herwegen, Farran, & Annaz, 2011). For both WS and DS, the proportion of each error type resembled that observed in typical development, and the developmental changes in the kinds of error types and the difficulty level of items were both typical in both groups. This demonstrated that the underlying mechanisms that support performance on this task in these groups, do not differ from typical development. This is an example where an analytical approach has enabled us to validate an association as a true association which reflects common strategy use across these groups. In contrast, for individuals with Autism, for whom a relative strength in performance is observed on the RSPM task, performance differs at the neural level, compared to TD controls. Performance in Autism is associated with increased activation in left occipital areas and decreased activation in medial posterior parietal and left lateral prefrontal areas. This is interpreted as reflecting a stronger reliance on visual processing, and reduced reliance on working memory (Soulieres et al., 2009) relative to TDs. Thus, for this group, the RSPM is tapping into underlying processes in an atypical manner, which should be taken into account when using this measure, particularly when used as a measure by which to match participant groups. These are important findings, given the frequent use of these tasks with these groups.

The example above regarding RSPM performance in Autism illustrates a multi-level approach in which dissociations are observed at the neural level that are not observed at the behavioural level. A similar approach was adopted by Sahyoun, Soulieres, Belliveau, Mottron, and Mody (2009), using a pictorial reasoning task. They demonstrated that despite an association at the behavioural level, the Autism group exhibited reduced activation of frontal and temporal language regions, and reduced neural connections within this language network, relative to TD controls. Instead, the Autism group relied on posterior occipito-temporal and ventral temporal regions to complete the task. This suggests group differences in task completion strategy, with visual mediation, rather than verbal mediation, being employed in Autism.

The future of associations and dissociations

Associations can often mask underlying dissociations. This is key to a fine-grained understanding of developmental differences across neurodevelopmental disorder groups. Analytic approaches which draw on error analyses, the practice of deriving sensitive measures, and the use of multi-level analysis, can often uncover that seemingly typical behaviour is being reached by alternative means. Multi-level analysis is becoming more accessible for use with neurodevelopmental disorders due to technological advances in genetic profiling as well as neural measurement techniques such as functional Near Infrared Spectroscopy (fNIRS), and wireless and portable EEG. With reference to behavioural measurement, standardised tasks can often lack sensitivity because they were not designed for the population that they are being tested on. The Benton Test of Facial Perception is one example; "normal" performance on this task can be achieved via a featural processing style (Duchaine & Weidenfeld, 2003). An understanding of whether a behaviour can be reached via alternative means enables us to understand the extent to which a phenotypic deficit is detrimental to performance versus whether it can be compensated for. This has implications for understanding both typical and atypical development.

Cross-syndrome comparison

> "... it is critical to raise cross-syndrome questions at multiple levels of analysis, such as: is this brain/cognitive/behavioural deficit syndrome-specific or is it syndrome-general, that is, is it characteristic of all atypical development where learning difficulties obtain or unique to a particular syndrome?"
>
> (Karmiloff-Smith & Farran, 2012)

Cross-syndrome comparison has the advantage of enabling one to determine which patterns of performance are syndrome-specific and which are artefacts of having learning difficulties. With reference to early predictors of later atypical development, it can also help to determine areas of developmental vulnerability versus disorder-specific risk factors.

Across a series of studies, some of which I discuss above, we used cross-syndrome comparison to investigate large-scale spatial navigation performance in individuals with WS and individuals with DS. These two groups present with contrasting cognitive profiles with respect to verbal and visuospatial ability. Furthermore, both groups also demonstrate atypicalities in the hippocampus

(Pinter et al., 2001; Meyer-Lindenberg et al., 2004), which is strongly associated with navigation performance in the typical population (Doeller, King, & Burgess, 2008). Coupled together, this suggested to us that navigation performance would be poor in both groups, but that performance would be associated with differing underlying mechanisms for each group.

Above, I describe one study in which the ability to learn a route from A to B was driven by different processes in DS, WS and TD groups (Purser et al., 2014), thus demonstrating that poor performance represented different limitations across these groups. In another study, participants were asked to learn two routes, A to B and A to C, within a grid environment (Farran et al., 2015; also see Courbois et al., 2013). Having learnt these routes, they were then asked to demonstrate a short-cut from B to C. Their ability to find the short-cut was our measure of configural knowledge (an understanding of the configural structure of the area; Siegel & White, 1975). We demonstrated that 59% of TD children could find the short-cut. This compared to only 10% of individuals with DS and 35% of individuals with WS. In fact, the DS group often used the strategy of simply adding the two known routes together, which requires no configural understanding at all. While configural knowledge was strongly related to non-verbal ability in the TD group, it was related to attention switching in the WS group (correlations were not useful in the DS group due to the low success rate). We suggest that the WS group are limited by their ability to switch perspectives from the egocentric first-person perspective that they experience in the environment itself (useful for gaining knowledge of a fixed route) to the allocentric viewer-independent perspective required to make decisions based on the configural structure of the environment. In contrast, the majority of individuals with DS are simply unable to develop configural knowledge; they do not have a cognitive map (also see Toffalini, Meneghetti, Carretti, & Lanfranchi, 2018). Our current ongoing research will provide insight into any further group difference. We are working with individuals with WS and individuals with DS to determine the benefits of providing an aerial view satellite-navigation style map (i.e., configural information) to participants while they are navigating from a first-person perspective. If enabling participants to map these two perspectives to one another is beneficial it might facilitate their ability to switch from an egocentric first-person perspective to an allocentric viewer-independent perspective. Based on the evidence above, this might facilitate performance for a WS group more than a DS group, and predicts a benefit to configural knowledge in the 'sat-nav' condition compared to the standard condition for the WS group, but not the DS group. This study has strong implications for navigation training. If our hypotheses are borne out, this would really have appealed to Annette with reference to its real-world application and the direct relevance to the families of individuals with neurodevelopmental disorders.

There are several neurodevelopmental disorders for whom deficits in functions associated with the dorsal visual stream have been reported. These include dyslexia, Autism, Developmental Coordination Disorder, Williams syndrome and Fragile X syndrome. Inspection of the data from individual disorder research suggests that the dorsal visual stream is developmentally vulnerable. However, the lack of cross-syndrome comparison means that it is currently not possible to differentiate between developmental vulnerability and phenotypically unique patterns of ability across the functions of the dorsal visual stream (see Grinter, Maybery, & Badcock, 2010). The former are relevant for understanding developmental characteristics of the brain, while the latter help us to understand the specifics of each disorder. This is a strong example of the potential benefit to the interpretation of research findings, that cross-syndrome comparisons would reap.

The future of cross-syndrome comparison

Cross-syndrome comparison enables one to determine whether a participant group characteristic reflects learning difficulties, developmental vulnerability or syndrome-specific phenotypic processing, that is, it provides a more subtle comparison than with TD controls alone.

Whilst the detection of the early markers of a disorder using "at risk" populations is a relatively new area of research, it will be important to determine the specificity of these markers via cross-syndrome comparison. Similar to the dorsal stream vulnerability example above, Levit-Binnun, Davidovitch, and Golland (2013) suggest that motor and sensory deficits reported across a number of neurodevelopmental disorders are a reflection of vulnerability in brain networks. Equally, there is a high level of co-occurrence across disorders such as Autism, ADHD and DCD, and there might be common early markers or risk factors for the development of these disorders. Johnson, Gliga, Jones, and Charman (2015) review the literature regarding early markers of Autism and ADHD to conclude that whilst each disorder demonstrates early deficits in sensory, attentional and motor domains, direct comparisons have not yet been made within a single study and comparisons across studies are difficult to make due to differences in the focus of studies across disorders. To my knowledge, while there are no published studies of cross-syndrome comparisons across these groups, labs often run parallel streams with common test protocols across, for example, Autism and ADHD streams, which allows for cross-syndrome comparison. This data will be vital if we are to further our understanding of developmental vulnerability, and I have little doubt that Annette will have sowed the seeds for cross-syndrome comparisons to be made with the data held at the CBCD.

Individual differences

"Consideration of individual variation at multiple levels opens a series of new questions... that remained hidden in studies at the... group level."

(Karmiloff-Smith et al., 2016)

The literature on individual differences within neurodevelopmental disorders has historically been scarce and, in published papers, can often only be inferred from looking at error bars. However, there is a growing impetus to acknowledge these differences and to use these to better understand the developmental processes which underpin disorder phenotypes.

Developmental trajectory analysis demonstrates individual differences. Data is plotted in the form of scatter plots, and thus the score of each individual is visible. Because data is plotted cross-sectionally against mental age, differences in rates of development across neurodevelopmental disorder groups can also reveal individual differences in the impact of an underlying process on a task. In Purser et al. (2014), discussed above, we demonstrated that stronger non-verbal ability was associated with lower errors on the route learning task for all groups, but that the developmental trajectory for the DS group was steeper than for the TD group. That is, lower levels of non-verbal ability were more detrimental to route learning ability in DS than in the TD group, but at higher levels of non-verbal ability, the DS trajectory had "caught up", and levels of route learning ability were commensurate with their non-verbal mental age. These group differences in the slope function demonstrate a cascading impact of individual differences; the steeper function of the DS group demonstrates broader individual differences in route learning ability in DS than one might expect for the range of non-verbal mental age. At a cognitive level, the group differences at the lower end of the trajectory might reflect an inability for some individuals (those with lowest level of non-verbal ability), but not others (those with relatively stronger non-verbal ability), to compensate for low non-verbal ability.

Individual differences in DS was the topic of one of Annette's most recent papers (Karmiloff-Smith et al., 2016). She describes broad heterogeneity in DS at many levels, including genetic, neural and cognitive. For example, at the genetic level, there are individual differences in the mechanisms which give rise to an extra copy of chromosome 21 (the most common form of DS), as well as individual differences in gene expression. This, along with individual differences in neural development and environmental factors, all contribute to how the phenotypical outcome of DS in any one individual is expressed. Taking this into consideration, individual differences hold strong promise in uncovering the developmental pathways to phenotypic outcome at a more fine-grained level than has hitherto been considered.

Broad individual differences are not only observed in DS. In my research (Farran et al., 2015) we have shown that navigation competence in both DS and WS ranged from those who were not able to learn a simple route from A to B and back again, to those who could learn the configuration of a new environment in sufficient detail to be able to determine a novel short-cut within the environment. Thus, while navigation is considered a relative weakness in these groups, this is not the case for all individuals. These differences might stem from differences in input from underlying processes (e.g., attention, memory), or influences at different levels of description (genetic, neural, environmental), which act as constraints or facilitators on cognitive outcome (in this case, navigation).

Jones et al. (2014) describe individual differences in early development, and the impact that this has on later heterogeneity. They describe a study by Landa, Gross, Stuart, and Bauman (2012) in which infants who went on to develop Autism could be categorised into four different subgroups with respect to their early characteristics. These ranged from those with general developmental delay to those with accelerated early development. This demonstration of heterogeneity from the infant start-state has broad implications for the understanding of how phenotypes develop. Jones et al. (2014) point out that this is not only important theoretically but also from a clinical standpoint with reference to the ability to predict the severity of a child's symptoms later in life, and parental treatment decisions.

The future of individual differences

Individual differences are inherent in any group of individuals. However, for neurodevelopmental disorders these individual differences are often broader than observed in the typical population. This can be detrimental to interpretation of results, as it is possible, particularly when small samples are involved, that sample characteristics are not representative of the broader population, leading to spurious findings. As researchers, we should be transparent in reporting individual differences. This can be as simple as representing group data as individual data points, for example by using scatter plots, violin plots or pirate plots, rather than presenting group means, to studies such as Landa et al. (2012), in which latent class analysis was used to categorise participants into subgroups. Furthermore, a multi-level approach of understanding input from genetic, neural and/or environmental factors will enable us to better understand the seemingly broad heterogeneity in neurodevelopmental disorder groups. Annette liked to include many parental questionnaires in testing batteries. This can be a cost-effective and time-efficient way of gathering environmental data, such as SES and maternal depression scores, which might have a strong impact on the outcome of the participant. An understanding of individual differences not only has implications for intervention but also asks questions more broadly regarding heterogeneity in phenotypic characteristics.

Realising neuroconstructivist principles and reproducible research

Throughout this chapter, I have illustrated neuroconstructivist principles that were strongly advocated by Annette. These principles are not mutually exclusive of course, and many of the examples illustrate more than one of the principles. Key considerations have surfaced, which can be used to both improve our understanding of typical and atypical development and improve the rigour of future research. Some of these considerations map onto recommendations that are easy to implement, while others require significant resources. First, we should strive to be transparent about heterogeneity within a group. Annette's most recent paper advocated a better understanding of individual differences; she would be thrilled that there is now a strong push to represent and understand individual differences in neurodevelopmental disorders. In its simplest form, we can plot individual data points rather than group means. This will enable the reader to better understand the range of performance of a group. And, beyond behaviour, Annette discussed individual differences at genetic and neural levels. Technological advances have and will make these kinds of individual differences more and more possible to investigate. Another example refers to the use of standardised tests. Standardised tests assume that the task is being completed using a typical strategy, and that the cognitive profile of the participant is uniform. We discussed above how performance in the normal range can be achieved on the Benton Test of Face Perception via an atypical strategy. Similarly, performance on the Test for Reception of Grammar (TROG; Bishop, 1983) is not fully representative of grammatical ability in WS on account of their poor performance on the spatial items in the task (Philips, Jarrold, Baddeley, Grant, & Karmiloff-Smith, 2004). I advocate taking an analytical approach to understanding task performance. While the use of standardised tasks has many advantages, detailed investigation of performance on such tasks (e.g., analysis of error types, looking performance, contributing mechanisms; e.g., Van Herwegen et al., 2011) is warranted in order to be certain that we understand what we are measuring.

The remaining considerations map onto recommendations that require broad resources in terms of time, money and geographical space. First, we need to study individuals longitudinally if we are to obtain a true picture of the dynamics of development (and Annette had begun to take this a step further, into the fetal brain, as part of her collaborations with Mary Rutherford – a legacy that continues via Michael Thomas and me). Cross-sectional data does not always map directly onto longitudinal developmental trajectories. To-date, with the exception of the cohorts of "at-risk" groups, there are few longitudinal studies of neurodevelopmental disorders. This is particularly true of genetically defined disorders. Second, we should be cautious in labelling a deficit as disorder-specific until cross-disorder comparison has been carried out. Currently, most of the longitudinal studies of "at-risk" groups are single disorder studies. This risks failing to answer questions that can be addressed via cross-disorder research, such as whether an identified antecedent of later diagnosis has specificity to that disorder. My final recommendation relates to sample size. This is a recognised issue when working with rare and vulnerable groups. Small sample sizes limit the kinds of analyses that we can employ (e.g., numerous studies, including many of my own, are designed to rely on determining differences in group means, simply because of the practical limitations in gaining the sufficient power required to employ designs such as associational designs or latent group/variable analyses). Furthermore, given that individual differences can be broader in an atypical group than observed in the typical populations, small samples also run an elevated risk of not being representative of the population. Acquiring adequately powered sample sizes is difficult. This and the recommendations above require collaboration. In an era where reproducible science is at the forefront of our minds, it is important that labs work together to agree common protocols and to collect data collectively such that participant numbers are sufficient to produce reproducible findings (the WISDom group is an emerging example). This also ensures that the analyses chosen are driven by the research question, rather than being driven by limitations in the availability of participants.

There are two standout legacies of Annette that come from a more personal level. In terms of advocacy, we can all also aim to be like Annette when interacting with families and charities. I am very fond of my WS participants and their families, and have worked with many of them for over 20 years. As a legacy to Annette, we should continue to take time to chat to the parents and the individuals themselves, and to help when we are able to (a letter to a school about the capacity of children with WS can work wonders). Do not aim to just collect data; seek to chat to the families (I have learnt so much from this), give your time at the family events, interact with the charities... and go and dance at the WS disco! The second legacy relates to collaboration. Annette was also a collaborator extraordinaire. She had time for so many people, and inputted to so many of our lives. As a legacy to Annette, we should push to collaborate as a method for improving the science of neurodevelopmental disorders. This will enable us to ask bigger questions and to be confident in the reproducibility of our findings. One of Annette's strongest pieces of advice was to "Ask the big questions".

References

Atkinson, J., Braddick, O., Anker, S., Curran, W., Andrew, R., Wattam-Bell, J., & Braddick, F. (2003). Neurobiological models of visuospatial cognition in children with Williams syndrome: Measures of dorsal-stream and frontal function. *Developmental Neuropsychology, 23*, 139–172.

Ansari, D., Donlan, C., Thomas, M., Ewing, S., Peen, T., & Karmiloff-Smith, A. (2003). What makes counting count? Verbal and visuo-spatial contributions to typical and atypical number development. *Journal of Experimental Child Psychology, 85*, 50–62.

Baddeley, A., Gathercole, S., & Papagno, C. (1998). The phonological loop as a language learning device. *Psychological Review, 105*, 158–173.

Bates, E., Benigni, L., Bretherton, I., Camaioni, I., & Volterra, V. (1979). *The emergence of symbols, cognition, and communication in infancy*. New York: Academic Press.

Bedford, R., Gliga, T., Shephard, E., Elsabbagh, M., Pickles, A., Charman, T., & Johnson, M. H. (2017). Neurocognitive and observational markers: prediction of autism spectrum disorder from infancy to mid-childhood. *Molecular Autism, 8*. doi:10.1186/s13229-017-0167-3

Bedford, R., Pickles, A., & Lord, C. (2016). Early gross motor skills predict the subsequent development of language in children with autism spectrum disorder. *Autism Research, 9*(9), 993–1001. doi:10.1002/aur.1587

Benton, A. L., Hamsher, K., Varney, N. R., & Spreen, O. (1983). *Contributions to neuropsychological assessment*. New York: Oxford University Press.

Bellugi, U., Sabo, H., & Vaid, J. (1988). Spatial deficits in children with Williams Syndrome. In J. Stiles-Davis, U. Kritchevshy, & U. Bellugi (Eds.), *Spatial Cognition: Brain Bases and Development* (pp. 273–297). Hillsdale, NJ: Lawrence Erlbaum.

Bishop, D. V. M. (1983). *Test For Reception of Grammar*. Manchester: Medical Research Council.

Bruner, J. S. (1966). *Toward a Theory of Instruction*. Cambridge: Harvard University Press.

Camp, J. S. (2014). The development of problem-solving abilities in typical and atypical development. Unpublished PhD Thesis.

Carney, D. P. J., Brown, J. H., & Henry, L. A. (2013). Executive function in Williams and Down syndromes. *Research in Developmental Disabilities, 34*(1), 46–55. doi:10.1016/j.ridd.2012.07.013

Courbois, Y., Farran, E. K., Lemahieu, A., Blades, B., Mengue-Topio, H., Sockeel, P. (2013). Wayfinding behaviour in Down Syndrome: A study with virtual environments. *Research in Developmental Disabilities, 34*, 1825–1831.

Clearfield, M. W. (2004). The role of crawling and walking experience in infant spatial memory. *Journal of Experimental Child Psychology, 89*(3), 214–241. doi:10.1016/j.jecp.2004.07.003

Deruelle, C., Mancini, J., Livet, M. O., Casse-Perrot, C., & deSchonen, S. (1999). Configural and local processing in Williams syndrome. *Brain and Cognition, 41*, 276–298.

Doeller, C. F., King, J. A., & Burgess, N. (2008). Parallel striatal and hippocampal systems for landmarks and boundaries in spatial memory. *Proceedings of the National Academy of Sciences USA, 105*, 5915–5920.

Duchaine, B. C., & Weidenfeld, A. (2003). An evaluation of two commonly used tests of unfamiliar face recognition. *Neuropsychologia, 41*, 713–720.

Elsabbagh, M., Fernandes, J., Webb, S. J., Dawson, G., Charman, T., Johnson, M. H., & British Autism Study Infant, S. (2013). Disengagement of visual attention in infancy is associated with emerging autism in toddlerhood. *Biological Psychiatry, 74*(3), 189–194. doi:10.1016/j.biopsych.2012.11.030

Ewing, L., Farran, E. K., Karmiloff-Smith, A., Smith, M. L. (2017). Understanding strategic information use during emotional expression judgments in Williams syndrome. *Developmental Neuropsychology, 42*, 323–335. doi.org/10.1080/87565641.2017.1353995

Facon, B., & Nuchadee, M. L. (2010). An item analysis of Raven's Colored Progressive Matrices among participants with Down syndrome. *Research in Developmental Disabilities, 31*, 243–249.

Farran, E. K., Brown, J. H., Cole, V. L., Houston-Price, C. & Karmiloff-Smith, A. (2008) A longitudinal study of perceptual grouping by proximity, luminance and shape in infants at two, four, six and eight months. *European Journal of Developmental Science, 2*, 353–369.

Farran, E. K. & Karmiloff-Smith, A. (Eds) (2012). *Neurodevelopmental Disorders Across the Lifespan: A Neuroconstructivist Approach*. Oxford University Press.

Farran, E. K., Bowler, A., D'Souza, H., Mayall, L., Hill, E., Karmiloff-Smith, A. (2018). Motor competence as a predictor of Large-scale spatial cognition in ADHD, Williams syndrome and Typical Development. Paper presented at Neurodevelopmental Disorder Annual Seminar series, Coventry, June 2018.

Farran, E. K., Blades, M., Boucher, J. & Tranter, L. J. (2010). How do individuals with Williams Syndrome learn a route in a real world environment? *Developmental Science, 13*, 454–468.

Farran, E. K., Courbois, Y., Van Herwegen, J., Blades, M. (2012a). How useful are landmarks when learning a route in a virtual environment? Evidence from typical development and Williams syndrome. *Journal of Experimental Child Psychology*, *111*, 571–586.

Farran, E. K., Courbois, Y., Van Herwegen, J., Cruickshank, A. G., Blades, M. (2012b). Colour as an environmental cue when learning a route in a virtual environment: Typical and atypical development. *Research in Developmental Disabilities*, *33*, 900–908.

Farran, E. K., Purser, H. R. M., Courbois, Y., Ballé, M., Sockeel, P., Mellier, D., Blades, M. (2015). Route knowledge and configural knowledge in typical and atypical development: A comparison of sparse and rich environments. *Journal of Neurodevelopmental Disorders*, *7*, 37. doi: 10.1186/s11689-015-9133-6

Farran, E. K., Broadbent, H., Atkinson, L. (2016). Impaired spatial category representations in Williams Syndrome; an investigation of the mechanistic contributions of non-verbal cognition and spatial language performance. *Frontiers in Psychology*. 7. doi: 10.3389/fpsyg.2016.01868

Gentner, D., & Goldin-Meadow, S. (Eds.). (2003). *Language in Mind: Advances in the study of language and thought*. MIT Press.

Gosselin, F., & Schyns, P. (2001). Bubbles: A technique to reveal the use of information in recognition tasks. *Vision Research*, *41*(17), 2261–2271.

Grinter, E. J., Maybery, M. T., & Badcock, R. D. (2010). Vision in developmental disorders: Is there a dorsal stream deficit? *Brain Research Bulletin*, *82*, 147–160.

Iverson, J. M. (2010). Developing language in a developing body: Therelationship between motor development and language development, *Journal of Child Language*, *37*, 229–261. doi: 10.1017/S0305000909990432

Jarrold, C., Baddeley, A. D., & Hewes, A. K. (1998). Genetically dissociated compnents of working memory: Evidence from Down's and Williams Syndrome. *Neuropsychologia*, *37*, 637–651.

Jarrold, C., Baddeley, A. D., Hewes, A. K., & Phillips, C. (2001). A longitudinal assessment of diverging verbal and non-verbal abilities in the Williams syndrome phenotype. *Cortex*, *37*, 423–431.

Johnson, M. H., Gliga, T., Jones, E., & Charman, T. (2015). Annual research review: Infant development, autism, and ADHD – early pathways to emerging disorders. *Journal of Child Psychology and Psychiatry*, *56*(3), 228–247. doi:10.1111/jcpp.12328

Jones, E. J. H., Gliga, T., Bedford, R., Charman, T., & Johnson, M. H. (2014). Developmental pathways to autism: A review of prospective studies of infants at risk. *Neuroscience and Biobehavioral Reviews*, *39*, 1–33. doi:10.1016/j.neubiorev.2013.12.001

Jones, E. J. H., Dawson, G., Kelly, J., Estes, A., & Webb, S. J. (2017). Parent-delivered early intervention in infants at tisk for ASD: Effects on electrophysiological and habituation measures of social attention. *Autism Research*, *10*(5), 961–972. doi:10.1002/aur.1754

Karmiloff-Smith, A., Al-Janabi, T., D'Souza, H., Groet, J., Massand, E., Mok, K.,… Strydom, A. (2016). The importance of understanding individual differences in Down syndrome. *F1000Research*, *5*. doi:10.12688/f1000research.7506.1

Karmiloff-Smith, A. (1998). Development itself is the key to understanding developmental disorders. *Trends in Cognitive Neuroscience*, *2*, 389–398.

Karmiloff-Smith, A. (2009). Nativism versus neuroconstructivism: Rethinking the study of developmental disorders. *Developmental Psychology*, *45*, 56–63.

Karmiloff-Smith, A., Thomas, M., Annaz, D., Humphreys, K., Ewing, S., Brace, N.,… Campbell, R. (2004). Exploring the Williams syndrome face processing debate: The importance of building developmental trajectories. *Journal of Child Psychology and Psychiatry*, *45*, 1258–1274.

Karmiloff-Smith, A., D'Souza, D., Dekker, T. M., Van Herwegen, J., Xu, F., Rodic, M., & Ansari, D. (2012). Genetic and environmental vulnerabilities in children with neurodevelopmental disorders. *Proceedings of the National Academy of Sciences of the United States of America*, *109*, 17261–17265. doi:10.1073/pnas.1121087109

Karmiloff-Smith, A. & Farran E. K. (2012). Conclusion: Future theoretical and empirical directions within a neuroconstructivist framework. In E. K. Farran & A. Karmiloff-Smith (Eds.), Neurodevelopmental Disorders Across the Lifespan: A Neuroconstructivist Approach (pp. 363–371). Oxford University Press.

Landa, R. J., Gross, A. L., Stuart, E. A., & Bauman, M. (2012). Latent class analysis of early developmental trajectory in baby siblings of children with autism. *Journal of Child Psychology and Psychiatry*, *53*(9), 986–996. doi:10.1111/j.1469-7610.2012.02558.x

Laing, E., Butterworth, G., Ansari, D., Gsodl, M., Longhi, E., Panagiota, G.,… Karmiloff-Smith, A. (2002). Atypical development of language and social communication in toddlers with Williams syndrome. *Developmental Science*, *5*(2), 233–246.

LeBarton, E. S. & Landa, R. J. (2019). Infant motor skill predicts later expressive language and autism spectrum disorder diagnosis, *Infant Behavior and Development*, *54*, 37–47. doi: 10.1016/j.infbeh.2018.11.003

Leonard, Hayley C. & Hill, Elisabeth L. (2014), Review: The impact of motor development on typical and atypical social cognition and language: a systematic review, *Child and Adolescent Mental Health*. doi: https://doi.org/10.1111/camh.12055

Levit-Binnun, N., Davidovitch, M., & Golland, Y. (2013). Sensory and motor secondary symptoms as indicators of brain vulnerability. *Journal of Neurodevelopmental Disorders*, *5*. doi:10.1186/1866-1955-5-26

Lombardi, C. M., Casey, B. M., Thomson, D., Nguyen, H. N., & Dearing, E. (2017). Maternal support of young children's planning and spatial concept learning as predictors of later math (and reading) achievement. *Early Childhood Research Quarterly*, *41*, 114–125. doi:10.1016/j.ecresq.2017.07.004

Newcombe, N. S. (2019). Navigation and the developing brain. *Journal of Experimental Biology*, *222*. doi:10.1242/jeb.186460

Philips, C., Jarrold, C., Baddeley, A., Grant, J., & Karmiloff-Smith, A. (2004). Comprehension of spatial language terms in Williams syndrome: Evidence for an interaction between domains of strength and weakness. *Cortex*, *40*, 85–101.

Pinter, J. D., Brown, W. E., Eliez, S., Schmitt, J. E., Capone, G. T., & Reiss, A. L. (2001). Amygdala and hippocampal volumes in children with Down syndrome: A high-resolution MRI study. *Neurology*, *56*(7), 972–974.

Meyer-Lindenberg, A., Kohn, P., Mervis, C. B., Kippenham, J. S., Olsen, R. A., Morris, C. A., & Faith Berman, K. (2004). Neural basis of genetically determined visuospatial construction deficit in Williams syndrome. *Neuron*, *43*, 623–631.

Pruden, S. M., Levine, S. C., & Huttenlocher, J. (2011) Children's spatial thinking: does talk about the spatial world matter? *Developmental Science*, *14*(6), 1417–1430. doi:10.1111/j.1467-7687.2011.01088.x.

Purser, P., Thomas, M., Farran, E. K., Karmiloff-Smith, A., & Van Herwegen, J. (2019). Comparing cross-sectional and longitudinal trajectories for vocabulary and nonverbal reasoning in Williams syndrome. Paper presented at Society for Research in Child Development Conference, Baltimore, March, 2019.

Purser, H. R. M., Farran, E. K., Courbois, Y., Lemahieu, A., Sockeel, P., Mellier, D., et al. (2014). The development of route learning in down syndrome, williams syndrome and typical development: Investigations with virtual environments. *Developmental Science*, *18*(4), 599–613. doi: 10.1111/desc.12236

Raven J. (1976). *Raven Progressive Matrices*. Toronto: The Psychological Corporation.

Raven, J. C. (1993). *Coloured Progressive Matrices*. Oxford, UK: Information Press Ltd.

Riby (2012). Face processing and social interaction. In E. K. Farran & A. Karmiloff-Smith (Eds.), *Neurodevelopmental Disorders Across the Lifespan: A Neuroconstructivist Approach* (pp. 225–246). Oxford University Press.

Sahyoun, C. P., Soulieres, I., Belliveau, J. W., Mottron, L., & Mody, M. (2009). Cognitive differences in pictorial reasoning between high-functioning autism and Asperger's syndrome. *Journal of Autism and Developmental Disorders*, *39*(7), 1014–1023. doi:10.1007/s10803-009-0712-9

Shallice, T. (1982). Specific impairments of planning. *Philosophical Transactions of the Royal Society of London, Series B, Biological Sciences*, *298*(1089), 199–209.

Siegel, A. W., & White, S. (1975). The development of spatial representation of large-scale environments. In Reese H. (Ed.), Advances in child development and behavior. Volume 10 (pp. 9–55). New York: Academic Press.

Soulieres, I., Dawson, M., Samson, F., Barbeau, E. B., Sahyoun, C. P., Strangman, G. E., … Mottron, L. (2009). Enhanced visual processing contributes to matrix reasoning in autism. *Human Brain Mapping*, *30*(12), 4082–4107. doi:10.1002/hbm.20831

Thomas, M. S. C., Annaz, D., Ansari, D., Scerif, G., Jarrold, C., & Karmiloff-Smith, A. (2009). Using developmental trajectories to understand genetic disorders. *Journal of Speech, Language, and Hearing Research*, *52*, 336–358.

Toffalini, E., Meneghetti, C., Carretti, B., & Lanfranchi, S. (2018). Environment learning from virtual exploration in individuals with down syndrome: the role of perspective and sketch maps. *Journal of Intellectual Disability Research*, *62*(1), 30–40. doi:10.1111/jir.12445

Van Herwegen, J., Farran, E. K., Annaz, D. (2011). Item and error analysis on Raven's coloured progressive matrices in Williams syndrome. *Research in Developmental Disabilities*, *32*, 93–99.

Young, A. W., Hellawell, D., & Hay, D. C. (1987). Configural information in face perception. *Perception*, *16*, 747–759.

Wynn, K. (1990). Children's understanding of counting. *Cognition*, *36*(2), 155–193.

15

AGE MATTERS

Yonata Levy

Annette, neuroconstructivism and me – a personal note

I first came across the name of Annette Karmiloff-Smith in the early 1980s when I was working on my dissertation on the acquisition of Hebrew gender. At that time, Annette's work that concerned French gender was among the few studies on the topic. Following an insightful correspondence, we met in 1988, when Annette came on a private visit to Israel. A year later, Rick Cromer invited me to give a talk at the MRC Cognitive Development Unit (CDU) in London. The talk was about the acquisition of Hebrew morphology, including gender. The Chomskyan revolution in linguistics, along with the modularity approach to cognition, were at their peak. Generative grammars had an enormous impact on theories of language acquisition. I was part of that school of thought, as was Cromer. I did not know who Annette sided with. After all, she worked with Piaget in Geneva!

After my talk, Annette and I had coffee in the small kitchen downstairs. I vividly recall that tiny room, below street level, with its narrow, rectangular table. Annette made coffee and was most concerned with the safety of the filters that were used for drinking water at the CDU. As for my talk, she said: "If I had heard you five years ago, I would have applauded. Now I am sceptical of the whole approach". What she had in mind of course was the direct transfer of models of mature cognitive states to account for development. With respect to language acquisition, the generative approach meant accounting for children's evolving knowledge in terms of missing bits of mature grammars. Furthermore, adhering to modularity in development amounted to the claim that language develops independent of the rest of cognition. Claims about innately specified rules and mechanisms of language acquisition, language-specific genetics and anatomy generated direct predictions with respect to developmental language pathology. The term 'neuroconstructivism' was not yet around.

A couple of years later, despite what Annette perceived as our theoretical differences, or perhaps because of it, Annette invited me to spend a sabbatical year at the CDU in London. The CDU, led by John Morton, was an acclaimed institution where exciting research on cognitive development and developmental pathology was conducted. Spending a sabbatical at the CDU was a real bonus and I was only too happy to accept. That year, 1994–1995, turned out to be the beginning of a new focus in my research interests. I was gradually becoming interested in neurodevelopmental disorders. From this perspective, one could not fail to see the importance of Annette's ideas, and the ideas of people she worked with.

In the years that followed, I spent three more sabbaticals in London along with several shorter visits. In between there were conferences and meetings and so Annette and I kept in touch. She was an important part of my professional life. Her intellectual influence has been immense. To an outside listener our professional exchanges may have seemed argumentative, yet none of it leaked into our personal conversations. We spoke about research. We discussed results. We talked about our projects. We talked about helping parents of children with neurodevelopmental disorders. We spoke about men, about husbands, about family, about aging, about clothes and how to choose them, about our children – my sons, her daughters. Grandchildren were not yet in sight, at least for me.

I am not sure if Annette realized that I was no longer a generativist and that I shared the theoretical convictions of neuroconstructivism. For those working with children with neurodevelopmental disorders, the lack of specificity of functional disorders especially in the early years is striking. The developing system simply cannot be described "as a normal brain with parts intact and parts impaired, as the popular view holds" (Karmiloff-Smith et al., 2012, p. 393). In referring to the 'then-popular view', Annette had in mind the view of the brain as innately modular and the neglect of the impact of development under atypical conditions upon the mature system, which she considered vital.

Regrettably, I have not told her that she had brought me around, that I considered neuroconstructivism as the right approach to the study of neurodevelopmental disorders. This article is my belated debt to Annette. It was unduly delayed. If only I knew that time would be so short.

Background

This chapter summarizes neurobiological evidence of the impact of timing on brain development and the behavioural consequences of developmental delay. It is argued that a re-evaluation of the role of timing on theories of functional development in health and in pathology is called for. I present results of a longitudinal study of the acquisition of Hebrew by typically developing children (TD), children with Williams syndrome and children with Down syndrome, matched on mean length of utterances (MLU). Results suggest that developmental trajectories were similar along most structural parameters, and age of language onset and pace of acquisition were major parameters differentiating among the groups

(Levy, & Eilam, 2013). These results suggest that along with the focus on structural properties of developmental trajectories advocated by neuroconstructivism, timing is of critical importance. I then present a short version of a review of neurobiological and behavioural research, suggesting that chronological age should be a key factor in our attempts to understand neurodevelopmental disorders (Levy, 2018).

The motivation to conduct a longitudinal study of language acquisition of children with these syndromes had its origin in a tantalizing problem. The neuroconstructivist approach to development rests on the view of the mind/brain as an interconnected network and the critical role of equilibrium among components. This view of the brain suggests that in cases of genetically atypical populations, it is unlikely that there will be parts intact and parts impaired. Rather, brain development will be affected across multiple levels and specific changes will have a stronger or a weaker impact on various systems (Karmiloff-Smith, 1998b). In Annette's words, "development itself plays a crucial role in phenotypical outcomes" (Karmiloff-Smith, 1998a, p. 397).

This view still leaves open the following question. Neurodevelopmental disorders differ in the impact of their unique genetics on brain development. Does a neuroconstructivist approach predict that the developmental course and the phenotypic outcome will be different for each disorder? Alternatively, does the nature of the problem-space dictate the developmental course? Consider language development – If the former hypothesis holds, then the prediction is that genetic syndromes will present with phonological errors, order of acquisition and types of syntactic and morphological errors that will uniquely characterize the developmental course of each syndrome. However, if the structural properties of human languages are such that there is a designated trajectory that the human brain *must* follow in acquiring language – any language – then at least in the early stages, when basic language is acquired, development will be similar across a variety of syndromes. Furthermore, this trajectory will be similar to the trajectory followed by TD children. If the disorder is such that the brain does not retain the capacity to learn along those lines, children will not develop complex language at all.

To test these hypotheses we compared the developmental course of basic grammar in TD children, children with WS and children with DS in a mixed longitudinal and cross-sectional design.

Longitudinal study of early language in children with Williams syndrome and children with Down syndrome

William syndrome (WS) is a rare genetic disorder with a prevalence of 1 in 20,000 live birth (Morris, Demsey, Leonard, Dilts, & Blackburn, 1988), although more recent studies suggest it is more common (1:7,500, Strømme, Bjørnstad, & Ramstad, 2002). The majority of individuals with WS have mild to moderate cognitive deficits within an IQ range of 50–70. Verbal IQ typically exceeds performance IQ and language is considered a relative strength. Deficits in performance on spatial-motor tasks are highly specific and universal in WS (Mervis,

Morris, Bertrand, & Robinson, 1999). Numerical skills are likewise poorer than predicted by general cognition (Paterson, Girelli, Butterworth, & Karmiloff-Smith, 2006). People with WS are verbose and typically show hypersociability.

WS individuals have smaller brains and early studies have shown that reductions are not uniformly distributed across brain regions (Jernigan & Bellugi, 1990). Gyrification abnormalities in the orbitofrontal cortex, Sylvian fissure, and superior parietal regions have been reported (Eckert et al., 2006; Gaser et al., 2006; Van Essen et al., 2006). There are some reports that amygdala volumes are disproportionally increased (Martens, Wilson, Dudgeon, & Reutens, 2009; Capitao et al., 2011, but see Meyer-Lindenberg et al., 2004; Meda, Pryweller, & Thornton-Wells, 2012). Convergent anomalies in structural and functional connectivity are characteristic of WS brains. Altered fractional anisotropy (FA) values in parieto-occipital regions were associated with increased connectivity in the antero-posterior pathways linking parieto-occipital with frontal regions. The analysis of resting-state data shows altered functional connectivity in the main brain networks (Gagliardi et al., 2018).

Down syndrome (DS) is the leading genetic cause of intellectual disability, with an overall prevalence of 1:700 live births (Mai et al., 2019). DS is associated with abnormalities in multiple organ systems and a characteristic phenotype that includes physical as well as behavioural features (Hazlett, Hammer, Hooper, & Kamphaus, 2011). Similar to people with WS, individuals with DS display mild to moderate cognitive handicaps with an IQ of 50–70 on average. Short-term auditory memory is an area of particular weakness relative to tasks measuring visuospatial short-term memory, or nonverbal mental age (MA; Seung, & Chapman, 2000; Jarrold, Baddeley, & Phillips, 2002; Miolo, Chapman, & Sindberg, 2005). Language, and more specifically syntax and morphology, present a recognized weakness in people with DS.

In a recent review article, Hamner, Udhnani, Osipowicz, and Lee (2018) state that relatively little is known about the developing brain in DS. There is evidence for smaller total brain volume, total grey matter and white matter, cortical lobar, hippocampal and cerebellar volumes. While frontal grey matter volume is negatively correlated with age in typical controls, such volume reduction is not seen in DS. Increased cortical thickness is probably derived by reduced tissue contrast between grey and white matter (Bletsch et al., 2018).

TD children, children with WS and children with DS learning Hebrew as their first language were followed longitudinally in their home environment. Hebrew has a rich inflectional and morpho-syntactic structure and a morphologically derived lexicon. The study focused on growth in MLU and correlations with correct productions and types of errors with respect to verb morphology (person, number, gender and tense) and noun morphology (number), sentence types, sentence structure, relative clauses and vocabulary size.

Results were clear cut. Language development over the entire span of data collection, MLU 1.2–4.4, revealed minor structural differences between children with WS, children with DS and TD controls of similar MLU. Thus, despite different brain atypicalities resulting in crucial phenotypic differences in language abilities in late childhood, the course of language development in the early stages was similar in both disorders. Furthermore, language development in children with DS and children with WS followed the course of typical development. Atypical errors were not recorded (Levy & Eilam, 2013).

The typicality of the structural properties of the developmental trajectories of children with WS and children with DS, despite very different profiles of later language skills, highlight the view of grammar as a computational system whose properties constrain development in unique ways, likely allowing a sole route to acquisition of basic grammatical skills. It supports the notion of Universal Grammar (UG) conceived as a set of constraints on possible grammars that limit the potential for alternative routes to mastery (e.g., Roeper, 2010).

From a neuroconstructivist perspective, similarity in order of acquisition and error types among the study groups was an unexpected result. On the face of it, it argues against a major tenet of neuroconstructivism since they point to similar developmental course with only minor structural differences among the study groups, despite major differences in developmental brain impairments. But what about the differences in age of language onset and developmental pace?

Importantly, while structural differences were minimal, age of language onset and pace of acquisition in the DS and WS groups departed significantly from typical timing. At the beginning of combinatorial language, MLU 1–1.5, average age was 22.8 months (SD = 2.6) for TD children, 46.8 months (SD = 8.8) for children with WS, and 54.7 months (SD = 10.5) for children with DS. Age of TD children at the end of the study, MLU 4.4, was 40.1 months (SD = 3.4). Within group variance in age of onset of word combinations and age at final stage of the study was relatively small. For the WS group, age at the final stage of the study was ~74 months (SD = 13). Age of the DS group was 89.7 months (SD = 8.03). Within group variance in the clinical groups was relatively large. Between-group differences in age of onset and age at final stage of the study were statistically significant.

Synchrony among variables, defined as developmental progress seen in language variables relative to growth in MLU, is yet another measure of developmental pace. Whereas development within one calendar year showed remarkable synchrony among the language variables, developmental pace was slow in the clinical groups, and even more so in the DS group than in the WS group.

Differences in age and pace of acquisition have been well documented in the clinic. Furthermore, disrupted developmental schedules, resulting in delays, are the hallmark of pathological populations. Thus, the fact that we documented it in our study was no surprise. This chapter addresses questions that concern the impact of delay on a neuroconstructivist viewpoint. Can structurally similar trajectories that differ in their time course be considered 'similar'? In the sections below, I suggest that the answer is 'no', since delay foretells developmental deviance.

Disorders of developmental timing

In typical development age is correlated with brain maturation and compatible environmental interaction. In atypical development however, age is not a reliable predictor of brain maturation. Area-specific brain maturation may be slower than normal (e.g., Wolff et al., 2012), or faster (e.g., Weinstein et al., 2011), or it may take an atypical course, which in some cases may nevertheless result in typical performance (e.g., Annaz, Karmiloff-Smith, Johnson, & Thomas, 2009). Genetic irregularities, disorders of timed gene expression and an environment that is not correlated with these atypicalities, and is therefore atypical for the child's developmental state, are liable to disrupt the correlation between age and maturation.

One attempt to maintain chronological age as a reference point when evaluating a child's performance is the concept of mental age (MA). MA in fact offers a substitute "age" for the study of disordered populations. Importantly, cognitive performance indexed by MA is typically attributed to intellectual potential, often at the cost of neglecting other age-related factors. Thus, in studies of populations with cognitive impairments participants are regularly compared to much younger controls, who, although matched on MA, differ on a variety of variables such as experience with age-appropriate environments, emotional development, social interactions and physical development. In an effort to further localize the behavioural deficit and hence the choice of controls, concepts such as language-age, or face processing age were introduced referring to performance in the area of interest, as distinct from the general cognitive functioning of the child (e.g., Rice, 2012 and see the discussion in Plante, Swisher, Kiernan, & Restrepo, 1993). Importantly, such concepts leave no room for considerations of the network properties of cognition and the effects that are the result of impairment of the equilibrium among functional components.

Whereas for the most part, work within the neuroconstructivist framework shows the expected focus on structural dissimilarities in the developmental course between disordered and typical populations (e.g., Karmiloff-Smith et al., 1997; Laing et al., 2002; Paterson et al. 2006; Thomas, & Karmiloff-Smith, 2003), there has been some concern with age and timing as well. In a précis of neuroconstructivism, Sirois et al. (2008) referred to chronotopy, stressing the fact that some patterns of gene expression occurred at specific developmental times, some aspects of neural development depended upon ordered events, and plasticity occurred at different ages in different parts of the developing system. In a similar vein, Thomas et al. (2009) stressed the fact that when developmental trajectories rather than data points were documented, age was not eliminated and could be considered in the comparison.

Current neurobiological research indeed suggests that age cannot be eliminated from the equation. A focus on the mind/brain as an interconnected network brings to the fore the critical role of equilibrium among components in brain development as well as in behavioural domains. Two parameters affecting such equilibrium are temporal dependencies, i.e., the fact that two or more brain areas or behavioural components mature at the same chronological age, and developmental co-variance, i.e., the fact that in different brain areas, similar morphological parameters mature in parallel. This is seen in age-dependent gene expression, parameters of cortical maturation, long-range connectivity and the interaction of the biological network with the environment (Lenroot & Giedd, 2011).

In what follows I summarize recent neurobiological research in typical and atypical populations that supports the conclusion that age-by-size trajectories of brain structures relate to functional properties more than absolute sizes (Giedd & Rapoport, 2010). It follows that disorders of developmental schedule disrupt development no less than structural brain atypicalities (for a more comprehensive review see, Levy, 2018).

Age-related effects in gene expression and neurobiological processes in typical and atypical populations

A key aspect of age-sensitive effects orchestrating synchronous development among brain networks is the timing of gene expression. Below are some central results related to this statement. Heritability of cortical thickness is regional and age-related (Schmitt et al., 2014). Minute disruptions of timing of synaptogenesis and pruning were seen to affect connectivity (Levitt, 2003). Studies in primates point to differences among functional areas in the time course of the formation and differentiation of tissues. The three-way interaction between age, genes and the environment and the way this interdependence shapes developmental trajectories suggests that the environment will have a differential effect on neuronal development in accordance with the time needed to form relevant connections (Rakic, 2006).

Age-related effects of different polymorphisms of the brain-derived neurotropic factor (BDNF) gene have mostly been studied in mice, but are also seen in humans (Casey et al., 2009). In mice as well as in human subjects, effects on hippocampal size as well as on memory and learning in response to stressful life events were shown to be pronounced during infancy but not during young adulthood. This is because BDNF is low during infancy, but high during adulthood. Furthermore, negative effects of low BDNF are restricted to those with the BDNFmet allele (BDNF methionine), but not evident in those with the BDNFval allele (BDNF valine). For example, in 4- to 12-year-old post-institutionalized children, who were adopted into families, those with BDNFmet allele, but not those with the val allele, had smaller cortical volumes relative to the non-institutionalized controls with the same allele (Choi, Zepp, Higgs, Weickert, & Webster, 2009).

Geschwind and Levitt (2007) stated what has become a truism in genetic research in general, and in research on autism in particular: "In neurodevelopmental disorders, the timing, location and degree to which gene expression is disrupted dictate the emergent phenotype" (p. 103). In a recent review article de la Torre-Ubieta, Won, Stein, and Geschwind (2016) citing Geschwind and Levitt, and Abrahams and Geschwind (2010), state that

... even small changes in timing will preferentially disrupt the connectivity of higher-order association areas that mediate social behavior, which include the frontal-parietal, frontal-temporal and frontal striatal circuits. Identification of the spatiotemporal dynamics of transcriptional and translational regulation and the subsequent changes in micro- and macro-circuit connectivity will be necessary to link synaptic dysfunction to complex behavioral traits in individuals with ASD. (ibid, p. 354)

In a similar vein, Courchesne, Campbell, and Solso (2011) argued that in ASD, gene expression abnormalities in the adolescent and adult brain were not expected to be the same as those in the prenatal or toddler years and delayed gene expression in the early years would have consequences for later development. Thus, results obtained with older children and adults were prone to reflect outcome associations rather than causal ones.

Note that differential effects of age were seen in the genetics of schizophrenia as well. Disruption of age-dependent changes in the expression of schizophrenia susceptibility genes in the prefrontal cortex occurring during the early years were shown to dispose the individual to schizophrenia (Choi et al., 2009). Thus, disrupted-in-scizophrenia-1 (DISC-1) gene allelic effects affected maturational schedule of fronto-cortical areas, which, in turn, increased risk levels for specific schizophrenia disease phenotypes. Furthermore, the age-dependent expression profiles of the neurotransmitter-related genes suggest that age affects responses to psychotropic medications.

Finally, Kendler, Gardner, and Lichtenstein (2008) reported genetic effects on symptoms of anxiety and depression, from middle childhood to young adulthood. The first genetic factor, which accounted for 72% of the variance in symptoms at ages 8–9, diminished in influence by ages 19–20, accounting for only 12% of the variance. New sets of genetic risk factors were emerging in adolescence and early adulthood.

Age-related effects and covariant developmental changes seen in imaging studies of typical and atypical populations

Age-related, temporal dependencies oversee long-range connectivity and covariance of brain morphological features such as increased or decreased cortical volume or degree of gyrification (i.e., the process of forming the folds of the cerebral cortex). Brain networks that are anatomically connected have been shown to co-vary in their morphological features and these correlations reflect similarities in timed maturational trajectories (Alexander-Bloch, Giedd, & Bullmore, 2013; Zielinski, Gennatas, Zhou, & Seeley, 2010). For example, posterior and anterior language areas in the left hemisphere co-vary in their cortical thickness (Lerch et al., 2006). Similarly, motor, auditory and visual systems show patterns of structural co-variance (Zielinski et al.).

Synchronized maturational change and structural covariance were reported for different regions within the same functional modules, suggesting that covariation in cortical anatomy characterizes functional organization (Alexander-Bloch et al., 2013). Interestingly, this is true not only for regions showing physical connectivity, but also for brain areas that are *not* anatomically connected, but do show functional connectivity (seen through synchronous neuronal firing). Maturation of brain structural co-variance manifests age-dependent linear as well as non-linear progressions, likely reflecting differences in functional maturation between different brain systems. Thus, the co-variant networks that tend to develop linearly include areas related to language and attention, whereas primary sensory and motor areas tend to expand more significantly at younger ages than later (Khundrakpam et al., 2013; Zielinski et al., 2010).

In a longitudinal study of cortical thickness and cortical surface area (i.e., the actual brain area that measures the folds – gyri – and the grooves – sulci), developmental changes were age-dependent and gender specific (Giedd et al. 2015; Raznahan et al. 2011). Tamnes et al. (2010) studied age-related changes in cortical thickness, regional white matter (WM) volume and diffusion characteristics (i.e., diffusion is the degree of freedom that characterizes the movement of water molecules along the axon. Neuronal maturation results in more restricted water flow). Results showed regional age-related cortical thinning, WM volume increases and changes in diffusion parameters. All measures showed unique associations with age, yet cortical thickness was the most strongly age-related parameter. Importantly, age-by-cortical thickness was more predictive of IQ than differences in thickness alone (Shaw et al., 2007). Forde et al. (2017) investigated the effect of age and sex on cortical development in healthy adolescents. Cortical thickness and degree of gyrification, but not cortical surface area, were inversely associated with age in all regions.

A longitudinal study of typically developing babies revealed a maturational, nonlinear developmental sequence of brain networks, that was age-dependent. The following age-defined, network-specific, developmental order was identified: primary sensorimotor/auditory areas, visual, attention/default mode and lastly control/executive function areas (Gao et al., 2015). Langeslag et al. (2013) performed resting-state functional magnetic resonance imaging (Rs-fMRI) to determine whether parietal-frontal functional connectivity was associated with intelligence in young children. Results were age-related, suggesting that higher intelligence in 6–8 years old was associated with increased connectivity of the right parietal region and the right frontal region. This association was stronger in girls than in boys.

Fewer studies have examined the effects of age in populations with neurodevelopmental disorders. In a review article, Courchesne et al. (2011) put forth a theory of age-specific anatomical abnormalities in autism spectrum disorder (ASD) that was developmental and age-specific. It was argued that an abnormally accelerated brain overgrowth was seen in children diagnosed with ASD between ages 2 and 4 years, followed by an abnormally slow or arrested growth between early and later childhood. An accelerated rate of decline in brain size was seen from adolescence to middle adulthood. Results reported in Raznahan et al. (2011) suggest that abnormal increase in cortical volume seen in 2- to 4-year-old males with autism could result from derailing maturational timing of surface area rather than from actual increase in thickness. Furthermore, the developmental trajectory of cortical surface area predicted that detrimental effects were likely to occur in males rather than in females, and in young children rather than later in development, thus supporting clinical observations in the autism population (Duvekot et al., 2017).

Fractional anisotropy (FA) reflects the extent to which diffusion of water molecules within the cell is restricted. Restricted diffusion is a measure of cell organization. Thus, increased FA suggests higher level of cell maturation. A recent study of brain WM in high-risk infant siblings of children with autism reported higher FA at six months, followed by slower changes in siblings who were later diagnosed with ASD, relative to siblings without ASD (Wolff et al., 2012). Consequently, by 24 months, those with ASD had lower FA values in 12 out of 15 fibre tracts examined. In contrast with these results, Weinstein et al. (2011) reported increased FA in the left superior longitudinal fasciculus and the body of the corpus callosum in 2- to 3-year olds, low-functioning children with autism. It was suggested that increased FA reflected premature maturation of the left-superior longitudinal fasciculus in autism, which could signal diminished plasticity for language.

A structural MRI study of children with ASD at two-time points (Time point I: mean age = 4.1 y; Time point II: mean age 6.6 y) revealed a lack of the expected age-related reduction of cortical thickness within areas related to language, social cognition and behaviour control. Age-related gains in expressive language correlated with gains in cortical thickness in the right hemisphere homolog areas, likely the result of compensatory processes (Smith et al., 2016). Resting-state fMRI showed age-related reduced connectivity between two higher-order cognitive networks in ASD (Bos et al., 2014). An interaction effect in the default mode network (i.e., the network that is active when the person is not focused on the outside world) seen in insula connectivity increased with age in ASD, whereas it decreased with age in TD children (Bos et al.).

As for children with attention deficit hyperactivity disorder (ADHD), developmental delays of trajectories of cortical thickness were noted specifically for the frontal lobes (Shaw et al., 2007). The median age by which 50% of the cortical points attained peak thickness was 7.5 years in TD children, and 10.5 years in children with ADHD. The greatest differences were seen in the middle prefrontal cortex, which reached peak thickness in typical children at 5.9 years, whereas it reached maximum thickness at 10.9 years in children with ADHD. Clearly, the behavioural effects of ADHD were not a function of optimal cortical thickness in itself, but referred to the age at which the desired values were attained . Posner, Park, and Wang (2014) review article of resting-state fMRI studies of connectivity within the default mode network and the interactions between the default mode network and the cognitive control network in people with ADHD supports the above MRI findings. Evidence for delayed neuromaturation was seen as well (Posner, et al.). The neuropsychological correlates of these delays remain sparsely explored.

In sum, timed gene expression is a key factor in processes related to pruning, myelination, brain plasticity and probably also the termination of sensitive periods in typical populations. Age-by-size trajectories of cortical thickness, cortical surface area and WM maturational processes are temporally correlated across functional trajectories. The conclusion from this work is that chronological age is a major parameter governing anatomical and functional brain development. Disruption of typical developmental schedule necessarily affects developmental equilibrium. It follows that developmental delays, whether global or specific, such as those that characterize clinical populations of children, are likely to take a long-lasting toll. Below I report evidence in support of this statement.

Behavioural studies reporting dependency relations between language and arithmetic, and the long-term impact of developmental language delay

Not many studies have considered developmental dependencies and synchrony among behavioural domains. Still, behavioural evidence pointing to correlations in performance across cognitive domains suggest that typical development is characterized by equilibrium within the network. Van de Cavey and Hartsuiker (2016) presented subjects with musical, mathematical or linguistic stimuli and discovered priming effects across these domains between sequences showing long- and short-distance dependencies. Houwen, Visser, and Vlaskamp (2016) present results indicating the impact of motor skills on language in children with and without intellectual disabilities. Correlations between motor skills and language were reported for children with autism as well (Choi, Leech, Tager-Flusberg, & Nelson, 2018). A relatively well-studied area relates to language delays and language disorders at an early age and the long-term effects of these early difficulties. Of relevance to the thesis of this chapter are TD children who were language delayed as toddlers or had language deficits during the early school years. Although such deficits may appear functionally specific and short-lived, results summarized below point to the long-term behavioural consequences of language delays and the interdependence between arithmetic and language in typical and atypical populations.

The co-occurrence of language problems, dyslexia and problems in numerical skills is a known fact. To cite just a few studies pointing to the interdependence between these domains, Moll, Snowling, Gobel, and Hulme (2015) reported that early language and executive skills predicted variation in later arithmetic skills. Studies of bilinguals suggest that they rely on different procedures in order to solve arithmetic problems in each of their languages (Van Rinsveld, Dricot, Guillaume, Rossion, & Schiltz, 2017). Scheepers and Sturt (2014) reported bidirectional priming effects between similar syntactic structures in language and in arithmetic. These results are supported by evidence of cross-domain, bi-directional neural activation (Nakai & Okanoya, 2018). Brain activation studies suggest that arithmetic and language are mediated by a common, but regionally differentiated, language-arithmetic network (Andin, Fransson, Ronnberg, & Rudner, 2015). A study of left-hemispheric asymmetry in sentence listening and reading and in mental arithmetic showed covariant collateralization of relevant areas. In view of the later lateralization for arithmetic relative to language, this may be evidence for the role of language acquisition in the organization of the arithmetic network (Pinel & Dehaene, 2010).

Late talkers present a relevant example to the effects of age of acquisition on later achievements. Toddlers who are late talkers in the absence of any other condition that may be associated with language delay are estimated at 5–8%. Delay is likely to resolve in three out of four children in the course of their preschool years (Bishop et al., 2012). Is language delay in these children indeed cost-free?

Late talkers with resolved delay consistently remain within the lower end of the normal distribution on most language measures at least into 5th grade. These results were first reported by Paul (1996) and replicated in numerous subsequent studies, as summarized in a review article by Rescorla (2011), and more recently in a study by Rescorla and Turner (2015). Longitudinal studies of teenagers who were late talkers showed significantly lower performance on grammaticality judgments, as well as on tasks involving ambiguous sentences (Rescorla, 2011).

Another group in which long-term effects of delayed acquisition have been reported are children with developmental language disorder (DLD). DLD is defined as a significant language delay in the absence of hearing loss, low-level non-verbal intelligence, neurological conditions or other clinically defined impairments. Most (though not all) 4- to 5-year-old children diagnosed with DLD were late talkers. While the majority of children with DLD achieve adequate everyday language in the early school years, school-age children with a history of DLD do not reach above 75% correct responses on judgments of grammaticality (Tomblin et al., 1997). A similar level of attainment was apparent in receptive vocabulary in adolescents with a history of DLD. Difficulties in literacy skills have often been noted in children with a history of DLD, including children who did not differ from controls on vocabulary and comprehension (Stothard, Snowling, Bishop, Chipchase, & Kaplan, 1998).

Durkin, Mok, and Conti-Ramsden (2013) examined numerical skills in children with DLD at age 7, and then again at age 8. Average scores were more than 1 SD below the population mean and language level correlated with number skills. This and similar studies have contributed to the ongoing debate about the specificity of DLD.

As for children with neurodevelopmental disorders, language delays characterize most of these populations. In view of the fact that children with neurodevelopmental disorders are often cognitively impaired, the long-term impact of language delays are hard to measure in these populations. It is worth noting however, that the extent of the delay, that is, age of language onset, developmental pace and children's ultimate achievements are syndrome specific and likely not determined by IQ level, as seen in the Levy and Eilam (2013) longitudinal study. This suggests that the time course of language acquisition is biologically controlled.

Final remarks

The work reviewed above underscores the conclusion that age-related, temporal dependencies and structural covariance are critical determinants of brain development. It supports the claim that age-by-size trajectories relate to functional properties more than absolute sizes (Giedd and Rapoport, 2010). Disrupted developmental schedules, characteristic of most neurodevelopmental disorders, are therefore likely to upset the network equilibrium, prompting functional consequences. Thus, even if the developmental course is structurally typical, such as was reported in Levy and Eilam (2013) with respect to early language acquisition in children with WS and DS, given the long delays in language onset and pace of acquisition, language development in these populations cannot be considered typical. In fact, referring to disorders of developmental schedule as 'delay' is misleading, as it implies that once children catch up, there will be no long-term consequences. The studies reviewed above suggest that this is not the case. Rather, delay predicts lower than average performance and has implications for interrelated domains.

Note that at least in theory, one can conceive of an overall, evenly distributed delay which will preserve the developmental synchrony among brain areas. Still, synchrony with physical development and the external environment is required for a typical, albeit slow, developmental course. It is rather unlikely that all of these factors will conspire in the face of a general developmental slow-down.

In sum, in line with the logic behind neuroconstructivism, typical development needs to be characterized not only structurally but temporally as well, stating temporal interactions among domains and modalities. Atypicalities in timing are the hallmark of neurodevelopmental disorders but the extent of the departure from typical schedule is unique to each syndrome. Consequently, atypical timing predicts functional deviance likely to result in syndrome-specific profile.

Whereas the correlation between cognitive domains has been studied to some extent, there is not much work on the **developmental interdependence** between domains that ultimately become interconnected. Especially lacking are behavioural studies that will address this question. Computational modelling might be away to go about it. It may be a first step toward a developmental model, which will reflect the fact that the whole of a healthy brain is more than the sum total of its parts. This is precisely the view of neuroconstructivism, with its focus on the development of the system as a whole, and its unique circumstances.

As this chapter approaches conclusion, my thoughts are with Annette. I first presented my ideas about the misguided use of the term 'delay' and the crucial role of chronological age in the way we think about neurodevelopmental disorders, in a talk at Birkbeck, London. It was in the winter of 2015. At that time my arguments were based solely on behavioural data. Annette was busy that afternoon and so we agreed to discuss the paper sometime in the future. We never did. What did she think about the results that I presented in that talk? What would she have thought about the neurobiological results that I reviewed in the current chapter? I hope she would have been persuaded. I am sure she would have had some intelligent, enriching comments along with some criticism. I miss her wisdom.

"Woe over those who are gone and no more to be found" Babylonian Talmud, Sanhedrin, 111a.

References

Abrahams, B. S., & Geschwind, D. H. (2010). Connecting genes to brain in the autism spectrum disorders. *Archives of Neurology, 67*, 395–399.

Alexander-Bloch, A., Giedd, J. N., & Bullmore, E. (2013). Imaging structural co-variance between human brain regions. *Nature Reviews Neuroscience, 14*, 322–336.

Andin, J., Fransson, P., Ronnberg, J., & Rudner, M. (2015). Phonology and arithmetic in the language calculation network. *Brain and Language, 143*, 97–105.

Annaz, D., Karmiloff-Smith, A., Johnson, M. H., & Thomas, M. S. C. (2009). A cross-syndrome study of the development of holistic face recognition in children with autism, Down syndrome, and Williams syndrome. *Journal of Experimental Child Psychology, 102*(4), 456–486.

Bishop, D. V. M., Holt, G., Line, E., McDonald, D., McDonald, S., & Watt, H. (2012). Parental phonological memory contributes to prediction of outcome of late talkers from 20 months to 4 years: A longitudinal study of precursors of specific language impairment. *Journal of Neurodevelopmental Disorders, 4*(3), 1–12.

Bletsch, A., Mann, C., Andrews, D. S., Daly, E., Tan, G. M. Y., Murphy, D. G. M., & Ecker, C. (2018). Down syndrome is accompanied by significantly reduced cortical grey-white matter tissue contrast. *Human Brain Mapping, 39*(10), 4043–4054.

Bos, D. J., van Raalten, T. R., Oranje, B., Smits, A. R., Kobussen, N. A., Belle, J. V., Rombouts, S. A., & Durston, S. (2014). Developmental differences in higher-order resting-state networks in Autism Spectrum Disorder. *Neuroimage Clinical, 14*(4), 820–827.

Capitao, L., Sampaio, A., Sampaio, C., Vasconcelos, C., Fernandez, M., Garayzabal, E., Shenton, M. E., & Goncalves, O. F. (2011). MRI amygdala volume in Williams Syndrome. *Research in Developmental Disabilities, 32*, 2767–2772.

Casey, B. J., Glatt, C. E., Tottenham, N., Soliman, F., Bath, K., Amso, D., Altemus, M., Pattwel, L. S., Jones, R., Levita, L., McEwen, B., Magariños, A. M., Gunnar, M., Thomas, K. M., Mezey, J., Clark, A. G., Hempstead, B. L., Lee, F. S. (2009). Brain-derived neurotrophic factor as a model system for examining gene by environment interactions across development. *Neuroscience, 164*, 108–120.

Choi, B., Leech, K. A., Tager-Flusberg, H., & Nelson, C. A. (2018). Development of fine motor skills is associated with expressive language outcome in infants at high and low risk for autism spectrum disorder. *Journal of Neurodevelopmental Disorders, 10*. https://doi.org/10.1186/s11689-018-9231-3

Choi, K. H., Zepp, M. E., Higgs, B. W., Weickert, C. S., & Webster, M. J. (2009). Expression profiles of schizophrenia susceptibility genes during human prefrontal cortical development. *Journal of Psychiatry and Neuroscience, 34*, 450–458.

Courchesne, E., Campbell, K., & Solso, S. (2011). Brain growth across the life span in autism: Age-specific changes in anatomical pathology. *Brain Research, 1380*, 138–145.

de la Torre-Ubieta, L., Won, H., Stein, J. L., & Geschwind, D. H. (2016). Advancing the understanding of autism disease mechanisms through genetics. *Nature Medicine, 22*(4), 345–361.

Durkin, K., Mok, P. L., & Conti-Ramsden, G. (2013). Severity of language impairment predicts delayed development in number skills. *Frontiers in Psychology*. doi: 10.3389/psyg.2013.00581.

Duvekot, J., van der Ende, J., Verhulst, F. C., Slappendel, G., van Daalen, E., Maras, A., & Greaves-Lord K. (2017). Factors influencing the probability of a diagnosis of autism spectrum disorder in girls versus boys. *Autism, 21*(6), 646–658.

Eckert, M. A., Galaburda, A. M., Karchemskiy, A., Liang, A., Thompson, P., Dutton, R. A., Lee, A. D., Bellugi, U., Korenberg, J. R., & Mills, D. (2006). Anomalous Sylvian fissure morphology in Williams syndrome. *NeuroImage, 33*, 39–45.

Forde, N. J., Ronan, L., Zwiers, M. P., Schweren, L. J., Alexander-Bloch, A. F., Franke, B., Faraone, S. V., Oosterlaan, J., Heslenfeld, D. J., Hartman, C. A., Buitelaar, J. K., Hoekstra, P. J. (2017). Healthy cortical development through adolescence and early adulthood. *Brain Structure and Function, 222*(8), 3653–3663.

Gagliardi, C., Arrigoni, F., Nordio, A.2, De Luca, A., Peruzzo, D., Decio, A., Leemans, A., & Borgatti, R. (2018). A different brain: Anomalies of functional and structural connections in Williams syndrome. *Frontiers in Neurology, 13*(9), 721. doi: 10.3389/fneur.2018.00721.

Gao, W., Alcauter, S., Elton, A., Hernandez-Castillo, C. R., Smith, J. K., Ramirez, J., & Lin, W. (2015). Functional network development during the first year: Relative sequence and socioeconomic correlations. *Cerebral Cortex, 25*(9), 2919–2928.

Gaser, C., Luders, E., Thompson, P. M., Lee, A. D., Dutton, R. A., Geaga, J. A., Hayashi, K. M., Bellugi, U., Galaburda, A. M., Korenberg, J. R., Mills, D. L., Toga, A. W., & Reiss, A. L. (2006). Increased local gyrification mapped in Williams syndrome. *NeuroImage, 33*, 46–54.

Geschwind, D. H., & Levitt, P. (2007). Autism spectrum disorders: developmental disconnection syndromes. *Current Opinion in Neurobiology, 17*(1), 103–111.

Giedd, J. N., & Rapoport, J. L. (2010). Structural MRI of pediatric brain development: What have we learned and where are we going? *Neuron, 67*(5), 728–734.

Giedd, J. N., Raznahan, A., Alexander-Bloch, A., Schmitt, E., Gogtay, N., & Rapoport, J. L. (2015). Child psychiatry branch of the National Institute of Mental Health longitudinal structural magnetic resonance imaging study of human brain development. *Neuropsychopharmacology, 40*(1), 43-

Hamner, T., Udhnani, M. D., Osipowicz, K. Z., & Lee, N. R. (2018). Pediatric brain development in down syndrome: A Field in its infancy. *Journal of the International Neuropsychological Society, 23*, 1–11.

Hazlett, H. C., Hammer, J., Hooper, S. R., & Kamphaus, R. W. (2011). Down syndrome. Goldstein, S., & Reynolds, C. R. (Eds.), *Handbook of neurodevelopmental and genetic disorders in children* (2nd ed.). New York, NY: Guilford Press, 362–381.

Houwen, S., Visser, L. van der Putten, & Vlaskamp, C. (2016). The interrelationships between motor, cognitive and language development in children with and without intellectual and developmental disabilities. *Research in Developmental Disabilities, 53–54*, 19–31.

Jarrold, C., Baddeley, A. D., & Phillips, C. (2002). Verbal short-term memory in Down syndrome: A problem of memory, audition, or speech? *Journal of Speech, Language, and Hearing Research, 45*, 531–544.

Jernigan, T. L., & Bellugi, U. (1990). Anomalous brain morphology on magnetic resonance images in Williams syndrome and Down syndrome. *Archives of Neurology, 47*, 529–533.

Karmiloff-Smith, A. (1998a). Development itself is the key to understanding developmental disorders. *Trends in Cognitive Sciences, 2*(10), 389–398.

Karmiloff-Smith, A. (1998b). Is atypical development necessarily a window on the normal mind/brain? The case of Williams syndrome. *Developmental Science, 1*(2), 273–277.

Karmiloff-Smith, A., D'Souza, D., Dekker, T. M., Van Herwegen, J., Xu, F., Rodic, M., Ansari, D. (2012). Genetic and environmental vulnerabilities in children with neurodevelopmental disorders. *Proceedings of the National Academy of Sciences USA, 109*(Suppl 2), 17261–17265.

Karmiloff-Smith, A., Grant, J., Berthoud, I., Davies, M., Howlin, P., & Udwin, O. (1997). Language and Williams syndrome: How intact is "intact"? *Child Development, 68*(2), 246–262.

Kendler, K. S., Gardner, C. O., & Lichtenstein, P. A. (2008). Developmental twin study of symptoms of anxiety and depression: Evidence for genetic innovation and attenuation. *Psychological Medicine, 38*(11), 1567–1575.

Khundrakpam, B. S., Reid, A., Brauer, J., Carbonell, F., Lewis, J., Ameis, S., Karama, S., Lee, J., Chen, Z., Das, S., & Evans, A. C. (2013). Developmental changes in organization of structural brain networks. *Cerebral Cortex, 23*(9), 2072–2085.

Laing, E., Butterworth, G., Ansari, D., Gsodl, M., Longhi, E., Panagiota, G., Paterson, S., & Karmiloff-Smith, A. (2002). Atypical development of language and social communication in toddlers with Williams syndrome. *Developmental Science, 5*(2), 233–246.

Langeslag, S. J. E., Schmidt, M., Ghassabian, A., Jaddoe, V. W., Hofman, A., van der Lugt, A., Verhulst, F. C., Tiemeier, H., & White, T. J. H. (2013). Functional connectivity between parietal and frontal brain regions and intelligence in young children: The Generation R study. *Human Brain Mapping, 34*(12), 3299–3307.

Lenroot, R. K., & Giedd, J. (2011). Annual research review: Developmental considerations of gene by environment interactions. *Journal of Child Psychology and Psychiatry, 52*(4), 429–441.

Lerch, J. P., Worsley, K., Shaw, W. P., Greenstein, D. K., Lenroot, R. K., Giedd, J., & Evans, A. C. (2006). Mapping anatomical correlations across cerebral cortex (MACACC) using cortical thickness from MRI. *NeuroImage, 31*(3), 993–1003.

Levitt, P. (2003). Structural and functional maturation of the developing primate brain. *Journal of Pediatrics, 143*, S35–S45.

Levy, Y. (2018). Developmental Delay reconsidered: The critical role of age-dependent, co-variant development. *Frontiers in Psychology, Conceptual Analysis*, 23 April 2018 doi: 10.3389/fpsyg.2018.00503.

Levy, Y., & Eilam A. (2013). Pathways to language: A naturalistic study of children with Williams syndrome and children with Down syndrome. *Journal of Child Language. Special Issue on Language Acquisition in Atypical Populations, 40*(1), 106–138.

Mai, C. T., Isenburg, J. L., Canfield, M. A., Meyer, R. E., Correa, A., Alverson, C. J., Lupo, P. J., Riehle-Colarusso, T., Cho, S. J., Aggarwal, D., & Kirby, R. S. (2019). National population-based estimates for major birth defects, 2010–2014. *Birth Defects Research, 111*(18), 1420–1435.

Martens, M. A., Wilson, S. J., Dudgeon, P., & Reutens, D. C., 2009. Approachability and the amygdala: insights from Williams syndrome. *Neuropsychologia, 47*(12), 2446–2453.

Meda, S. A., Pryweller, J. R., & Thornton-Wells, T. A. (2012). Regional brain differences in cortical thickness, surface area and subcortical volume in individuals with Williams syndrome. *PLoS One, 7*, e31913.

Mervis, C. B., Morris, C. A., Bertrand, J., & Robinson, B. F. (1999). Williams syndrome: Findings from an integrated program of research. In: Tager-Flusberg, H. (Ed.) *Neurodevelopmental Disorders*. Cambridge, MA: MIT Press, 65–110.

Meyer-Lindenberg, A., Kohn, P., Mervis, C. B., Kippenhan, J. S., Olsen, R. K., Morris, C. A., & Berman, K. F. (2004). Neural basis of genetically determined visuospatial construction deficit in Williams syndrome. *Neuron, 43*, 623–631.

Miolo, G., Chapman, R. S., & Sindberg, H. A. (2005). Sentence comprehension in adolescents with Down syndrome and typically developing children: Role of sentence voice, visual context, and auditory-verbal short-term memory. *Journal of Speech, Language, and Hearing Research, 48*(1), 172–188.

Moll, K., Snowling, M. J., Gobel, S. M., & Hulme, C. (2015). Early language and executive skills predict variation in number and arithmetic skills in children at family risk for dyslexia and typically developing controls. *Learning and Instruction, 38*, 53–62.

Morris, C. A., Demsey, S. A., Leonard, C. O., Dilts, C., & Blackburn, B. L. (1988). Natural history of Williams syndrome: Physical characteristics. *Journal of Pediatrics, 113*, 318–326.

Nakai, T., & Okanoya, K. (2018). Neural evidence of cross-domain structural interaction between language and arithmetic. *Science Report, 8*(1). Published online Aug 7. doi: 10.1038/s41598-018-31279-8.

Paterson, S. J., Girelli, L., Butterworth, B., & Karmiloff-Smith, A. (2006). Are numerical impairments syndrome specific? Evidence from Williams syndrome and Down's syndrome. *Journal of Child Psychology and Psychiatry, 47*(2), 190–204.

Paul, R. (1996). Clinical implications of the natural history of slow expressive language development. *American Journal of Speech-Language Pathology, 5*(2), 5–21.

Pinel, P., & Dehaene, S. (2010). Beyond hemispheric dominance: Brain regions underlying the joint lateralization of language and arithmetic to the left hemisphere. *Journal of Cognitive Neuroscience, 22*(1), 48–66.

Plante, E., Swisher, L., Kiernan, B., & Restrepo, M. A. (1993). Language matches: Illuminating or confounding? *Journal of Speech & Hearing Research, 36*(4), 772–776

Posner, J., Park, C., & Wang, Z. (2014). Connecting the dots: A review of resting connectivity MRI studies in attention-deficit/hyperactivity disorder. *Neuropsychology Review, 24*(1), 3–15.

Rakic, P. (2006). A century of progress in corticoneurogenesis: From silver impregnation to genetic engineering. *Cerebral Cortex 16*, i3–i17.

Rescorla, L. (2011). Late talkers: Do good predictors of outcome exist? *Developmental Disabilities Research Reviews, 17*(2), 141–150.

Rescorla, L., & Turner, H. (2015). Morphology and syntax in late talkers at age 5. *Journal of Speech, Language, and Hearing Research, 58*(2), 434–444.

Roeper, T. (2010). Interference, frequency and the primary linguistic data. *Lingua. International review of general linguistics. Revue internationale de linguistique generale, 120*, 2538–2545.

Raznahan, A., Shaw, P., Lalonde, F., Stockman, M., Wallace, G. L., Greenstein, D., Clasen, L., Gogtay, N., & Giedd, J. N. (2011). How does your cortex grow? *Journal of Neuroscience, 31*(19), 7174–7177.

Rice, M. (2012). Toward epigenetic and gene regulation models of specific language impairment: looking for links among growth, genes, and impairments. *Journal of Neurodevelopmental Disorders, 4*(27), 1–14.

Scheepers, C., & Sturt, P. (2014). Bidirectional syntactic priming across cognitive domains: From arithmetic to language and back. *The Quarterly Journal of Experimental Psychology, 67*(8), 1643–1654.

Seung, H.-K., & Chapman, R. S. (2000). Digit span in individuals with Down syndrome and typically developing children: Temporal aspects. *Journal of Speech, Language, and Hearing Research, 43*, 609–620.

Shaw, P., Eckstrand, K., Sharp, W., Blumenthal, J., Lerch, J. P., Greenstien, D., Clasen L., Evans A., Giedd J., & Rapoport, J. L. (2007). Attention deficit hyperactivity disorder is characterized by a delay in cortical maturation. *Proceedings of the National Academy of Sciences USA, 104*, 19649–19654.

Smith, E., Thurm, A., Greenstein, D., Farmer, C., Swedo, S., Giedd, J., & Raznahan, A. (2016). Cortical thickness change in autism during early childhood. *Human Brain Mapping*, *37*(7), 2616–2629.

Stothard, S. E., Snowling, M. J., Bishop, D. V. M., Chipchase, B. B., & Kaplan, C. A. (1998). Language-impaired preschoolers: A follow-up into adolescence. *Journal of Speech, Language, and Hearing Research*, *41*(2), 401–418.

Strømme, P., Bjørnstad, P. G., & Ramstad, K. (2002). Prevalence estimation of Williams syndrome. *Journal of Child Neurology*, *17*, 269–271.

Schmitt, J. E., Neale, M. C., Fassassi, B., Perez, J., Lenroot, R. K., Wells, E. M., & Giedd, J. N. (2014). The dynamic role of genetics on cortical patterning during childhood and adolescence. *Proceedings of the National Academy of Science USA*, *111*(18), 6774–6779.

Sirois, S., Spratling, M., Thomas, M. S. C., Westermann, G., Mareschal, D., & Johnson, M. H. (2008). Précis of neuroconstructivism: How the brain constructs cognition. *Behavioral and Brain Sciences*, *31*(3), 321–331.

Tamnes, C. K., Ostby, Y., Fjell, A. M., Westlye, L. T., Due-Tønnessen, P., & Walhovd, K. B. (2010). Brain maturation in adolescence and young adulthood: Regional age-related changes in cortical thickness and white matter volume and microstructure. *Cerebral Cortex*, *20*(3), 534–548.

Thomas, M. S. C., Annaz, D., Ansari, D. Serif, G., Jarrod, C., & Karmiloff-Smith, A. (2009). Using developmental trajectories to understand developmental disorders. *Journal of Speech, Language, and Hearing Research*, *52*, 336–358.

Thomas, M. S. C., & Karmiloff-Smith, A. (2003). Modeling language acquisition in atypical phenotypes. *Psychological Review*, *110*(4), 647–682.

Tomblin, J. B., Records, N. L., Buckwalter, P., Zhang, X., Smith, E., & O'Brien M. (1997). The prevalence of specific language impairment in kindergarten children. *Journal of Speech and Hearing Research*, *40*, 1245–1260.

Van de Cavey, J., & Hartsuiker, R. J. (2016). Is there a domain general cognitive structuring system? Evidence from structural priming across music, math, action description and language. *Cognition*, *146*, 172–184.

Van Essen, D. C., Dierker, D., Snyder, A. Z., Raichle, M. E., Reiss, A. L., & Korenberg, J. (2006). Symmetry of cortical folding abnormalities in Williams syndrome revealed by surface based analyses. *Journal of Neuroscience*, *26*, 5470–5483.

Van Rinsveld, A., Dricot, L., Guillaume, M., Rossion, B., & Schiltz, C. (2017). Mental arithmetic in the bilingual brain: language matters. *Neuropsychologia*, *101*, 17–29.

Weinstein, M., Ben-Sira, L., Levy, Y., Zachor, D. A., Ben Itzhak, E., Artzi, M., & Ben Bashat, D. (2011). Abnormal white matter integrity in young children with autism. *Human Brain Mapping*, *32*(4), 534–543.

Wolff, J. J., Gu, H., Gerig, G., Elison, J. T., Styner, M., Gouttard, S., Botteron, K. N., Dager, S. R., Dawson, G., Estes, A. M., Evans, A. C., Hazlett, H. C., Kostopoulos, P., McKinstry, R. C., Paterson, S. J., Schultz, R. T., Zwaigenbaum, L., Piven, J., & The IBIS Network. (2012). Differences in white matter fiber tract development present from 6 to 24 months in infants with autism. *The American Journal of Psychiatry*, *169*(6), 589–600.

Zielinski, B. A., Gennatas, E. D., Zhou, J., & Seeley, W. W. (2010). Network-level structural covariance in the developing brain. *Proceedings of the Natural Academy of Science U. S. A.* *107*(42), 18191–18196.

16

ALIGNING COGNITIVE STUDIES IN MOUSE MODELS AND HUMAN INFANTS/TODDLERS: THE CASE OF DOWN SYNDROME

Hana D'Souza, Daniel Brady, Frances K. Wiseman, Mark A. Good, Michael S.C. Thomas, and The LonDownS Consortium

Professor Annette Karmiloff-Smith and LonDownS: Aligning mouse and human phenotyping

Of Annette Karmiloff-Smith's many contributions to developmental psychology, the thrust of her work in developmental disorders can be summarised as follows: in order to truly understand neurodevelopmental disorders, we need to study development across levels of description, across disorders and across species. This approach was evident in how she led the Infant stream within the *London Down Syndrome* (LonDownS) *Consortium*. LonDownS is a large interdisciplinary collaboration between human geneticists, cellular biologists, psychiatrists, psychologists, neuroscientists and mouse geneticists, whose aim is to understand the link between Down syndrome (DS) and Alzheimer's disease (AD), and to identify protective and risk factors that could inform interventions. One of Annette's goals within LonDownS was to nurture cross-talk between mouse and human phenotyping. In particular, Annette focused on one key issue – aligning the designs of memory tasks for human infants/toddlers with DS with behavioural tasks that are used with mice.

In this chapter, we discuss how mouse models can deepen our understanding of neurodevelopmental disorders more generally, before focusing on a specific example of an attempt to align mouse model designs with human infant/toddler studies in DS within LonDownS. We present this cross-species study to illustrate Annette's thinking on the subject. While the design is not fully complete, it serves to demonstrate what is required to make cross-species comparisons scientifically valuable. We begin by considering mouse models of neurodevelopmental disorders.

How can mouse models deepen our understanding of neurodevelopmental disorders?

The mouse is a commonly used model organism in neurodevelopmental disorder research (Sukoff Rizzo, & Crawley, 2017). The advantage of the mouse model is that it permits a high degree of intervention in the system to establish causality, while human studies are understandably usually confined to natural experiments and correlations. In mouse studies, genetic background as well as environment can be tightly controlled and manipulated in order to establish causal pathways between genotype and phenotype. Any tissue can be accessed at any stage of development. This makes it possible to study where and when particular genes in the mouse are expressed. Furthermore, the lifespan of the mouse is short (one mouse year equals about 40 human years) and hence development is rapid (e.g., the time from birth to reproductive age is around 8 weeks) (Sukoff Rizzo, & Crawley, 2017). This allows researchers to repeat longitudinal experiments multiple times within a relatively short period of time and several generations can be observed. Finally, mouse models present an opportunity to test potential therapeutic interventions.

The value of each mouse model of a neurodevelopmental disorder depends on two sorts of alignment: (1) How well the genetics and physiology of the mouse align with the human; and (2) How well the cognitive and behavioural phenotyping of the mouse maps to the human.

Genetic and physiological alignment

Using the mouse as a model organism is particularly advantageous as the genes conserved between mouse and human are well mapped, and the mouse genome can be relatively easily manipulated. The mouse and human genomes share many similarities, including large syntenic genomic regions that directly map between the two species (Breschi, Gingeras, & Guigó, 2017). Evolutionary divergences vary between functional gene clusters, with genes involved in immunology and reproductive organs being particularly divergent, and regulation of gene expression and splicing often varying between the two species (Brawand et al., 2011; Yue et al., 2014). However, many biochemical and physiological processes are conserved in humans and mice, allowing us to draw conclusions about the function of conserved genes and genetic networks across these species.

Cognitive and behavioural phenotyping alignment

When comparing the cognitive processes and behaviours of mice and humans, it is important to use cross-species tasks that measure the same variable. Here, it is crucial to ascertain that task analogues across species not only look similar from a behavioural perspective, but also that they have the same cognitive demands and neural underpinnings (Edgin, Mason, Spanò, Fernández, & Nadel, 2012; Karmiloff-Smith et al., 2012). Certain variables may be easier to align tasks on than others. Particularly, it may not be feasible to conduct cross-species studies on higher-level cognitive functions, such as language. This is a significant limitation, because such functions are often the core phenotypic difficulties described across many neurodevelopmental disorders, including DS.

In Annette's presentations and writings, she would often highlight the need for better alignment of tasks across species by using an example from spatial cognition (e.g., Karmiloff-Smith, 2007). To study this domain in rodent research, the Morris water maze is commonly used (see Figure 16.1a; Morris, 1981). In this task, a rodent is placed in a water basin with opaque water, which contains a platform just below the surface of the water. Learning of location of this platform is assessed across multiple trials. In contrast, in humans, spatial abilities are often assessed using tests which involve the manipulation of spatial relations between objects while the participant remains seated at a table (e.g., the Block design subtest from the Wechsler Intelligence Scale for Children [WISC, Wechsler, 2014]; see Figure 16.1b). Even though both the rodent and human tasks are supposed to measure spatial abilities, it is difficult to directly compare these two tasks as a number of differences exist between them. For example, in the case of the Morris water maze, the rodent needs to constantly update its changing body position in space, in contrast to the Block design where the body remains static while the human needs to represent the changing relations between objects (Karmiloff-Smith, 2007). These are viewed as different types of spatial cognition even within humans (Uttal et al., 2013). To improve the alignment of tasks across species, a Morris-type virtual water maze task for humans was developed in which human participants navigate through space on a computer (e.g., Kallai, Makany, Karadi, & Jacobs, 2005). An even closer analogue to the rodent task has been used in humans – a spatial test where humans have to find targets in an arena (Kalová, Vlček, Jarolímová, & Bureš, 2005; Laczó et al., 2017; Smith, Gilchrist, Hood, Tassabehji, & Karmiloff-Smith, 2009). Even though this may seem like a close analogue to the rodent task in terms of body position changes, these tasks still differ on a number of dimensions. For example, most of the spatial cognition tasks used in humans rely on understanding verbal instructions, something that is not possible to implement with either mice or young children with/without neurodevelopmental disorders.

Annette argued that we need to think about how to better align tasks across species and age groups. In her own research, she would try to come as close as ethically possible to a Morris water maze with children by using a ball pit with hidden treasure (see Figure 16.1c; Westermann, Thomas, & Karmiloff-Smith, 2011). However, even with this task, Annette would argue that better

FIGURE 16.1 Comparisons of rodent–human analogues: (a) Morris water maze used with rodents to study spatial cognition (adapted from Hamilton, & Rhodes, 2015); (b) Block design used with humans to study spatial cognition; (c) Ball pit with hidden treasure used with children

alignment is necessary because motivation remains different across the species: fear in the case of the rodent to escape drowning, and reward in the case of the child to find the treasure. Here, ethical considerations prevent closer alignment from human to rodent. Therefore, it would be necessary to adjust the design in the opposite direction, with a version of the animal task that emphasises reward (e.g., the Barnes Maze [Barnes, 1979], Radial Arm Maze [Olton, & Samuelson, 1976]). To optimise alignment, researchers working with the different species therefore need to collaborate at the design stage.

We now turn to consider mouse models in the specific context of Down syndrome, the disorder that Annette focused on in her final work within LonDownS.

Focus on Down syndrome

Down syndrome (DS) is the most prevalent neurodevelopmental disorder of known genetic origin (trisomy 21) associated with intellectual disability, occurring approximately in one in 1,000 live births (Wu, & Morris, 2013). One of the core difficulties described in DS is memory (for review, see Godfrey, & Lee, 2018). This is of particular interest as individuals with DS show higher prevalence of

Alzheimer's disease (AD) than typically developing (TD) individuals (Zigman, & Lott, 2007), and the defining clinical feature of AD is an acquired memory deficit (Carlesimo, & Oscar-Berman, 1992). By the age of 40 years, individuals with DS show universal neuropathology associated with AD, including depositions of plaques and tangles, likely due to trisomy of the amyloid precursor protein (*APP*) gene which lies on chromosome 21. Ninety-seven percent of people with full trisomy of chromosome 21 develop dementia by the age of 80 years (McCarron et al., 2017; Zigman, & Lott, 2007). However, despite the elevated risk, a great deal of variation is observed in the age of onset of dementia in people with DS (McCarron et al., 2017). The cause of these differences in disease course is not clear, and both genetic and environmental factors are likely to play a role. Understanding these factors is not only of interest for understanding AD in DS, but also AD in the general population.

Mouse models of Down syndrome

Mouse models present a useful tool for unpacking mechanisms through which trisomy 21 contributes to AD in DS, as well as to memory difficulties in DS more broadly. Humans have 23 pairs of chromosomes, while mice have 20 pairs. Human chromosome 21 is the smallest human autosome (non-sex chromosome), representing about 1.5% of the total DNA (Hattori et al., 2000). It contains 222 protein-coding genes and 325 non-protein-coding genes (Gupta, Dhanasekaran, & Gardiner, 2016). Out of the protein-coding genes, 158 are conserved in the mouse but are distributed across three different mouse chromosomes. Most of these genes (102) lie on mouse chromosome 16, and a few on mouse chromosome 10 and 17 (37 and 19, respectively) (Gupta et al., 2016). Out of the non-coding genes, 75 are distributed across these three mouse chromosomes (Gupta et al., 2016). The distribution of conserved genetic material across the three different chromosomes, as well as some genes which are not conserved across the species, makes it challenging to create a complete mouse model of trisomy 21 (Antonarakis, Lyle, Dermitzakis, Reymond, & Deutsch, 2004). Nevertheless, a number of mouse models of DS have been developed (see Figure 16.2), which greatly advance our understanding of how trisomy of chromosome 21 relates to the DS phenotype (e.g., Herault et al., 2017; Ruparelia, Pearn, & Mobley, 2013).

Comprehensive reviews of the extent to which findings from mouse models of DS map onto findings in humans with DS have been conducted elsewhere (e.g., Gupta et al., 2016; Herault et al., 2017; Zhao, & Bhattacharyya, 2018). Here, we focus in detail on a specific attempt led by Annette to design memory tasks to be used with infants/toddlers with DS based on memory tasks used with the trisomy 21 mouse model – the Tc1 mouse (O'Doherty et al., 2005; see Figure 16.2). The Tc1 mouse is a transchromosomic model of DS – it carries a freely segregating copy of human chromosome 21 (with approximately 75% of gene content; Gribble et al., 2013). However, unlike individuals with DS, Tc1 mice are not

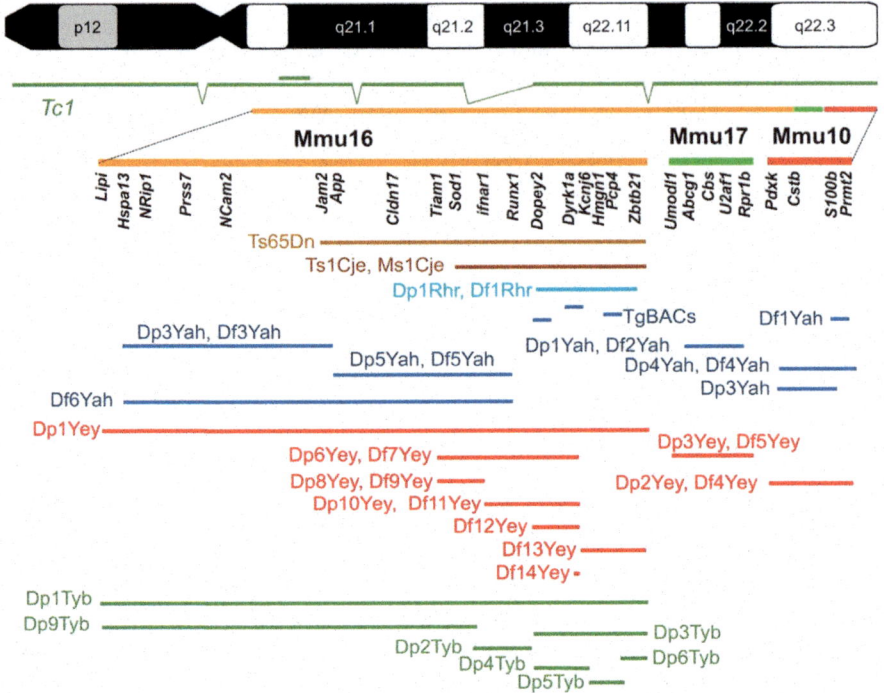

FIGURE 16.2   Mouse models of DS. Human chromosome 21 is depicted at the top of the figure; Mmu10 = mouse chromosome 10; Mmu16 = mouse chromosome 16; Mmu17 = mouse chromosome 17. Tc1 mouse which is of interest in the current chapter is shown in dark green under the human chromosome, with deletions and a duplication (double bar) depicted. Other codes and lines correspond to other mouse models of DS. Adapted from Herault et al. (2017)

trisomic for the *APP* gene, the gene thought to cause early onset AD (Gribble et al., 2013). Therefore, the Tc1 mouse allows us to investigate the contribution of chromosome 21 genes to memory, isolating it from contributions caused by the overexpressed *APP* gene.

Memory phenotyping of Tc1 mouse

Hall et al. (2016) investigated memory function in the Tc1 mouse across four different tasks: (1) novel object recognition task, (2) object-in-place memory task, (3) object location memory task, and (4) novel odour recognition task (Figure 16.3). All tasks were administered using the same apparatus – a square arena (60 × 60 × 40 cm) with a pale grey floor and white walls. An overhead camera was used to record the mouse's object exploration. This was defined as the time spent actively attending to (sniffing or interacting with) the object at a distance no greater than 1 cm.

(a) **Novel object recognition task**

(b) **Object-in-place task**

(c) **Object location task**

(d) **Novel odour recognition task**

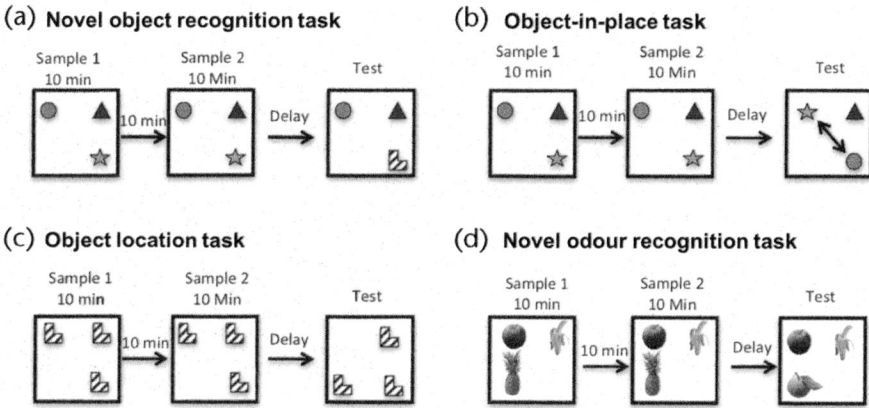

FIGURE 16.3 Design of memory tasks used with mice in Hall et al. (2016): (a) Novel object recognition task; (b) Object-in-place task; (c) Object location task; (d) Novel odour recognition task. Modified from Hall et al. (2016)

In the *novel object recognition task* (Figure 16.3a), a mouse was placed in the centre of the arena and presented with three different objects, each in a different corner. The mouse was allowed to explore for 10 minutes and was then removed from the arena for 10 minutes, followed by another 10-minute sample phase and removal from the arena. After these two sample phases, the mouse was returned to the arena either immediately (within approximately 30 seconds), after a short-term delay (10 minutes), or after a long-term delay (24 hours) for the test phase. In this test, one of the objects was replaced with a novel object. The objects and the arena were wiped down between sample phases and test phase to prevent a mouse from using olfactory cues. Increased exploration of the novel object in the test phase was taken as evidence of novel object recognition memory. Compared to non-genetically modified (wild type [WT]) mice, Tc1 mice showed intact immediate, impaired short-term, and intact long-term object recognition memory behaviour (for the summary of results, see Figure 16.4).

To test whether immediate and long-term memory behaviour were also intact in Tc1 mice for spatial organisation of objects, the *object-in-place memory task* was administered (Figure 16.3b). The procedure was identical to the novel object recognition task described above except that in the test phase, two of the objects swapped their spatial locations instead of a novel object appearing. As with the novel object recognition task, Tc1 mice showed sensitivity to the object-in-place change (i.e., increased exploration of the objects in a different location) in both immediate and long-term memory delay, although the short-term condition was not run here (for the summary of results, see Figure 16.4).

As described above, in the novel object recognition task, Tc1 mice demonstrated an impaired ability to detect the novel object after a 10-minute delay. However, it was unclear whether this impairment was specific to novel object detection only or broadly present across other types of memory. Therefore,

Mouse model – Memory task		Immediate memory	Short-term memory	Long-term memory
Tc1 – Novel object		Intact	Impaired	Intact
Tc1 – Object-in-place		Intact	NA	Intact
Tc1 – Object location		NA	Intact	NA
Tc1 – Novel odour		NA	Impaired	Intact
Tg2576 – Novel object		NA	Intact	Impaired

FIGURE 16.4 Summary of results from Hall et al. (2016). Immediate memory = approximately 30-second delay; Short-term memory = 10-minute delay; Long-term memory = 24-hour delay; NA = not administered. Images modified from Hall et al. (2016)

sensitivity to change in location of a familiar object was tested in Tc1 mice after a 10-minute delay using the *object location memory task* (Figure 16.3c). In this task, the procedure was similar to the two tasks above, except that three identical objects were present in three corners of the arena during the sample phase. In the test phase, one of the three objects was moved to the previously empty corner of the arena. Increased exploration of the object in the novel location in the test phase was taken as evidence of memory for the locations occupied by objects in the sample phase. Tc1 mice showed sensitivity to the location change of the familiar object after a 10-minute delay (for the summary of results, see Figure 16.4). Therefore, the difficulty in detecting object novelty following a short-term delay was not due to a general problem with either detecting novelty or modifying exploratory behaviour after this delay.

To assess the extent to which the impairment in object recognition was part of a more general recognition memory deficit in Tc1 mice, an olfactory version of this task (the *novel odour recognition task*) was administered (Figure 16.3d). Here, a mouse was presented with three visually identical plastic cubes each containing a different scent, each in a different corner of the arena. After the sample phase, following a 10-minute or 24-hour delay, one of the odour cubes was replaced with a novel odour cube for the test phase. As with the novel object recognition task, the Tc1 mouse showed impaired short-term but intact long-term recognition memory behaviour for odours (for the summary of results, see Figure 16.4). Therefore, the short-term impairment in recognition memory was not specific to a single sensory modality, but generalisable across different sensory domains.

To ascertain whether the pattern of memory differences in Tc1 mice was specific to human chromosome 21 expression and not a general consequence of the expression of human genes, Tg2576 mice were also tested. Tg2576 mice express a human Swedish *APP* mutation associated with early onset AD (Hsiao et al., 1996). The overexpression of *APP* is absent in Tc1 mice, unlike in individuals with DS, because the Tc1 mouse model does not contain an additional copy of *APP*. Therefore, it is theoretically interesting to compare these two mouse models. Tg2576 mice were administered the novel object recognition task. The pattern of results was opposite to what was found in Tc1 mice: Tg2576 showed intact performance following a short-term memory delay but impairment after long-term delay (for the summary of results, see Figure 16.4). This suggests that the short-term recognition deficit in the Tc1 mouse was not simply a non-specific consequence of the expression of human genes.

As summarised in Figure 16.4, Tc1 mice showed delay-dependent difficulties in recognition memory. Specifically, they showed intact immediate (30 seconds), impaired short-term (10 minutes), and intact long-term (24 hours) memory behaviour for novel object/odour recognition. No deficit was present in the ability to detect spatial novelty. The pattern suggests a memory consolidation rate deficit (such that memories are not consolidated by the end of the short-term delay, but are consolidated by the end of the long-term delay) in the object/odour memory system but not the spatial memory system.

Building analogues of mouse memory tasks for infants/toddlers with Down syndrome

The results from the Tc1 mice described above are broadly consistent with reports of short-term memory difficulties in individuals with DS (Godfrey, & Lee, 2018). However, any meaningful comparison would require a test of object and spatial memory using similar procedures in humans. There is currently only a limited number of memory studies on infants/toddlers with DS (for a review, see Godfrey, & Lee, 2018). Their designs are rather distant from the designs of the mouse memory tasks described above, predominantly focusing on imitation abilities (e.g., Milojevich, & Lukowski, 2016; Rast, & Meltzoff, 1995). Therefore, novel tasks are needed.

In Annette's presentations, she would emphasise several points important for establishing genotype/phenotype correlations across species when designing new tasks. Human/animal models must compare "like with like". We need to develop tasks that can be used across different species (mouse, human) and age groups (infants/toddlers, children, adults, elderly). This means that the tasks have to be non-verbal: they cannot rely on the ability to understand or produce language. Furthermore, the tasks must impose comparable cognitive and neural demands.

Clearly, when designing analogues of the mouse memory tasks for infants/toddlers with DS, a number of factors need to be considered. Unfortunately, many decisions have to be made with little empirical guidance, as there is a lack of systematic investigation of these factors across species (and often, even within species). Some of the factors will be discussed below.

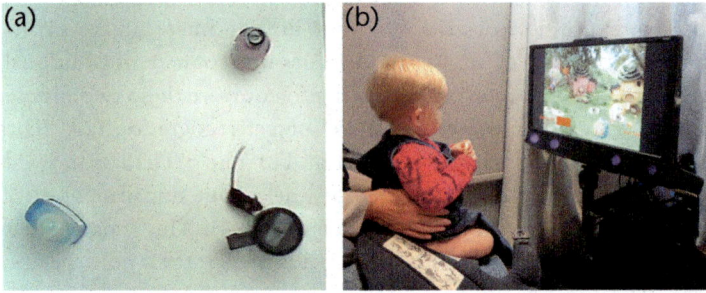

FIGURE 16.5 (a) Hall et al.'s procedure with mice relies on gross motor abilities during exploratory behaviour; (b) Remote eye-tracking as a data collection technique suitable across different ages in humans, minimising the need for whole-body or limb motor skills

Measure of exploration

Hall et al.'s procedure with mice relies on gross motor abilities (Figure 16.5a), a domain in which many infants/toddlers with DS show delays (D'Souza, Karmiloff-Smith, Mareschal, & Thomas, 2020; Fidler, Hepburn, & Rogers, 2006; Will, Caravella, Hahn, Fidler, & Roberts, 2018). Even though virtually all children with DS eventually learn to walk, some of them do not start walking until they are 5 years of age (Palisano et al., 2001). Therefore, any measure relying on locomotor abilities is unsuitable for infants/toddlers with DS. Similarly, manual exploration has been found to be delayed (de Campos, da Costa, Savelsbergh, & Rocha, 2013).

Thus, comparing "like with like" across species may not always mean using an identical method of exploration, but instead one should consider what the most efficient mode of exploration is for a particular species and age group. Visual exploration has been used in a number of studies with infants/toddlers with DS as an indicator of cognitive abilities (D'Souza, D'Souza, Johnson, & Karmiloff-Smith, 2015; D'Souza, D'Souza, Johnson, & Karmiloff-Smith, 2016; Fidler, Schworer, Will, Patel, & Daunhauer, 2019). In the past, it was necessary to video record an infant's eyes and tediously hand-code changes in eye gaze frame-by-frame. However, nowadays eye movements can be detected and coded automatically, using infant-friendly remote eye-tracking technology (see Figure 16.5b; Gredebäck, Johnson, & von Hofsten, 2009). Therefore, ease of administration and analysis made remote eye-tracking the method of choice for mouse task analogues in this study.

Stimuli

Another decision to be made when analogues of mouse and human studies are designed is what stimuli should be used. Here again, comparing "like with like" may not always mean using identical stimuli across species. Rather the concern should be about using stimuli which engage interest (i.e., attentional processes) across species. The stimuli used by Hall et al. (2016) consisted of everyday objects made of non-porous materials, selected on the basis of differences in shape and pattern (based on pilot work), as mice do not have particularly good colour vision (Jacobs, 1993).

Selecting engaging stimuli for use in human infant/toddler studies is of utmost importance for retention of these participants in eye-tracking paradigms. The stimuli selected by the LonDownS Infant stream involved photos of everyday objects or child friendly pictures (see below in Section 4) that infants/toddlers showed interest in during the pilot study.

Number of trials and timing

Another set of considerations when aligning mouse and human studies relates to differences in the timing of cognitive processes across species. This is crucial to consider when thinking about the number of trials needed for learning, as well as the duration of familiarisation and test trials. What is the human infant/toddler equivalent of a 10-minute exposure to objects in the mouse? What is the human infant/toddler equivalent of a 10-minute delay in the mouse? Hall et al. (2016) presented mice with two familiarisation trials, each lasting 10 minutes. For any study with infants/toddlers relying on visual exploration, this would be a very long time frame. We know from widely used habituation designs in human infant testing, which rely on the infant's diminished interest in a stimulus over time, that human infants lose interest in novel visual stimuli after just a few seconds (e.g., Bremner, & Fogel, 2004). A lack of interest in experimental stimuli would increase the chance of premature termination of the testing paradigm before any test trials could be administered, as infants/toddlers are more likely to become bored and/or upset. Therefore, it is necessary to think carefully about the minimum exposure needed to tap into similar cognitive processes in the mouse and human. Again, there is a lack of systematic comparative studies on rate of learning in mice and human infants. Therefore, the convention was borrowed from human infant studies, where the length of trials used with human infants is usually in the order of seconds. The exact length of trials (see Section 4 below) was determined by piloting.

LonDownS memory task design for human infants/ toddlers with Down syndrome

In this section, we describe the specific task design constructed in an attempt to align mouse models of DS with human infant/toddler studies. The LonDownS Infant stream focused on two memory tasks administered by Hall et al. (2016): *novel object recognition* and *object-in-place* (described in detail above), with the aim of creating analogues in human infants/toddlers with DS for the immediate time point. In the human infant/toddler equivalent of the task, a number of objects were presented to the child on a screen. Each memory task consisted of two

(a) **Novel object recognition task**

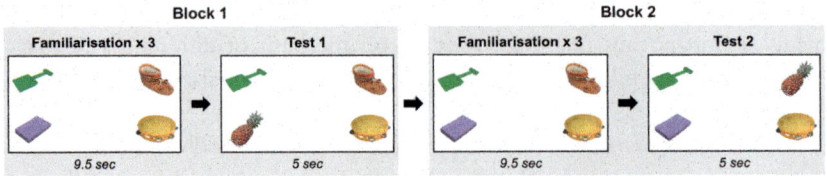

(b) **Object-in-place task**

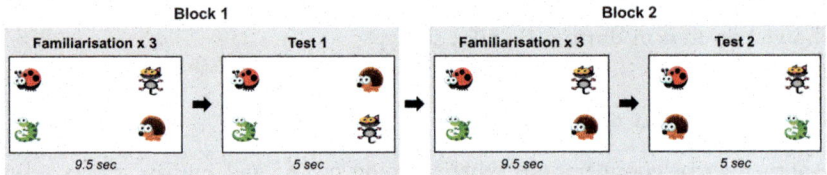

FIGURE 16.6 The layout of the stimuli on the screen in the human analogues of the mouse tasks: (a) Novel object recognition task; (b) Object–in–place task

blocks. In each block, three familiarisation trials were followed by a test trial. In both memory tasks, each familiarisation trial contained four stimuli onscreen. In the test trial of the *novel object recognition task* (Figure 16.6a), one of the familiar stimuli was replaced with a novel stimulus. In the test trial of the *object-in-place memory task* (Figure 16.6b), two of the familiar stimuli exchanged positions. To make the task more interesting for children, the images in the familiarisation trials expanded and contracted, and the tasks were accompanied by child-friendly music. Furthermore, to increase data quality, the pace of the stimuli presentation was tailored to each child, in that each trial was manually initiated by the experimenter only when the child was looking at the screen.

An eye-tracker was used to record eye gaze. Change in looking was computed by subtracting the proportion of looking at each quadrant in the familiarisation trial that immediately preceded the test trial, from the proportion of looking at each quadrant in the test trial. A positive change in looking indicates an increase in looking towards the quadrant after the change.

Participants

Data were collected from 85 infants and toddlers with a clinical diagnosis of DS and 63 typically developing (TD) children at Birkbeck, University of London as part of the LonDownS Consortium Infant stream protocol[1]. The DS group did not significantly differ from the TD group on chronological age or gender (see Table 16.1). A further 12 young children with DS and two TD children were tested but did not yield usable data for diverse reasons including fatigue, fussiness and technical difficulties.

TABLE 16.1 Participant characteristics

		Group		
		DS	TD	Comparison
Number		85	63	NA
Age (months)		6.13–51.50	4.67–52.47	$t(146) = 1.27, p = .207$
		$M = 26.76, SD$	$M = 24.11, SD$	
		$= 12.35$	$= 12.93$	
Gender	Females	39 (45.9%)	34 (54.0%)	$\chi^2(1) = 0.95, p = .331$
	Males	46 (54.1%)	29 (46.0%)	

Note. DS = children with Down syndrome; TD = typically developing children; NA = not applicable.

The participants were recruited via existing participant databases and support groups. Ethical approval was obtained from the North West Wales National Health Service (NHS) Research Ethics Committee (13/WA/0194) and Birkbeck Psychological Sciences Ethics Committee (121373). Prior to testing, informed consent was obtained from parents. Participants were given a small gift (e.g., a T-shirt) in return for their participation.

Equipment

A Tobii TX300 remote eye-tracker was used to capture information on moment-to-moment point of gaze, with measurement accuracy of approximately 0.5° and spatial resolution of approximately 0.05°. The tracking equipment and stimulus presentation were controlled using customised Matlab scripts (Mathworks Inc., Natick, MA, U.S.). The screen was 58.42 cm with a resolution of 1920 × 1080 pixels. Participants were seated approximately 65 cm from the screen. A camera mounted directly above the horizontal midpoint of the screen was used to monitor and record the child's behaviour. Auditory stimuli were delivered via two speakers centrally positioned below the screen.

Results

Novel object recognition task

As illustrated in Figure 16.7, both TD and DS groups showed an increase in looking to the quadrant in which the novel object appeared – for both the first test trial (bottom left quadrant, one sample t-test against 0, TD: $t(53) = 10.52$, $p < .001$, $d = 1.43$; DS: $t(77) = 6.50$, $p < .001$, $d = 0.74$) and second test trial (top right quadrant, one sample t-test against 0, TD: $t(53) = 4.90$, $p < .001$, $d = 0.67$; DS: $t(76) = 2.58$, $p = .012$, $d = 0.29$).

To compare looking behaviour in the quadrant of change across groups and blocks, as well as examine the effect of chronological age, we conducted a 2 × 2 × 2

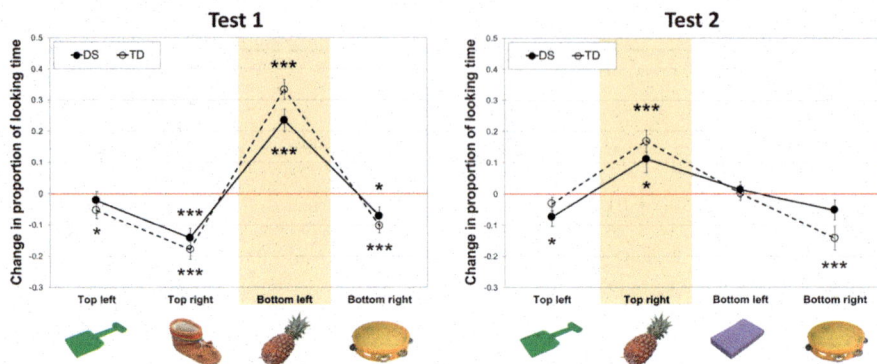

FIGURE 16.7 Change in proportion of looking time from the last familiarisation trial to the test trial in the novel object recognition task in typically developing children (TD) and children with Down syndrome (DS). The quadrant in which a new object appeared is highlighted in yellow. Stars indicate that the values significantly differed from 0 (no change, red line), analysed using one-sample t-tests. Error bars show ± 1 SE; $\star p < .050$, $\star\star p < .010$, $\star\star\star p < .001$

mixed ANCOVA. This analysis included trial type (last familiarisation trial, test trial) and block (first, second) as the within-subject factors, group (TD, DS) as the between-subject factor, and chronological age as a covariate. There was a significant main effect of block, $F(1,126) = 15.82$, $p < .001$, $\eta_p^2 = .11$. This was unexpected and is consistent with a location/object bias. Importantly, there was a significant main effect of trial, $F(1,126) = 30.63$, $p < .001$, $\eta_p^2 = .20$. This suggests that irrespective of the group, children showed a significant increase in looking in the quadrant of change in the test trial compared to the last familiarisation trial. This pattern did not differ between blocks, as none of the interactions including trial x block were significant.

Object-in-place memory task

As illustrated in Figure 16.8, both TD and DS groups showed an increase in looking to the quadrants in which the objects changed their position in the first test trial. As indicated by one-sample t-tests against 0, the TD group increased their looking to the bottom right quadrant ($t(56) = 3.03$, $p = .004$, $d = 0.40$) and not to the top right quadrant ($t(56) = -1.57$, $p = .121$); the opposite was true for the DS group (bottom right quadrant: $t(75) = -1.84$, $p = .070$; top right quadrant: $t(75) = 2.35$, $p = .021$, $d = 0.27$). For the second test trial, both TD and DS increased their looking to the bottom left quadrant (TD: $t(54) = 2.12$, $p = .039$, $d = 0.29$; DS: $t(77) = 4.90$, $p < .001$, $d = 0.55$).

To compare looking behaviour in the quadrant of change across groups and blocks, and to examine the effect of chronological age, we conducted a 2 × 2 × 2 mixed ANCOVA. This analysis included trial type (last familiarisation trial, test trial) and block (first, second) as the within-subject factors, group (TD, DS) as the between-subject factor, and chronological age as a covariate. The quadrant of change for the analysis was selected based on the above presented one sample

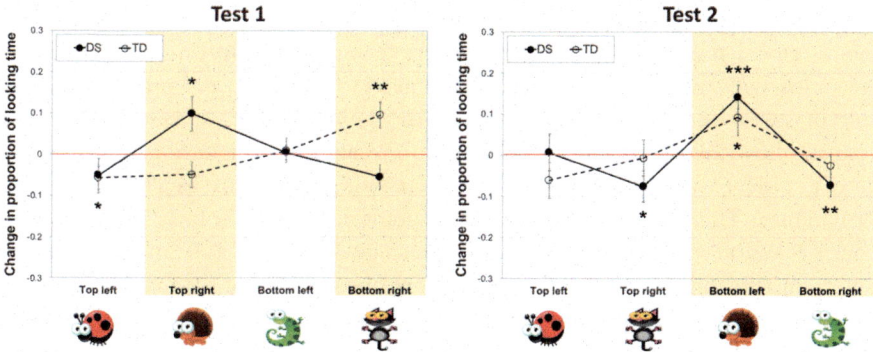

FIGURE 16.8 Change in proportion of looking time from the last familiarisation trial to the test trial in the object-in-place memory task in typically developing children (TD) and children with Down syndrome (DS). The quadrants in which the two objects changed location are highlighted in yellow. Stars indicate that the values significantly differed from 0 (no change, red line), analysed using one-sample t-tests. Error bars show ± 1 SE; $\star p < .050$, $\star\star p < .010$, $\star\star\star p < .001$

t-tests analysis (i.e., in the first block, the bottom right quadrant for TD and the top right quadrant for DS; for the second block, the bottom left quadrant for both groups). There was a significant main effect of block ($F(1,125) = 5.06$, $p = .026$, $\eta_p^2 = .04$) and a significant interaction between block and group ($F(1,125) = 16.55$, $p < .001$, $\eta_p^2 = .12$). This was unexpected and is consistent with a location/object bias. Importantly, there was a significant main effect of trial ($F(1,125) = 15.12$, $p < .001$, $\eta_p^2 = .11$). This suggests that irrespective of the group, children showed a significant increase in looking in the quadrant of change in the test trial compared to the last familiarisation trial. This pattern did not differ between blocks, as none of the interactions including trial x block were significant.

Summary

In sum, like TD children of the same chronological age, infants/toddlers with DS showed looking behaviour consistent with an ability to recognise the change in the novel object recognition task as well as in the object-in-place memory task at the immediate test. This is consistent with behaviour shown in the Tc1 mouse.

Improving alignment between mouse and human studies in the current study

More tasks, more delayed timepoints, and converging evidence

The current memory study with human infants/toddlers with DS focused only on the novel object recognition task and object-in-place memory task. We present the data here not because the design is complete, but because it illustrates the kinds of decisions that need to be made in attempting to align human infant/toddler and mouse model paradigms. For a more comprehensive comparison, other memory tasks used with mice (as introduced above) need to be included. Furthermore, the current study tested only the immediate timepoint. Future studies need to investigate how infants/toddlers with DS perform after a short-term and long-term delay in order to ascertain the extent to which their performance parallels that of the Tc1 mouse. The short-term delayed time point can be easily built into the eye-tracking session, though it needs to be determined whether a 10-minute delayed time point in mice taps into similar cognitive processes as a 10-minute delay in humans. Here, combining behavioural and electrophysiological/neuroimaging measures could help ascertain the progression of various cognitive processes, as well as confirm that the underlying neural processes do not differ across species. Matching time courses of different cognitive processes is of importance as the crucial finding in the Tc1 mouse was a specific deficit in short-term memory delay (10 minutes), consistent with a consolidation rate deficit (the impairment was not manifested at either the immediate or long-term timepoints). Ideally, one would like to test infants/toddlers and mice at incrementally increasing delays to map and compare memory functioning across time. Of course, this would be a rather challenging endeavour, as a large sample size would be needed: different participants would have to be tested at different delays, because participants would be likely to habituate to the paradigm even if numerous interesting objects were included.

Re-centering infant/toddler attention and the use of a mask between trials

The test trials in the current study followed directly from the familiarisation trials, without re-centring the infant's attention in between. For future studies, the infant's attention needs to be brought back to the centre of the screen at the beginning of each test trial in parallel to the mice being placed in the centre of the arena. Furthermore, a mask between trials should be considered to flush out any retinal image in the human infants/toddlers, similar to the arena being wiped down between trials in the mouse task to control for potentially confounding factors.

Number and positioning of stimuli

In the mouse memory tasks, only three objects were used, while four images were presented in the case of human infants/toddlers. It is difficult to know whether these created a similar memory load across species. Furthermore, the duration of the familiarisation trials substantially differed across the two species (2 × 10 minutes for the mice, 2 × 28.5 seconds for the human infants/toddlers). The effects of the number of stimuli and duration of familiarisation on memory across species require systematic investigation. The current study with humans suggests that processing four stimuli within the familiarisation window is possible for both young children with DS and TD children of the same chronological age.

The stimuli across the human and mouse studies were positioned in the corners of their respective "arenas". In the case of the mouse, that meant in three corners of the square arena; in the case of the human infant/toddler, in four corners of the rectangular screen. This means that while objects were evenly distributed along the perimeter in the case of the mouse, this was not the case in the human study. Future studies should consider whether this could affect task performance. Generally, it is advisable to keep distances between objects constant in order to control for any confounding effect of object distance between object pairs. Furthermore, it is advisable to introduce counterbalancing in order to disentangle object bias from quadrant bias. Finally, using only three objects for future infant/ toddler designs would allow researchers to more easily include an object location memory task which requires one of the corners to be kept empty (for one of the objects to move to a new location). Keeping the number of stimuli consistent across the different tasks would allow for a more direct comparison (at least within species).

Analogous developmental stages across species

One key challenge is how to match the ages of human participants to the animals in the model species. Hall et al. (2016) tested adult mice (4–7 months of age), while the LonDownS Infant stream focused on human infants/toddlers (4–52 months). Comparison of different developmental stages complicates comparisons across species, as some of the memory impairments may emerge later in development or be present early and then disappear. The cognitive profile in individuals with DS changes across the lifespan, showing not only general slowing over development, but also changes in strengths and weaknesses (D'Souza, Karmiloff-Smith, et al., 2020; Grieco, Pulsifer, Seligsohn, Skotko, & Schwartz, 2015). Therefore, taking a snapshot of development in infancy/toddlerhood in humans and in adulthood in mice is unlikely to provide us with insights about such developmental changes. As Annette ferociously argued throughout her career, we need to place development at the very core of our studies (D'Souza, D'Souza, & Karmiloff-Smith, 2017; D'Souza, & Karmiloff-Smith, 2011; D'Souza, & Karmiloff-Smith, 2017; Karmiloff-Smith, 1981, 1998, 2010; Karmiloff-Smith, Scerif, & Thomas, 2002). Therefore, future studies should conduct systematic longitudinal comparisons across species. Understanding the development of memory in infancy/toddlerhood in DS may help to identify precursors of memory decline in middle and later adulthood, which are of interest due to highly comorbid AD (Wiseman et al., 2015), as well as factors that exaggerate or attenuate the risk of memory decline (D'Souza, Mason, et al., 2020; Thomas et al., 2020).

Future focus: Understanding individual differences in mouse models and humans

Large individual differences have been described in DS (Karmiloff-Smith et al., 2016). For example, even though the median IQ in DS is around 50, some individuals have an IQ as low as 30 or as high as around 90 (D'Ardhuy et al., 2015). Furthermore, although all individuals with DS exhibit the neuropathology associated with AD, the onset of AD is highly variable (McCarron et al., 2017; Zigman, & Lott, 2007). Individual differences are already present early in development, and understanding these individual differences across time has been a focus of the LonDownS Infant stream (e.g., D'Souza, Karmiloff-Smith, et al., 2020; D'Souza, Lathan, Karmiloff-Smith, & Mareschal, 2020; D'Souza, Mason, et al., 2020; Thomas et al., 2020). Following this line of research, it would be of interest to see whether individual differences on the eye-tracking memory tasks are predictive of developmental outcomes in DS. However, the current design is unlikely to permit such investigations. Although the low number of test trials in the current study was sufficient to analyse group performance, it may be psychometrically too weak to enable us to reliably measure individual performance. With the small number of trials, even though the mean performance of the groups was above chance, there was large variability in looking responses. Furthermore, no relationship with chronological age was observed either in infants/toddlers with DS or TD infants/toddlers (for illustration, see Figure 16.9). This may reflect lack of predictive validity due to the low number of trials (Siegelman, Bogaerts, & Frost, 2017).

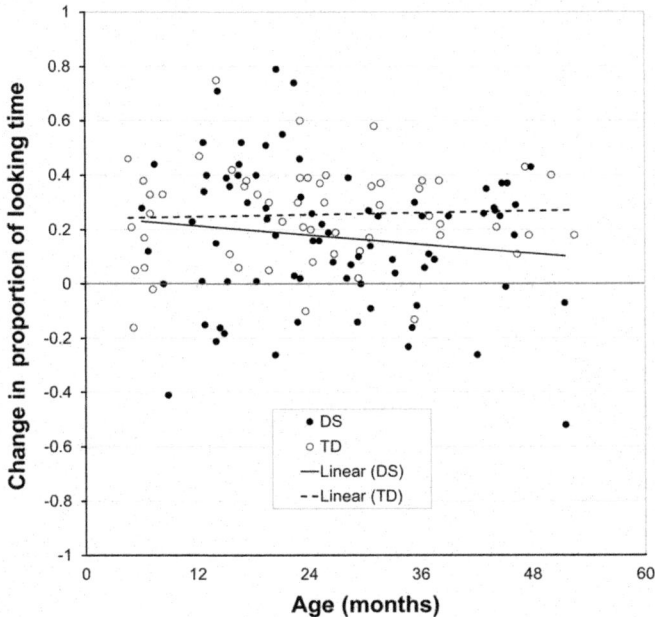

FIGURE 16.9 Change in proportion of looking time from the last familiarisation trial to the test trial averaged over the two blocks of the novel object recognition task, plotted against chronological age in months. The red line indicates no change (0). There is no correlation with age in either typically developing children (TD; $r(55) = .04$, $p = .753$) or children with Down syndrome (DS; $r(79) = -.14$, $p = .230$)

Future studies need to increase the number of trials. There is in fact a need for systematic investigation of how many trials are sufficient to learn about individual differences. While the number of trials can be easily increased in adult participants who are compliant, in infants/toddlers an increase of trials will increase the risk of boredom. This will have a negative impact on retention and limit how many different tasks the infants/toddlers can be administered within one testing session. Therefore, in order to reliably capture individual differences in infants/toddlers across a range of tasks, future projects will likely have to take place over multiple sessions. Developing tasks that can reliably measure individual differences is of utmost importance as there is an interest in pharmacological interventions early in development (even prenatally, Guidi et al., 2014), yet there is a lack of outcome measures which could be reliably used at this stage. Keeping the designs across mice and humans consistent would make interventions developed in mice easier to translate to humans.

The individual differences in infants/toddlers with DS are likely to be an outcome of different genetic backgrounds interacting with various environments (Karmiloff-Smith et al., 2016). These complex interactions between genes and environment are difficult to disentangle in humans, as there is limited scope for experimentally controlling factors contributing to development – something which is possible to do more comprehensively in mice. However, even though mouse models have provided us with important insights into possible mechanistic pathways, their explanatory power for individual differences is rather limited. This is partly due to careful control of mouse research to limit unknown genetic variation by control of genetic background (Finlay, 2019; Herault et al., 2017). However, the choice of genetic background can have a drastic impact on the outcome, even determining the offspring's viability (Sanford, Kallapur, Ormsby, & Doetschman, 2001; Threadgill et al., 1995). The influence of variability in the genetic background can also be detected in phenotypic outcomes when modelling neurodevelopmental and psychiatric disorders (Arbogast et al., 2016; Sittig et al., 2016).

Control of the genetic background of mouse models, and hence the limited range of combination of alleles studied in research, is accompanied by lack of variability in the environment, since mice are raised in controlled, stable environments. Yet, it is known that environment can influence phenotypic outcomes. For example, variability in maternal care has been shown to impact the phenotype of the offspring in both rodents and humans (Watamura, & Roth, 2019; Weaver et al., 2004). Furthermore, environmental enrichment has been shown to affect a phenotypic outcome in a mouse model of DS (Ts65Dn;Martínez-Cué et al., 2002; Martínez-Cué et al., 2005).

In sum, it is challenging to untangle the complex interactions between genes and environment across development in humans as there is very little space for the experimental manipulation of human development. Here, mouse models present promising future avenues because genetic background and environment can be systematically manipulated and measured across developmental time.

Conclusions

Mouse models hold great promise for advancing our understanding of the ae-tiology of cognitive profiles in humans with DS and may pave the way for more targeted interventions. However, the validity of each mouse model critically de-pends on how well the tasks used with the mice map onto cognitive processes of interest in humans. In this chapter, we have outlined some of the first attempts, led by Annette, to design tasks for infants/toddlers with DS based on those used with mouse models. Going forward, it is essential to keep communication lines be-tween different disciplines open in order to design tasks that would enable us to directly compare emerging phenotypes across species and developmental time.

Annette addressed head on the methodological challenges of employing mouse models to understand behavioural phenotypes in neurodevelopmental disorders, as was her style. It was part of her broader commitment to a multidisciplinary, multi-level approach to investigating cognitive development, where the very complexity of the problems she was facing seemed to inspire Annette to new levels of methodological and theoretical innovation. This pioneering spirit will be much missed.

Acknowledgements

We would like to thank all the families in this study for their time. This work would also not have been possible without the help of George Ball, Joanne Ball, Nazanin Biabani, Amelia Channon, Charlotte Dennison, Dean D'Souza, Maria Eriksson, Jennifer Glennon, Kate Hughes, Malin Karstens, Amanda Lathan, Esha Massand, Kate Mee, Beejal Mehta, Lidija Nikolova, Isabel Quiroz, Irati Rodriguez Saez de Urabain, Jessica Schulz, Laura Simmons, Rebecca Thoburn Pallant, Margaux Trombert, Kate Whitaker, and many others. We would parti-cularly like to highlight the contribution of George Ball and Esha Massand in designing the eye-tracking tasks, and Irati Rodriguez Saez de Urabain for pro-gramming these. We would also like to thank Tamara Al-Janabi for managing the LonDownS Consortium as a whole. This work was funded by a Wellcome Trust Strategic Award (grant number: 098330/Z/12/Z) conferred upon the LonDownS Consortium. Our work was further supported by the Waterloo Foundation and the Baily Thomas Charitable Fund. HD is the Beatrice Mary Dale Research Fellow supported by Newnham College (University of Cambridge) and the Isaac Newton Trust. FKW was supported by the UK Dementia Research Institute which receives its funding from DRI Ltd, funded by the UK Medical Research Council, Alzheimer's Society and Alzheimer's Research UK, and by an Alzheimer's Research UK Senior Research Fellowship.

The LonDownS Consortium principal investigators are Andre Strydom (chief investigator), Department of Forensic and Neurodevelopmental Sciences, Institute of Psychiatry, Psychology and Neuroscience, King's College London, London, UK, and Division of Psychiatry, University College London, London, UK; Elizabeth

Fisher, Department of Neurodegenerative Disease, Institute of Neurology, University College London, London, UK; Dean Nizetic, Blizard Institute, Barts and the London School of Medicine, Queen Mary University of London, London, UK, and Lee Kong Chian School of Medicine, Nanyang Technological University, Singapore, Singapore; John Hardy, Reta Lila Weston Institute, Institute of Neurology, University College London, London, UK, and UK Dementia Research Institute at UCL, London, UK; Victor Tybulewicz, Francis Crick Institute, London, UK, and Department of Medicine, Imperial College London, London, UK; Annette Karmiloff-Smith, Birkbeck University of London, London, UK (deceased); Michael Thomas, Birkbeck University of London, London, UK; and Denis Mareschal, Birkbeck University of London, London, UK.

Notes

1 For detailed demographics and health information on the children with DS in the LonDownS Consortium, see Startin et al. (2020).

References

Antonarakis, S. E., Lyle, R., Dermitzakis, E. T., Reymond, A., & Deutsch, S. (2004). Chromosome 21 and Down syndrome: From genomics to pathophysiology. *Nature Reviews Genetics, 5*(10), 725–738. https://doi.org/10.1038/nrg1448

Arbogast, T., Ouagazzal, A.-M., Chevalier, C., Kopanitsa, M., Afinowi, N., Migliavacca, E., Cowling, B. S., Birling, M.-C., Champy, M.-F., Reymond, A., & Herault, Y. (2016). Reciprocal effects on neurocognitive and metabolic phenotypes in mouse models of 16p11.2 deletion and duplication syndromes. *PLoS Genetics, 12*(2), e1005709. https://doi.org/10.1371/journal.pgen.1005709

Barnes, C. A. (1979). Memory deficits associated with senescence: A neurophysiological and behavioral study in the rat. *Journal of Comparative and Physiological Psychology, 93*(1), 74–104. https://doi.org/10.1037/h0077579

Brawand, D., Soumillon, M., Necsulea, A., Julien, P., Csárdi, G., Harrigan, P., Weier, M., Liechti, A., Aximu-Petri, A., Kircher, M., Albert, F. W., Zeller, U., Khaitovich, P., Grützner, F., Bergmann, S., Nielsen, R., Pääbo, S., & Kaessmann, H. (2011). The evolution of gene expression levels in mammalian organs. *Nature, 478*(7369), 343–348. https://doi.org/10.1038/nature10532

Bremner, J. G., & Fogel, A. (2004). *Blackwell Handbook of Infant Development.* Blackwell Publishing Ltd. https://doi.org/10.1002/9780470996348

Breschi, A., Gingeras, T. R., & Guigó, R. (2017). Comparative transcriptomics in human and mouse. *Nature Reviews Genetics, 18*(7), 425–440. https://doi.org/10.1038/nrg.2017.19

Carlesimo, G. A., & Oscar-Berman, M. (1992). Memory deficits in Alzheimer's patients: A comprehensive review. *Neuropsychology Review, 3*(2), 119–169. https://doi.org/10.1007/bf01108841

D'Ardhuy, X. L., Edgin, J. O., Bouis, C., de Sola, S., Goeldner, C., Kishnani, P., Nöldeke, J., Rice, S., Sacco, S., Squassante, L., Spiridigliozzi, G., Visootsak, J., Heller, J., & Khwaja, O. (2015). Assessment of cognitive scales to examine memory, executive function and language in individuals with Down syndrome: Implications of a 6-month observational study. *Frontiers in Behavioral Neuroscience, 9*, 300. https://doi.org/10.3389/fnbeh.2015.00300

D'Souza, D., D'Souza, H., Johnson, M. H., & Karmiloff-Smith, A. (2015). Concurrent relations between face scanning and language: A cross-syndrome infant study. *PLoS One, 10*(10), e0139319. https://doi.org/10.1371/journal.pone.0139319

D'Souza, D., D'Souza, H., Johnson, M. H., & Karmiloff-Smith, A. (2016). Audio-visual speech perception in infants and toddlers with Down syndrome, fragile X syndrome, and Williams syndrome. *Infant Behavior and Development, 44*, 249–262. https://doi.org/10.1016/j.infbeh.2016.07.002

D'Souza, D., D'Souza, H., & Karmiloff-Smith, A. (2017). Precursors to language development in typically and atypically developing infants and toddlers: The importance of embracing complexity. *Journal of Child Language, 44*(3), 591–627. https://doi.org/10.1017/S030500091700006X

D'Souza, D., & Karmiloff-Smith, A. (2011). When modularization fails to occur: A developmental perspective. *Cognitive Neuropsychology, 28*(3–4), 276–287. https://doi.org/10.1080/02643294.2011.614939

D'Souza, H., & Karmiloff-Smith, A. (2017). Neurodevelopmental disorders. *Wiley Interdisciplinary Reviews: Cognitive Science, 8*(1–2), e1398. https://doi.org/10.1002/wcs.1398

D'Souza, H., Karmiloff-Smith, A., Mareschal, D., & Thomas, M. S. C. (2020). The Down syndrome profile emerges gradually over the first years of life. *In Preparation.*

D'Souza, H., Lathan, A., Karmiloff-Smith, A., & Mareschal, D. (2020). Down syndrome and parental depression: A double hit on early expressive language development. *Research in Developmental Disabilities, 100*, 103613. https://doi.org/10.1016/j.ridd.2020.103613

D'Souza, H., Mason, L., Mok, K. Y., Startin, C. M., Hamburg, S., Hithersay, R., Baksh, R. A., Hardy, J., Strydom, A., Thomas, M. S. C., & the LonDownS Consortium (2020). Differential associations of apolipoprotein E ε4 genotype with attentional abilities across the life span of individuals with Down syndrome. *JAMA Network Open, 3*(9), e2018221. https://doi.org/10.1001/jamanetworkopen.2020.18221

de Campos, A. C., da Costa, C. S. N., Savelsbergh, G. J. P., & Rocha, N. A. C. F. (2013). Infants with Down syndrome and their interactions with objects: Development of exploratory actions after reaching onset. *Research in Developmental Disabilities, 34*(6), 1906–1916. https://doi.org/10.1016/j.ridd.2013.03.001

Edgin, J. O., Mason, G. M., Spanò, G., Fernández, A., & Nadel, L. (2012). Human and mouse model cognitive phenotypes in Down syndrome: Implications for assessment. In M. Dierssen & R. de la Torre (Eds.), *Progress in Brain Research* (Vol. 197, pp. 123–151). Elsevier. https://doi.org/10.1016/B978-0-444-54299-1.00007-8

Fidler, D. J., Hepburn, S., & Rogers, S. (2006). Early learning and adaptive behaviour in toddlers with Down syndrome: Evidence for an emerging behavioural phenotype? *Down Syndrome Research and Practice, 9*(3), 37–44. https://doi.org/10.3104/reports.297

Fidler, D. J., Schworer, E., Will, E. A., Patel, L., & Daunhauer, L. A. (2019). Correlates of early cognition in infants with Down syndrome. *Journal of Intellectual Disability Research, 63*(3), 205–214. https://doi.org/10.1111/jir.12566

Finlay, B. L. (2019). Generic *Homo sapiens* and unique *Mus musculus*: Establishing the typicality of the modeled and the model species. *Brain, Behavior and Evolution, 93*(2–3), 122–136. https://doi.org/10.1159/000500111

Godfrey, M., & Lee, N. R. (2018). Memory profiles in Down syndrome across development: A review of memory abilities through the lifespan. *Journal of Neurodevelopmental Disorders, 10*(1), 5. https://doi.org/10.1186/s11689-017-9220-y

Gredebäck, G., Johnson, S., & von Hofsten, C. (2009). Eye tracking in infancy research. *Developmental Neuropsychology, 35*(1), 1–19. https://doi.org/10.1080/87565640903325758

Gribble, S. M., Wiseman, F. K., Clayton, S., Prigmore, E., Langley, E., Yang, F., Maguire, S., Fu, B., Rajan, D., Sheppard, O., Scott, C., Hauser, H., Stephens, P. J., Stebbings, L. A., Ng, B. L., Fitzgerald, T., Quail, M. A., Banerjee, R., Rothkamm, K., … Carter, N. P. (2013). Massively parallel sequencing reveals the complex structure of an irradiated human chromosome on a mouse background in the Tc1 model of Down syndrome. *PLoS One, 8*(4), e60482. https://doi.org/10.1371/journal.pone.0060482

Grieco, J., Pulsifer, M., Seligsohn, K., Skotko, B., & Schwartz, A. (2015). Down syndrome: Cognitive and behavioral functioning across the lifespan. *American Journal of Medical Genetics, Part C: Seminars in Medical Genetics, 169*(2), 135–149. https://doi.org/10.1002/ajmg.c.31439

Guidi, S., Stagni, F., Bianchi, P., Ciani, E., Giacomini, A., De Franceschi, M., Moldrich, R., Kurniawan, N., Mardon, K., Giuliani, A., Calzà, L., & Bartesaghi, R. (2014). Prenatal pharmacotherapy rescues brain development in a Down's syndrome mouse model. *Brain, 137*(2), 380–401. https://doi.org/10.1093/brain/awt340

Gupta, M., Dhanasekaran, A. R., & Gardiner, K. J. (2016). Mouse models of Down syndrome: Gene content and consequences. *Mammalian Genome, 27*(11–12), 538–555. https://doi.org/10.1007/s00335-016-9661-8

Hall, J. H., Wiseman, F. K., Fisher, E. M. C., Tybulewicz, V. L. J., Harwood, J. L., & Good, M. A. (2016). Tc1 mouse model of trisomy-21 dissociates properties of short- and long-term recognition memory. *Neurobiology of Learning and Memory, 130*, 118–128. https://doi.org/10.1016/j.nlm.2016.02.002

Hamilton, G. F., & Rhodes, J. S. (2015). Exercise regulation of cognitive function and neuroplasticity in the healthy and diseased brain. *Progress in Molecular Biology and Translational Science, 135*, 381–406. https://doi.org/10.1016/BS.PMBTS.2015.07.004

Hattori, M., Fujiyama, A., Taylor, T. D., Watanabe, H., Yada, T., Park, H.-S., Toyoda, A., Ishii, K., Totoki, Y., Choi, D.-K., Soeda, E., Ohki, M., Takagi, T., Sakaki, Y., Taudien, S., Blechschmidt, K., Polley, A., Menzel, U., Delabar, J., … Yaspo, M.-L. (2000). The DNA sequence of human chromosome 21. *Nature, 405*(6784), 311–319. https://doi.org/10.1038/35012518

Herault, Y., Delabar, J. M., Fisher, E. M. C., Tybulewicz, V. L. J., Yu, E., & Brault, V. (2017). Rodent models in Down syndrome research: Impact and future opportunities. *Disease Models & Mechanisms, 10*(10), 1165–1186. https://doi.org/10.1242/dmm.029728

Hsiao, K., Chapman, P., Nilsen, S., Eckman, C., Harigaya, Y., Younkin, S., Yang, F., & Cole, G. (1996). Correlative memory deficits, Aβ elevation, and amyloid plaques in transgenic mice. *Science, 274*(5284), 99–102. https://doi.org/10.1126/science.274.5284.99

Jacobs, G. H. (1993). The distribution and nature of colour vision among the mammals. *Biological Reviews of the Cambridge Philosophical Society, 68*(3), 413–471. https://doi.org/10.1111/j.1469-185x.1993.tb00738.x

Kallai, J., Makany, T., Karadi, K., & Jacobs, W. J. (2005). Spatial orientation strategies in Morris-type virtual water task for humans. *Behavioural Brain Research, 159*(2), 187–196. https://doi.org/10.1016/j.bbr.2004.10.015

Kalová, E., Vlček, K., Jarolímová, E., & Bureš, J. (2005). Allothetic orientation and sequential ordering of places is impaired in early stages of Alzheimer's disease: Corresponding results in real space tests and computer tests. *Behavioural Brain Research, 159*(2), 175–186. https://doi.org/10.1016/J.BBR.2004.10.016

Karmiloff-Smith, A. (1981). Getting developmental differences or studying child development? *Cognition, 10*(1–3), 151–158. https://doi.org/10.1016/0010-0277(81)90039-1

Karmiloff-Smith, A. (1998). Development itself is the key to understanding developmental disorders. *Trends in Cognitive Sciences, 2*(10), 389–398. https://doi.org/10.1016/S1364-6613(98)01230-3

Karmiloff-Smith, A. (2007). Williams syndrome. *Current Biology, 17*(24), R1035–R1036. https://doi.org/10.1016/j.cub.2007.09.037

Karmiloff-Smith, A. (2010). Neuroimaging of the developing brain: Taking "developing" seriously. *Human Brain Mapping, 31*(6), 934–941. https://doi.org/10.1002/hbm.21074

Karmiloff-Smith, A., Al-Janabi, T., D'Souza, H., Groet, J., Massand, E., Mok, K., Startin, C., Fisher, E., Hardy, J., Nizetic, D., Tybulewicz, V., & Strydom, A. (2016). The importance of understanding individual differences in Down syndrome. *F1000Research, 5*, 389. https://doi.org/10.12688/f1000research.7506.1

Karmiloff-Smith, A., Broadbent, H., Farran, E. K., Longhi, E., D'Souza, D., Metcalfe, K., Tassabehji, M., Wu, R., Senju, A., Happé, F., Turnpenny, P., & Sansbury, F. (2012). Social cognition in Williams syndrome: Genotype/phenotype insights from partial deletion patients. *Frontiers in Psychology, 3*, 168. https://doi.org/10.3389/fpsyg.2012.00168

Karmiloff-Smith, A., Scerif, G., & Thomas, M. (2002). Different approaches to relating genotype to phenotype in developmental disorders. *Developmental Psychobiology, 40*(3), 311–322. https://doi.org/10.1002/dev.10035

Laczó, J., Markova, H., Lobellova, V., Gazova, I., Parizkova, M., Cerman, J., Nekovarova, T., Vales, K., Klovrzova, S., Harrison, J., Windisch, M., Vlcek, K., Svoboda, J., Hort, J., & Stuchlik, A. (2017). Scopolamine disrupts place navigation in rats and humans: A translational validation of the Hidden Goal Task in the Morris water maze and a real maze for humans. *Psychopharmacology, 234*(4), 535–547. https://doi.org/10.1007/s00213-016-4488-2

Martínez-Cué, C., Baamonde, C., Lumbreras, M., Paz, J., Davisson, M. T., Schmidt, C., Dierssen, M., & Flórez, J. (2002). Differential effects of environmental enrichment on behavior and learning of male and female Ts65Dn mice, a model for Down syndrome. *Behavioural Brain Research, 134*(1–2), 185–200. https://doi.org/10.1016/S0166-4328(02)00026-8

Martínez-Cué, C., Rueda, N., García, E., Davisson, M. T., Schmidt, C., & Flórez, J. (2005). Behavioral, cognitive and biochemical responses to different environmental conditions in male Ts65Dn mice, a model of Down syndrome. *Behavioural Brain Research, 163*(2), 174–185. https://doi.org/10.1016/j.bbr.2005.04.016

McCarron, M., McCallion, P., Reilly, E., Dunne, P., Carroll, R., & Mulryan, N. (2017). A prospective 20-year longitudinal follow-up of dementia in persons with Down syndrome. *Journal of Intellectual Disability Research, 61*(9), 843–852. https://doi.org/10.1111/jir.12390

Milojevich, H., & Lukowski, A. (2016). Recall memory in children with Down syndrome and typically developing peers matched on developmental age. *Journal of Intellectual Disability Research, 60*(1), 89–100. https://doi.org/10.1111/jir.12242

Morris, R. G. M. (1981). Spatial localization does not require the presence of local cues. *Learning and Motivation, 12*(2), 239–260. https://doi.org/10.1016/0023-9690(81)90020-5

O'Doherty, A., Ruf, S., Mulligan, C., Hildreth, V., Errington, M. L., Cooke, S., Sesay, A., Modino, S., Vanes, L., Hernandez, D., Linehan, J. M., Sharpe, P. T., Brandner, S., Bliss, T. V. P., Henderson, D. J., Nizetic, D., Tybulewicz, V. L. J., & Fisher, E. M. C. (2005).

An aneuploid mouse strain carrying human chromosome 21 with Down syndrome phenotypes. *Science*, *309*(5743), 2033–2037. https://doi.org/10.1126/science.1114535

Olton, D. S., & Samuelson, R. J. (1976). Remembrance of places passed: Spatial memory in rats. *Journal of Experimental Psychology: Animal Behavior Processes*, *2*(2), 97–116. https://doi.org/10.1037/0097-7403.2.2.97

Palisano, R. J., Walter, S. D., Russell, D. J., Rosenbaum, P. L., Gémus, M., Galuppi, B. E., & Cunningham, L. (2001). Gross motor function of children with Down syndrome: Creation of motor growth curves. *Archives of Physical Medicine and Rehabilitation*, *82*(4), 494–500. https://doi.org/10.1053/apmr.2001.21956

Rast, M., & Meltzoff, A. N. (1995). Memory and representation in young children with Down syndrome: Exploring deferred imitation and object permanence. *Development and Psychopathology*, *7*(3), 393–407. https://doi.org/10.1017/S0954579400006593

Ruparelia, A., Pearn, M. L., & Mobley, W. C. (2013). Aging and intellectual disability: Insights from mouse models of Down syndrome. *Developmental Disabilities Research Reviews*, *18*(1), 43–50. https://doi.org/10.1002/ddrr.1127

Sanford, L. P., Kallapur, S., Ormsby, I., & Doetschman, T. (2001). Influence of genetic background on knockout mouse phenotypes. In *Gene Knockout Protocols* (pp. 217–225). Humana Press. https://doi.org/10.1385/1-59259-220-1:217

Siegelman, N., Bogaerts, L., & Frost, R. (2017). Measuring individual differences in statistical learning: Current pitfalls and possible solutions. *Behavior Research Methods*, *49*(2), 418–432. https://doi.org/10.3758/s13428-016-0719-z

Sittig, L. J., Carbonetto, P., Engel, K. A., Krauss, K. S., Barrios-Camacho, C. M., & Palmer, A. A. (2016). Genetic background limits generalizability of genotype-phenotype relationships. *Neuron*, *91*(6), 1253–1259. https://doi.org/10.1016/J.NEURON.2016.08.013

Smith, A. D., Gilchrist, I. D., Hood, B., Tassabehji, M., & Karmiloff-Smith, A. (2009). Inefficient search of large-scale space in Williams syndrome: Further insights on the role of LIMK1 deletion in deficits of spatial cognition. *Perception*, *38*(5), 694–701. https://doi.org/10.1068/p6050

Startin, C. M., D'Souza, H., Ball, G., Hamburg, S., Hithersay, R., Hughes, K. M. O., Massand, E., Karmiloff-Smith, A., Thomas, M. S. C., LonDownS Consortium, & Strydom, A. (2020). Health comorbidities and cognitive abilities across the lifespan in Down syndrome. *Journal of Neurodevelopmental Disorders*, *12*(1), 4. https://doi.org/10.1186/s11689-019-9306-9

Sukoff Rizzo, S. J., & Crawley, J. N. (2017). Behavioral phenotyping assays for genetic mouse models of neurodevelopmental, neurodegenerative, and psychiatric disorders. *Annual Review of Animal Biosciences*, *5*(1), 371–389. https://doi.org/10.1146/annurev-animal-022516-022754

Thomas, M. S. C., Ojinaga Alfageme, O., D'Souza, H., Patkee, P. A., Rutherford, M. A., Mok, K. Y., Hardy, J., Karmiloff-Smith, A., & the LonDownS Consortium (2020). A multi-level developmental approach to exploring individual differences in Down syndrome: Genes, brain, behaviour, and environment. *Research in Developmental Disorders*, *104*, 103638. https://doi.org/10.1016/j.ridd.2020.103638

Threadgill, D. W., Dlugosz, A. A., Hansen, L. A., Tennenbaum, T., Lichti, U., Yee, D., LaMantia, C., Mourton, T., Herrup, K., Harris, R., Barnard, J. A., Yuspa, S. H., Coffey, R. J., & Magnuson, T. (1995). Targeted disruption of mouse EGF receptor: Effect of genetic background on mutant phenotype. *Science*, *269*(5221), 230–234. https://doi.org/10.1126/science.7618084

Uttal, D. H., Meadow, N. G., Tipton, E., Hand, L. L., Alden, A. R., Warren, C., & Newcombe, N. S. (2013). The malleability of spatial skills: A meta-analysis of training studies. *Psychological Bulletin*, *139*(2), 352–402. https://doi.org/10.1037/a0028446

Watamura, S. E., & Roth, T. L. (2019). Looking back and moving forward: Evaluating and advancing translation from animal models to human studies of early life stress and DNA methylation. *Developmental Psychobiology*, *61*(3), 323–340. https://doi.org/10.1002/dev.21796

Weaver, I. C. G., Cervoni, N., Champagne, F. A., D'Alessio, A. C., Sharma, S., Seckl, J. R., Dymov, S., Szyf, M., & Meaney, M. J. (2004). Epigenetic programming by maternal behavior. *Nature Neuroscience*, *7*(8), 847–854. https://doi.org/10.1038/nn1276

Wechsler, D. (2014). *Wechsler Intelligence Scale for Children—Fifth Edition*. NCS Pearson.

Westermann, G., Thomas, M. S. C., & Karmiloff-Smith, A. (2011). Neuroconstructivism. In U. Goswami (Ed.), *The Wiley-Blackwell Handbook of Childhood Cognitive Development* (2nd ed., pp. 723–748). Wiley-Blackwell.

Will, E. A., Caravella, K. E., Hahn, L. J., Fidler, D. J., & Roberts, J. E. (2018). Adaptive behavior in infants and toddlers with Down syndrome and fragile X syndrome. *American Journal of Medical Genetics, Part B: Neuropsychiatric Genetics*, *177*(3), 358–368. https://doi.org/10.1002/ajmg.b.32619

Wiseman, F. K., Al-Janabi, T., Hardy, J., Karmiloff-Smith, A., Nizetic, D., Tybulewicz, V. L. J., Fisher, E. M. C., & Strydom, A. (2015). A genetic cause of Alzheimer disease: Mechanistic insights from Down syndrome. *Nature Reviews Neuroscience*, *16*(9), 564–574. https://doi.org/10.1038/nrn3983

Wu, J., & Morris, J. K. (2013). The population prevalence of Down's syndrome in England and Wales in 2011. *European Journal of Human Genetics*, *21*(9), 1016–1019. https://doi.org/10.1038/ejhg.2012.294

Yue, F., Cheng, Y., Breschi, A., Vierstra, J., Wu, W., Ryba, T., Sandstrom, R., Ma, Z., Davis, C., Pope, B. D., Shen, Y., Pervouchine, D. D., Djebali, S., Thurman, R. E., Kaul, R., Rynes, E., Kirilusha, A., Marinov, G. K., Williams, B. A., ... The Mouse ENCODE Consortium. (2014). A comparative encyclopedia of DNA elements in the mouse genome. *Nature*, *515*(7527), 355–364. https://doi.org/10.1038/nature13992

Zhao, X., & Bhattacharyya, A. (2018). Human models are needed for studying human neurodevelopmental disorders. *The American Journal of Human Genetics*, *103*(6), 829–857. https://doi.org/10.1016/j.ajhg.2018.10.009

Zigman, W. B., & Lott, I. T. (2007). Alzheimer's disease in Down syndrome: Neurobiology and risk. *Mental Retardation and Developmental Disabilities Research Reviews*, *13*(3), 237–246. https://doi.org/10.1002/mrdd.20163

17

SLEEP TO REMEMBER: TYPICAL AND ATYPICAL SLEEP AND CONSTRAINTS ON REPRESENTATIONAL DEVELOPMENT

Katharine Hughes and Jamie Edgin

The contributors to this volume make evident the extensive and long-lasting influences Annette Karmiloff-Smith has had on many fields of scientific inquiry. These fields include developmental psychology, developmental cognitive neuroscience and even paediatrics, through her extensive investigation of developmental disorders and more recent focus on sleep. The work reviewed in this chapter represents our unfinished scientific conversations with Annette; we provide the overview to advance an interest in the science of sleep, and particularly sleep's role for developmental disorders, an interest that we all shared. Dr. Katharine Hughes was one of Annette's final graduate students, studying memory development in Down syndrome (DS) in the context of the LonDowns consortium, and Professor Jamie Edgin began a personal friendship and science collaboration with Annette when she initiated her final studies of DS in 2012. Both of us were in communication with Annette about science, life, sleep, and DS in the last months of her life, and at that point in time she continued to be keenly focused on scientific endeavours and mentoring. She placed great effort into advancing our careers through her mentorship, even across oceans, and for that we are exceptionally grateful.

The work overviewed in this chapter represents some of what we wish we could have discussed with her in more detail. We focus on sleep's role in development as it was a constant topic in our conversations, and one that we know Annette would have felt deserved a greater audience outside of our private communications. The three of us understood the importance of sleep for the ongoing construction of representational change and memory formation, but have been surprised to find it relatively understudied as a line of inquiry until recent years. Annette's neuroconstructivist approach has clear implications for how we view the role of sleep in development and developmental disorders. Annette argued against developmental models and theories built on static neuropsychological processes and deficits. Instead of mapping adult-derived mechanisms onto developing children or those with disorders, she believed that it is imperative to examine developmental trajectories, and we suggest here and in previous writings that the role and function of sleep develops and changes over the course of childhood (Gómez & Edgin, 2015). Further, we cannot assume that the functions of sleep are equivalent in those with and without developmental disorders, and we overview several studies that highlight how developmental disorders may reveal differing functions of sleep, and as such, inform our theories of these learning processes in general.

In Karmiloff-Smith's Representational Redescription Hypothesis, she high-lighted two parallel processes of how representations evolve across development: (1) the progressive emergence of adult cognitive specializations and (2) emerging conscious access to knowledge through representational redescription (Karmiloff-Smith, 1992). Central to this theory is that adult end-state specialized cognitive functions emerge over time through an interactive process. This involves the system's inherent domain-relevant processing biases and neurocompetition between regions as they develop, connect and process input from the external environment. In accordance with this, atypical development may result from subtle differences in initial states that cause altered developmental trajectories and variations in cognitive representations. Annette's seminal work on developmental disorders suggested early subtle differences in infant internal states, such as patterns of attention, may dramatically alter those trajectories. Here we argue that aberrant sleep could also interfere with typical cognitive attainments and representation. We argue that healthy development requires the infant to have well-coordinated states of alertness and rest, a "Yin and Yang" of developing cognition. As the field advances, it is becoming increasingly clear that offline states, such as sleep and rest, may be just as important as online behaviours, and the smooth coordination of these internal states is a fundamental developmental achievement. In a paper in 2015, Annette added to her original theoretical account in "Beyond Modularity" when she addressed the possibility that sleep may be one key driver of representational development (Karmiloff-Smith, 1992; Karmiloff-Smith, 2015):

> "However, to allow for conscious access to the same knowledge, I argued for a second, parallel process: representational redescription, i.e., representations embedded in brain processing that are implicit to the organism are redescribed, and thereby become knowledge to the brain, with the redescriptions from one domain becoming transportable to other domains (see details and supporting developmental data in Karmiloff-Smith, 1992). I would now argue that this process is likely to occur via brain activity during **sleep**."

Therefore, in this chapter we discuss sleep, its role in constraining development and, in particular, if sleep may provide one mechanism of representational redescription as originally suggested by Annette. We highlight new data on sleep and learning in developmental disorders, including some of our mutual research group's recent work, and we highlight how the field must study both typical as well as disordered sleep to best inform our understanding of sleep's capacity for affecting developmental change.

We conclude with a discussion regarding how the mechanisms underlying sleep and development suggests that the hippocampus and its associated memory networks should have greater consideration in developmental theory in general.

Theories of sleep and memory

Early in the philosophical exploration of sleep, the question was asked, what could possibly be the evolutionary function of this unconscious and vulnerable human state? Given the nature of early humans' environments, sleep, especially deep sleep, where we are not conscious of our surroundings and potential hazards, must have some exceptional benefit to be worth the added risk. As early as Ancient Greece, it was theorized that dreaming may involve processing information in a manner beneficial to our brain and cognitive function (Aristotle's text from 350BCE, "*On Sleep and Sleeplessness*"). Sleep has been shown to be correlated with daytime behaviour and attention, academic achievement and numerous other developmental attainments important for daily function in children and adults (Curcio, Ferrara, & De Gennaro, 2006; Fallone, Acebo, Arnedt, Seifer, & Carskadon, 2001; Zimmerman, & Aloia, 2012). But can brain processes during sleep actually change neural and cognitive representations? In recent decades, researchers have examined the more nuanced features of sleep, and discovered sleep-driven activity associated with brain regions implicated in memory function, and thus potentially representational development. Convincing data also suggest sleep contributes to functions of physiological maintenance and repair, including the clearance of neuropathologies associated with atypical ageing and Alzheimer's disease (Holth et al., 2019), but in this paper we focus on sleep's role for re-presentational development given Annette's particular interest in this aspect of sleep.

Sleep is a universal human experience, but the structure of sleep changes extensively across development. Some of those changes seem to mirror changes in neurodevelopment and are hypothesized to drive some developmental brain changes across childhood. Data from animal models suggest that sleep may play a particularly important role in driving cortical plasticity, including the development of ocular dominance columns (Aton et al., 2009; Hoffman, & McNaughton, 2002). Indeed, organisms that are exposed to extensive early experience show increased sleep need following that experience (Ganguly-Fitzgerald, Donlea, & Shaw, 2006), perhaps similar to the increased sleep need in early childhood. In early infancy, humans show polyphasic sleep up to 18 hours in a 24-hour period, with multiple daytime sleep episodes, but by adulthood 7–9 hours of sleep is sufficient (Ohayon, Carskadon, Guilleminault, & Vitiello, 2004). Infants do not consolidate their sleep into a night-time period completely until ~3.5 years, when they cease habitual napping. Some evidence suggests that daytime naps may be essential for memory retention and related to increased developmental attainments, including vocabulary acquisition (Hupbach, Gómez, Bootzin, & Nadel, 2009; Horváth, & Plunkett, 2018).

Sleep is divided into two main types: rapid eye movement (REM) and non-REM (NREM). REM shows the most "wake-like" higher frequency neural activity, and is associated with muscle paralysis and vivid dreaming, whereas NREM sleep has stages 1 to 3, increasing in the "depth" of sleep associated across these periods. During the night, NREM and REM stages are cycled through in ~60–90 minute cycles (varying across age), with each NREM sleep stage associated with different neurophysiological features, such as slow-wave activity (SWA or "delta"; electroencephalogram (EEG) slow oscillations of 0.5-5 Hz), sleep spindles (~10-16 Hz), and hippocampal associated sharp-wave ripples (SWR; high-frequency hippocampal bursts >100 Hz). Much of the work on memory formation has been related to the role of slow-wave sleep and its relationship with memory neurophysiological features, but more recent data have also suggested a role for REM sleep and combined NREM-REM periods (Spanò et al. 2018). By 6 months of age, human infants show many of the features of adult sleep-states, including increasing consolidation into a night-time sleep period which includes measurable REM and NREM states. By 9 months, active sleep (similar to REM) decreases to 30% of the sleep period, and there is an increase in the amplitude of SWA, which is theorized to reflect an increase in long-range synaptic connectivity (Scholle et al., 2011; Galland, Taylor, Elder, & Herbison, 2012). Children have proportionally more SWA and stage 3 NREM than adults until this begins to decline aged 10–12 years (Kurth et al., 2010; Ohayon et al., 2004). As slow-wave and REM sleep have been linked to memory retention, it seems likely that these shifts would predict differences in the role of sleep across development.

Current theories of sleep's role in memory retention suggest that sleep provides active as well as passive functions to modify memory representations, including integration with existing knowledge. The predominant account for sleep-dependent changes in memory is the Active System Consolidation Hypothesis (Diekelmann, & Born, 2010), which expands beyond the complementary learning systems theory of McClelland, McNaughton, and O'reilly (1995). These theories posit that long-term storage requires integration into cortical stores through processes of transfer from a precise code in hippocampus to integration with pre-existing knowledge in the cortex. Because of the potential for extensive interference within cortical networks during initial learning, information must be coded in a separate store (i.e., hippocampus) to be accessed later. Buzsáki (1998) expanded on this theory to suggest that sleep was pivotal to the active consolidation process. Along these lines, newly learned information is coded in the hippocampus and through a series of temporally coupled neurophysiological events occurring during sleep, information is thought to be replayed and integrated into cortical stores and existing knowledge structures. It is during slow-wave sleep that this process is thought to occur. Spindles are correlated with long-term potentiation in neocortical synapses, providing a functional link between hippocampal reactivation and strengthening of information in long-term stores

(Rosanova, & Ulrich, 2005). The finding that spindle activity is seen more in previously activated synapses and networks provides some support for the theory that these spindles are associated with memory consolidation (Werk, Harbour, & Chapman, 2005). Disrupting hippocampal sharp-wave ripple activity in rat models also impairs the consolidation and recall of spatial memory (Girardeau, Benchenane, Wiener, Buzsáki, & Zugaro, 2009), and Buzsáki (1998) suggested that random patterns of this replay (or functionally immature patterns) could actually serve to "spoil" memory patterns obtained over wake. Much of this work suggests that both hippocampal ripples (which can only be measured intracranially) and spindles are beneficial to the consolidation of memory traces and may be part of a sequence of events leading to active processes of memory consolidation during sleep.

Another theory of the memory functions of sleep has suggested a passive role of sleep in memory retention: The Synaptic Homeostasis Hypothesis (Tononi & Cirelli, 2014). In this view, there is a global downscaling of synaptic connectivity across slow-wave periods, which would preserve the strongest associations encountered across wake, while irrelevant or weakly encoded details may be forgotten. Indeed, it has been suggested that sleep may play an equally important role in facilitating some aspects of forgetting (Poe, 2017). Evidence for this theory comes from animal models, in particular the fruit fly, which shows fluctuations in protein levels at the synapse based on sleep and wake states, suggesting homeostatic regulation via sleep (Gilestro, Tononi, & Cirelli, 2009). Another recent theory, the Contextual Binding Theory, has emphasized the importance of considering the state-dependent aspects of sleep for memory consolidation, and particularly that sleep may reflect a context change which could primarily reduce effects of memory interference, without any active process at all (Yonelinas, Ranganath, Ekstrom, & Wiltgen, 2019). Therefore, sleep-dependent learning benefits could be due to either active or passive states, and one cannot assume that these changes necessarily require hippocampal involvement or active systems consolidation.

Sleep as a constraint on representational development

In adults, there is clear evidence that sleep affects memory retention, resulting in shifts in neural representations. Comparison of sleep and wake intervals has shown a benefit of sleep on declarative memory retention in adults (Mednick, Nakayama, & Stickgold, 2003; Tamminen, Payne, Stickgold, Wamsley, & Gaskell, 2010; Gais, Lucas, & Born, 2006; Noack, Schick, Mallot, & Born, 2017). Neural representations of memories also seem to be modified after sleep in adults, but these changes may not be immediately evident and continue to occur across days to months, with some studies suggesting hippocampal involvement until 48 hours after encoding, and a shift to medial prefrontal activation as much as 6 months later (Gais et al., 2007; Takashima et al., 2006; Takashima et al., 2009). Sleep-dependent neural changes in adults also seem to reflect task demands. In a study assessing neural function using fMRI, three-day retention of spatial navigation following sleep or sleep deprivation found that although there was comparable hippocampal-neocortical activity in both groups, the sleep group had increased striatal activity at delay (Orban et al, 2006). In another example of representational change across sleep in adults, Dumay and Gaskell (2007) found that recall of novel words was significantly better after a period of sleep but not after wake delays, and the novel words only engaged lexical competition effects with existing words after sleep. Specifically, lexical competition demonstrates information integration when newly learned competitor words are incorporated into a lexical representation and will cause competition, or a slowed detection process, for previously learned words (e.g., we are slower to detect "dolphin" if we just learned the novel word "dolpheg"). These studies provide evidence that sleep promotes both the consolidation and integration of novel information into human memory in adults, resulting in both neural and cognitive representational change.

But does sleep act in the same way in children and adults, or are there developmental shifts in sleep's role? Previous studies examining correlational data suggest that sleep patterns may relate to specific cognitive attainments across childhood, including the development of expressive vocabulary, attentional control, and memory (Gómez, Newman-Smith, Breslin, & Bootzin, 2011; Horváth, & Plunkett, 2018). However, findings have been mixed regarding sleep's relation to global developmental milestones (Mindell & Moore, 2018). Therefore, sleep may not be a *domain-general mechanism*, affecting developmental globally, but one that acts in a *domain-relevant manner*, or influencing specific skills depending on the developmental constraints at certain points in time (Karmiloff-Smith, 1992). Experimental data have been beneficial in teasing apart the changing functions of sleep, and a picture is beginning to emerge suggesting that a developmental approach is necessary in interpreting sleep's role in memory formation.

While data suggest that sleep benefits retention and generalization in children (Kurdziel, Duclos, & Spencer, 2013; Gómez, Bootzin, & Nadel, 2006), few studies have examined how cognitive representations are specifically modified by sleep. In work conducted by Karmiloff-Smith and colleagues, Annette and her associates directly tested whether or not sleep may act on explicit or procedural (implicit problem solving) memories, thus probing sleep's impact on some of the levels of representations addressed in *Beyond Modularity*. Specifically, Annette proposed that knowledge was first encoded procedurally and then an offline process (e.g., sleep) integrated information into explicit representations. In Ashworth's study designed to test these processing levels, typically developing (TD) children aged 6 to 12 years performed significantly better on **both** explicit word-learning and implicit problem-solving tasks following intervals containing sleep compared to wake (Ashworth, Hill, Karmiloff-Smith, & Dimitriou, 2014). This finding suggested that children's sleep can act to enhance both implicit and explicit memory representations, although Karmiloff- Smith and colleagues noted that explicit aspects of the Tower of Hanoi task might have been affected, as opposed to a purely implicit effect. Other group's research has shown some further differentiation in sleep's effects on implicit and explicit learning. Giganti et al. (2014) used a within-subjects design to demonstrate that naps benefit explicit memory recall but not implicit priming in preschool-age children. Similarly, in Wilhelm's study employing a finger-sequencing task, children gained more explicit knowledge of the task demands after sleep than adults, with the childhood groups' ability level correlating to higher BOLD response in the left posterior hippocampus (Wilhelm et al., 2013). These findings emphasize that we should consider how sleep operates across domains and at different representational levels, as the effects are likely to differ.

In school-age children, sleep does seem to modify cognitive and neural representations. Similar to the results of lexical competition effects measured in adults, children also demonstrate such integration effects with sleep (Henderson, Weighall, & Gaskell, 2013). Urbain et al. (2016) examined the shifts of neural representations in a group of school-age children roughly half of which napped (a 90-minute polysomnography nap) following a task requiring memory for the associations of novel object and their functions. Learning-dependent changes were observed in hippocampal and para-hippocampal regions in both groups, with sleep-dependent changes appearing in the prefrontal cortex. The findings of prefrontal activation in the sleep group suggest a shift from hippocampal involvement to cortical consolidation. No changes were observed over wakeful rest. Memory performance was unchanged in the wake and sleep groups across visits, suggesting a change in neural representation that could not be detected at the behavioural level. This study suggested that cortical integration via sleep may commence rapidly in children (i.e., over one nap period), but that sleep may affect neural representations over longer time scales in adults (Gais et al., 2007, Takashima et al., 2009, Takashima et al., 2006).

In addition to benefits in veridical memory demonstrated in older children (Kurdziel *et al.* 2013), several studies suggested that naps might benefit infants and young children's ability to generalize new concepts. In the first demonstration of these sleep-dependent generalization effects in infants, 15-month olds generalized artificial grammar associations to unheard stimuli if they napped in the 4-hours after learning (Gómez *et al.* 2006). If the infants stayed awake they listened longer to sentences that contained the same stimuli as encountered in training, evidencing no generalization of the stimuli. Hupbach et al. (2009) replicated this effect and suggested that naps were required to elicit learning, as at 24-hours the infants again generalized the grammar rules, but only if they napped in the 4 hours after learning.

Simon et al. (2017) demonstrated that naps facilitate generalization of statistical language learning in 6.5-month infants, and that these effects related to global power differences in alpha and theta as well as frontal SWA. However, these memories were very fragile and highly susceptible to interference. This pattern would suggest some process allowing for temporary maintenance at this age, instead of cortical integration. In another study examining both veridical memory and category generalization of object-label mappings using evoked response potentials, Friedrich, Wilhelm, Born, and Friederici (2015) showed that 9- to 16-month-old napping infants demonstrated sleep benefits for both veridical memory and generalization, with evoked response potentials (ERPs) which indicated deficits in specific object retention as well as integration. Sleep spindles correlated with the ERP components suggesting generalization, which was not apparent in infants who did not nap. These findings propose that sleep affects generalization across much of childhood, although the processes leading to these patterns require further investigation. For instance, forgetting may lead to loss of detail across repeated exemplars that might allow the child to extract commonalities, but this forgetting may differ from more mature processes of generalization in which both veridical memory and generalizations are mutually maintained. In fact, in *Beyond Modularity*, Karmiloff-Smith suggested that a representation's original details would be stored but that there would be a transcription process allowing for the extraction of commonalities. Is this what the young infant is able to maintain with sleep? Studies aimed at a better understanding of the paths to and nature of sleep-related generalization could be a fruitful area of new research.

A developmental approach to sleep's role

Across the last ten years, studies have shown that sleep can impact memory retention, integration and generalization in children. It might appear that sleep's effects are constant across development, but as often emphasized by Annette, a developmental perspective is required to fully understand sleep's function. On the whole, studies suggest that young infants can generalize new learning after sleep, but that early memories are quite fragile and will not be retained without sleep periods close to learning opportunities. Gómez and Edgin (2015) theorized that the role of sleep may change across development as the hippocampus gains greater functionality and precision in its representations. The hippocampus shows developmental change into late adulthood, and memory for precise representations is not fully mature in the preschool years (Lavenex, & Lavenex, 2013; Keresztes, Ngo, Lindenberger, Werkle-Bergner, & Newcombe, 2018). Therefore, in young infants, sleep serves to facilitate generalization through processes of forgetting irrelevant detail and extraction of key features, which could be due to non-specific or passive processes of sleep. Evidence for shifting functionality of sleep comes from work suggesting that sleep may interfere with generalization, and toddlers may generalize better after wake periods than nap periods under certain conditions. For example, Werchan and Gómez (2014) found that 2.5-year olds who learned objects that varied in colour and background were better at generalizing those mappings to new surface details and contexts after wake than sleep. In this case, sleep may have helped the child to form complex veridical associations that would have made generalization difficult. Wake, on the other hand, facilitated forgetting, and aided generalization. At 2.5 years toddlers were able to store neural representations with the support of more mature hippocampal representation. In total, a picture is emerging that sleep is needed to retain most information in infants because of fragile memory formation, but that the function of sleep shifts as the infant gains more precision in representation and storage of those specific details.

In contrast to the developmental theory of Gómez and Edgin, recent data has suggested that very young infants retain memories better after naps, which has been attributed to declarative memory formation and active sleep-dependent consolidation mechanisms of the hippocampus (Horváth, & Plunkett, 2018). In one study, the authors habituated 3-month-old infants to a line drawing of a human figure, and then tested the infant's looking patterns to the learned stimuli versus a distractor item. Infants that napped showed a looking preference to the new item, but only on the first trial, suggesting the memory was susceptible to interference at recall. Another recent study tested 6- to 12-month-old infants on the retention of deferred imitation sequences after exposure to naps (Konrad, Seehagen, Schneider, & Herbert, 2016). Again, naps seemed to benefit infants, allowing them to recall about half the actions in a three-action sequence which included removing a puppet's mitten, shaking the mitten and replacing the mitten, at 4 hours and after a 24-hour test interval. However, in these studies the infants had multiple exposures to the stimuli, a procedure which may tap cortical learning and allow sleep to benefit by simply maintaining strongly encoded associations. Furthermore, these studies do not measure initial memory retention across conditions. Without this outcome, it is difficult to determine whether or not sleep played an active role in consolidation or if sleep served to protect the memory from interference from external stimuli. All in all, no study has demonstrated precise memory in infants that includes the explicit retention of distinct associations, and in these studies the findings could be generated by passive sleep mechanisms alone.

Therefore, while on the surface it may appear that sleep can be of benefit across development and into adulthood, more research must be conducted to determine the nature of these sleep-dependent memory effects. At what level of representation are they acting? Are they a mere enhancement, do they allow for subsequent cross-modal integration, or do they guide behaviours and explicit access? These are some of the key developmental shifts in knowledge representation suggested by Karmiloff-Smith in *Beyond Modularity* (1992). Few studies in infants or adults have gone on to determine the long-term fate of newly acquired knowledge structures across intervals including sleep, so the field is lacking direct evidence for how these memories may evolve and begin to influence cognitive representation.

Direct comparisons of sleep effects between young infants and individuals at differing developing stages can be exceedingly difficult. However, the examination of children with developmental disorders can help to determine whether sleep functions may be continuous or whether they differ in those who have altered trajectories of sleep and neural, and particularly hippocampal, development. We make some strong predictions in this regard. If on the one hand sleep only serves to eliminate interference, or is simply a result of contextual change, sleep should show benefits across all age ranges and also in developmental disorders, although these benefits may be reduced in those with sleep impairment. These benefits would directly relate to the forgetting curves of the domain-specific memories at those ages. Therefore, sleep would shield the organism from the external stimuli of wake, and would reduce forgetting. Memory preservation would be greater for strongly encoded items that may not suffer as much from any downscaling process. However, if on the other hand, sleep serves an active process, and one that engages hippocampus, there should be increases in retention over and beyond initial learning levels with concomitant changes in cognitive or neural representations. Furthermore, there should be points in development at which developmental disorders of the hippocampus show sleep-dependent decrements in veridical memory as a result of poor translation of hippocampal representations. Karmiloff-Smith and colleagues were among the first researchers to begin to add to the body of experimental data on sleep-dependent learning in developmental disorders, helping to highlight the role of sleep in development in general and providing some data to address these predictions.

Sleep in developmental disorders: Annette Karmiloff-Smith's influence

In 2007, Karmiloff-Smith and colleagues reviewed the importance of considering sleep in development and disorders in a paper aimed to influence both clinicians and researchers (Hill, Hogan, & Karmiloff-Smith, 2007). Annette's group then conducted a series of studies central to the field of sleep and developmental disorders, including some of the first experimental studies to examine sleep across intellectual and developmental disability syndromes. Her most recent workrelated infant longitudinal memory development to sleep profiles in TD infants, showing that higher levels of night-time wakings related to delayed trajectories of memory development (Pisch, Wiesemann, & Karmiloff-Smith, 2019).

Annette conducted a series of studies examining sleep in Williams syndrome (WS) in collaboration with colleagues including Anna Joyce (Ashworth), Dagmara Dimitriou and Cathy Hill. They demonstrated higher sleep difficulties in WS than in age-matched controls (Annaz, Hill, Ashworth, Holley, & Karmiloff-Smith, 2011). An additional cross-syndrome study including a combination of parental report and objective sleep measures (e.g., actigraphy) with school-aged children with Down syndrome (DS), WS and TD children found that although sleep problems occurred in both atypical populations, the profiles were substantially different (Ashworth, Hill, Karmiloff-Smith, & Dimitriou, 2013). Children with DS had more frequent and longer night wakings and restlessness whereas children with WS had delayed sleep onset. The cause of these profiles of disordered sleep is still to a large degree unknown, although physiological aspects of DS are proposed to account for some percentage of issues (i.e., obstructive sleep apnea, Marcus, Keens, Bautista, von Pechmann, & Ward, 1991). This study highlights the importance of not taking a one-size-fits-all approach to atypical development and obtaining thorough and precise characterizations of disorder-specific sleep profiles.

As demonstrated by the cross-syndrome studies conducted by Annette and her colleagues, DS shows particularly frequent sleep fragmentation, and is the clearest example of a developmental disorder with impaired sleep-dependent memory consolidation. Specifically, Karmiloff-Smith's collaborative study with Joyce, Hill and Dimitriou provided the first demonstration that memory consolidation may be impaired across sleep intervals in children with DS. This study looked at the effect of sleep on explicit word pair consolidation in children with WS and DS across multiple test sessions (e.g., learning a novel name for a cartoon animal) and found that school-aged children with DS did not improve following sleep compared to when awake, or overall, whereas children with WS improved across time, but not in a sleep-dependent manner (Ashworth, Hill, Karmiloff-Smith, & Dimitriou, 2015; See Figure 17.1). Individuals with DS performed better overall when they were taught information at the beginning versus the end of the day, indicating some timing induced differences of sleep's effects. In typical infancy, data suggest sleeping shortly after learning is necessary to retain memories, but here these findings seemed to suggest the opposite pattern in DS. Ashworth et al. (2015) was the first sleep-dependent memory study conducted in DS, and these findings motivated later work from the Edgin lab in this area.

Annette's work similarly inspired other research groups studying developmental disorders to conduct studies with sleep-dependent learning designs to determine patterns of memory consolidation in relation to disordered sleep profiles (for a recent review of sleep-dependent learning across developmental disorders see

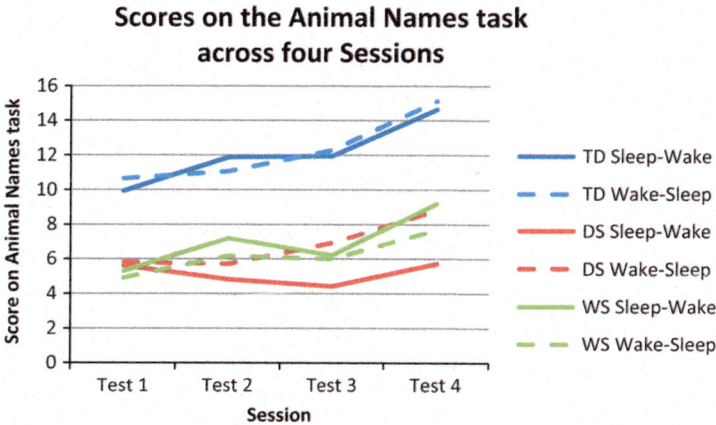

FIGURE 17.1 In an important study of sleep-dependent learning in Down syndrome (DS) and William's syndrome, Karmiloff-Smith and colleagues (Ashworth et al.) found that school-age children with Down syndrome showed divergent patterns of name-animal associations when learning occurred directly before sleep vs. training that occurred in the morning. This study also suggested sleep-related consolidation deficits in DS

Luongo et al., in press). In a small sample of children with ADHD, Prehn-Kristensen et al. (2011) showed a reduction is SWA early in the night and a reduced benefit of sleep on declarative memory consolidation, although both children with ADHD and controls showed greater recognition accuracy following wake versus sleep intervals. There is also evidence that sleep-dependent memory consolidation is altered in autism, where studies show sleep quality deficits, including delayed sleep onset, reduced total sleep time and increased REM latency (Miano et al., 2007). Given the variation in cognitive and language function in this group, sleep may provide one level of explanation for the diversity in learning deficits. Henderson, Powell, Gareth Gaskell, and Norbury (2014) demonstrated that children with autism showed effects of lexical integration immediately after learning that were not maintained across sleep intervals, suggesting the group did not integrate new words into the lexicon under the same time course as controls. In another recent study, children with autism showed impairments in visual recognition memory across sleep and wake, but enhanced gist memory after sleep, including increased semantic errors of commission on a list recall task (i.e., misremembering one heard "sleep" in a list of words all related to that content area; Kurz et al., 2019). These findings suggest that patterns of sleep-related forgetting may be present in autism, potentially affected by the profile of altered sleep.

In regard to DS, a syndrome with documented hippocampal impairment (Nadel, 2003), Edgin's lab found that naps are beneficial to habitually napping toddler's retention of novel object-label associations, but actually increased the deficits in veridical word-learning in DS (Spanò et al., 2018; See Figure 17.2). Similar to Karmiloff-Smith and colleagues, we found that if children with DS sleep immediately after learning, naps have a substantial negative impact on their patterns of long-term retention. As is shown in Figure 17.2, this study had a within-subjects design, allowing us to demonstrate that sleep benefited retention in typical children even though there was rapid memory decay after 5 minutes. In typical children, the nap benefit related to the proportion of REM sleep, but REM was significantly impaired in DS. These results suggest that sleeping immediately after learning may be particularly problematic for children with DS, whereas in early typical development, naps support memory retention and help counteract the rapid memory decay across hours without a nap. Further, these data suggest that the sleep in naps actively influences memory formation at these ages, because nap retention exceeds the retention level at a 5-minute delay. Given the relative resilience to interference in the DS group, we would not expect such a substantial performance loss across sleep if sleep solely consists of a passive process. The loss measured in the DS group suggests an active process, but one that has gone awry, and it is possible that sleep replays memories in an erroneous way in DS, thus erasing what is learned before the nap. While future research is required to expand on these findings, these results do suggest that work in developmental disorders can highlight how sleep's role may shift across different states of neural development.

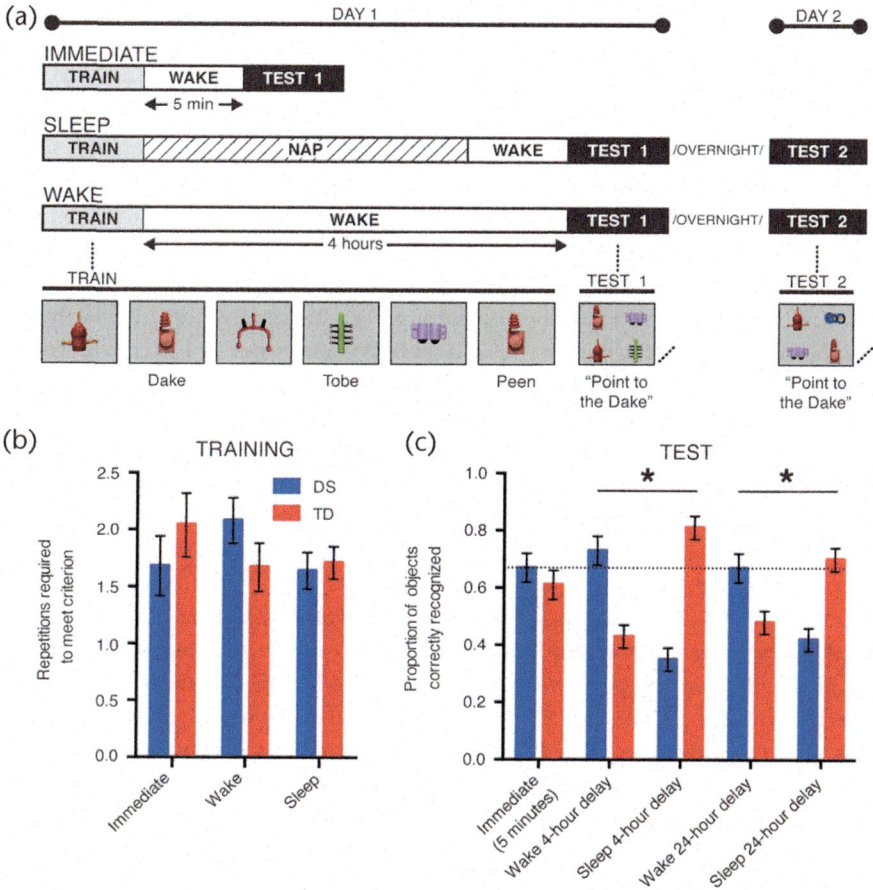

FIGURE 17.2 Experimental procedure and findings from Spanò et al., 2018 showing sleep-dependent learning benefits in typical children and deficits across naps in toddlers with Down syndrome

Summary, conclusions, and directions for future research and theory

Does sleep have a developmental trajectory?

Based on behavioural data in TD children alone, it would appear that sleep confers a benefit from early infancy onward, relating to patterns of generalization and veridical memory. However, some studies have suggested uneven trajectories of sleep's effects on memory, with sleep actually impairing generalization or veridical memory in certain developmental states and disorders. In this chapter we highlight exceptions that suggest developmental shifts in these processes. Further, in developmental disorders, dissociations in sleep-dependent learning are evident, suggesting non-linear changes in these processes.

One of the limitations in this field is the different methods that were used across previous studies, making comparison of results difficult, especially across different age groups. Further complicating the picture is how sleep interacts with the strength of the memory and preexisting knowledge framework of the participant, which in itself differs extensively from childhood to adulthood. For instance, it has been suggested that offline consolidation may benefit children to a greater degree than adults because adults may rely more on semantic integration strategies and prior knowledge. James, Gaskell, Weighall, and Henderson (2017) conducted a meta-analysis suggesting that children rely on SWS to support learning and neural reorganization, whereas adults benefit from pre-existing knowledge that facilitates consolidation. How can future research address these issues to better understand the developmental course of these functions? First, future research may try to adopt a more detailed analysis of the underlying changes in memory representations by using stimuli that equate prior knowledge, by measuring long-term retention of memories across representational demands (i.e., explicit and implicit), and in studies in which multiple pre-sleep encoding levels are examined to determine if sleep allows for the strengthening of weaker memories or merely supports stronger memories by protecting them from interference. The examination of the longitudinal trajectory of sleep-related learning processes in typical and atypical development would also help to answer this question in greater detail. Regarding this point we offer another prediction – if the generalization effects in infancy are due to the passive processes of sleep, we should see that sleep also allows for generalization effects due to forgetting in individuals with DS, but that sleep-dependent memory impairments are only evident under conditions tapping the sleep and memory functions of the hippocampus (as measured by our mutual research groups in Spanò et al., 2018 and Ashworth et al., 2015).

Does sleep modify neural and cognitive representations?

Karmiloff-Smith recently suggested that sleep may play an active role in representational change (Karmiloff-Smith, 2015). In her theory of Representational Redescription, she described four levels of cognitive representation, each allowing for greater integration, explicitation conscious and verbal access to information. At its first level, knowledge is procedural and implicit and then undergoes an offline process, which involves the integration of information and loss of specific sensory-motor details. Expanding beyond these levels, information then becomes consciously accessible and verbalizable. While most studies have not examined sleep's role in affecting different representational access, those that have examined implicit and explicit memory have suggested that adult's and older children's sleep may serve as an offline function that integrates knowledge in a way that it becomes accessible explicitly, influencing behaviour (Wilhelm et al., 2013). Studies of shifts in neural representations across sleep suggest shifts from hippocampal to cortical representation in both children and adults, although with different regional distributions depending on task demands, and more rapid integration in children.

However, few studies have directly addressed this question, and much more work is required to understand the nature of sleep-related changes in neural representations, particularly in infants and young children. In much of this work researchers have not measured "conscious" or "verbal" access, which could be one direction for future research.

Additional theoretical implications

In the final years of her career, Annette Karmiloff-Smith clearly communicated the importance of sleep for development. Theories of developmental change should follow-on from her insight to include these processes as one potential mechanism of representational development. Additionally, given the relation between sleep's functions and the neural mechanisms of hippocampus (i.e., representational replay through sharp-wave ripples), it is clear that developmental theory must also incorporate a more complete understanding of the role of the developing hippocampus into our explanations for cognitive development. The hippocampus has been relatively unexplored as a significant player in most developmental theory, primarily because some previous studies suggested its functions were early to emerge and did not show a developmental trajectory (as reviewed in Keresztes et al., 2018). However, there is mounting evidence in opposition to this view, and future work should incorporate the increased multi-modal representational capacity of this structure across development as another level of explanation for developmental process. Indeed, non-linear developmental trajectories of memory formation are evident and are related to the emergence of the adult-like functional capacity of this structure (Zeamer, Heuer, & Bachevalier, 2010; Edgin, Spanò, Kawa, & Nadel, 2014).

In summary, the work highlighted here helps to support a better understanding of how the typical and atypical development of sleep and its associated memory functions may be incorporated into future research. The results presented herein suggest that sleep may be as important a function as attention, and even as important as the functions of play emphasized by Piaget. The "offline" behavior of sleep relates to the persistence of memories in infancy, and sleep deficits may constrain trajectories in developmental disorders in ways that we are just beginning to understand. While the mechanisms underlying these changes require further experimental data, these findings bring to light the intriguing possibility that sleep may play a key role in representational change and cognitive development as noted by our friend and mentor Annette Karmiloff-Smith. This possibility warrants future conversation and research.

Acknowledgement

The work on this chapter was funded by grants from the Fondation Jérôme Lejeune and the National Institutes of Health (R01HD088409 to JOE).

References

Annaz, D., Hill, C. M., Ashworth, A., Holley, S., & Karmiloff-Smith, A. (2011). Characterisation of sleep problems in children with Williams syndrome. *Research in Developmental Disabilities, 32*(1), 164–169.

Ashworth, A., Hill, C. M., Karmiloff-Smith, A., & Dimitriou, D. (2013). Cross syndrome comparison of sleep problems in children with DS and Williams syndrome. *Research in Developmental Disabilities, 34*(5), 1572–1580.

Ashworth, A., Hill, C. M., Karmiloff-Smith, A., & Dimitriou, D. (2014). Sleep enhances memory consolidation in children. *Journal of Sleep Research, 23*(3), 304–310.

Ashworth, A., Hill, C. M., Karmiloff-Smith, A., & Dimitriou, D. (2015). The importance of sleep: attentional problems in school-aged children with DS and Williams syndrome. *Behavioral Sleep Medicine, 13*(6), 455–471.

Aton, S. J., Seibt, J., Dumoulin, M., Jha, S. K., Steinmetz, N., Coleman, T.,... & Frank, M. G. (2009). Mechanisms of sleep-dependent consolidation of cortical plasticity. *Neuron, 61*(3), 454–466.

Buzsáki, G. (1998). Memory consolidation during sleep: A neurophysiological perspective. *Journal of Sleep Research, 7*(S1), 17–23.

Curcio, G., Ferrara, M., & De Gennaro, L. (2006). Sleep loss, learning capacity and academic performance. *Sleep Medicine Reviews, 10*(5), 323–337.

Diekelmann, S., & Born, J. (2010). The memory function of sleep. *Nature Reviews Neuroscience, 11*(2), 114.

Dumay, N., & Gaskell, M. G. (2007). Sleep-associated changes in the mental representation of spoken words. *Psychological Science, 18*(1), 35–39.

Edgin, J. O., Spanò, G., Kawa, K., & Nadel, L. (2014). Remembering things without context: development matters. *Child Development, 85*(4), 1491–1502.

Fallone, G., Acebo, C., Arnedt, J. T., Seifer, R., & Carskadon, M. A. (2001). Effects of acute sleep restriction on behavior, sustained attention, and response inhibition in children. *Perceptual and Motor Skills, 93*(1), 213–229.

Friedrich, M., Wilhelm, I., Born, J., & Friederici, A. D. (2015). Generalization of word meanings during infant sleep. *Nature Communications, 6*, 6004.

Gais, S., Albouy, G., Boly, M., Dang-Vu, T. T., Darsaud, A., Desseilles, M.,... & Maquet, P. (2007). Sleep transforms the cerebral trace of declarative memories. *Proceedings of the National Academy of Sciences, 104*(47), 18778–18783.

Gais, S., Lucas, B., & Born, J. (2006). Sleep after learning aids memory recall. *Learning & Memory, 13*(3), 259–262.

Galland B. C., Taylor B. J., Elder D. E., Herbison P. (2012) Normal sleep patterns in infants and children: a systematic review of observational studies. *Sleep Medicine Reviews, 16*(3), 213–222.

Ganguly-Fitzgerald, I., Donlea, J., & Shaw, P. J. (2006). Waking experience affects sleep need in Drosophila. *Science, 313*(5794), 1775–1781.

Giganti, F., Arzilli, C., Conte, F., Toselli, M., Viggiano, M. P., & Ficca, G. (2014). The effect of a daytime nap on priming and recognition tasks in preschool children. *Sleep, 37*(6), 1087–1093.

Gilestro, G. F., Tononi, G., & Cirelli, C. (2009). Widespread changes in synaptic markers as a function of sleep and wakefulness in Drosophila. *Science, 324*(5923), 109–112.

Girardeau, G., Benchenane, K., Wiener, S. I., Buzsáki, G., & Zugaro, M. B. (2009). Selective suppression of hippocampal ripples impairs spatial memory. *Nature Neuroscience, 12*(10), 1222.

Gómez, R. L., Bootzin, R. R., & Nadel, L. (2006). Naps promote abstraction in language-learning infants. *Psychological Science, 17*(8), 670–674.

Gómez, R. L., & Edgin, J. O. (2015). Sleep as a window into early neural development: Shifts in sleep-dependent learning effects across early childhood. *Child Development Perspectives, 9*(3), 183–189.

Gómez, R. L., Newman-Smith, K. C., Breslin, J. H., & Bootzin, R. R. (2011). Learning, memory, and sleep in children. *Sleep Medicine Clinics, 6*(1), 45–57.

Henderson, L., Powell, A., Gareth Gaskell, M., & Norbury, C. (2014). Learning and consolidation of new spoken words in autism spectrum disorder. *Developmental Science, 17*(6), 858–871.

Henderson, L., Weighall, A., & Gaskell, G. (2013). Learning new vocabulary during childhood: Effects of semantic training on lexical consolidation and integration. *Journal of Experimental Child Psychology, 116*(3), 572–592.

Hill, C. M., Hogan, A. M., & Karmiloff-Smith, A. (2007). To sleep, perchance to enrich learning?. *Archives of Disease in Childhood, 92*(7), 637–643.

Hoffman, K. L., & McNaughton, B. L. (2002). Sleep on it: cortical reorganization after-the-fact. *Trends in Neurosciences, 25*(1), 1–2.

Holth, J. K., Fritschi, S. K., Wang, C., Pedersen, N. P., Cirrito, J. R., Mahan, T. E.,... & Lucey, B. P. (2019). The sleep-wake cycle regulates brain interstitial fluid tau in mice and CSF tau in humans. *Science, 363*(6429), 880–884.

Horváth, K., Hannon, B., Ujma, P. P., Gombos, F., & Plunkett, K. (2018). Memory in 3-month-old infants benefits from a short nap. *Developmental Science, 21*(3), e12587.

Horváth, K., & Plunkett, K. (2018). Spotlight on daytime napping during early childhood. *Nature and Science of Sleep, 10,* 97.

Hupbach, A., Gómez, R. L., Bootzin, R. R., & Nadel, L. (2009). Nap-dependent learning in infants. *Developmental Science, 12*(6), 1007–1012.

James, E., Gaskell, M. G., Weighall, A., & Henderson, L. (2017). Consolidation of vocabulary during sleep: The rich get richer?. *Neuroscience & Biobehavioral Reviews, 77,* 1–13.

Karmiloff-Smith A. (1992) *Beyond Modularity: A Developmental Perspective on Cognitive Science.* Cambridge, Mass: MIT Press/Bradford Books.

Karmiloff-Smith, A. (2015). An alternative to domain-general or domain-specific frameworks for theorizing about human evolution and ontogenesis. *AIMS Neuroscience, 2*(2), 91.

Keresztes, A., Ngo, C. T., Lindenberger, U., Werkle-Bergner, M., & Newcombe, N. S. (2018). Hippocampal Maturation Drives Memory from Generalization to Specificity. *Trends in Cognitive Sciences, 22*(8), 676–686.

Konrad, C., Seehagen, S., Schneider, S., & Herbert, J. S. (2016). Naps promote flexible memory retrieval in 12-month-old infants. *Developmental Psychobiology, 58*(7), 866–874.

Kurdziel, L., Duclos, K., & Spencer, R. M. (2013). Sleep spindles in midday naps enhance learning in preschool children. *Proceedings of the National Academy of Sciences, 110*(43), 17267–17272.

Kurth, S., Ringli, M., Geiger, A., LeBourgeois, M., Jenni, O. G., & Huber, R. (2010). Mapping of cortical activity in the first two decades of life: A high-density sleep electroencephalogram study. *Journal of Neuroscience, 30*(40), 13211–13219.

Kurz, E. M., Conzelmann, A., Barth, G. M., Hepp, L., Schenk, D., Renner, T. J.,... & Zinke, K. (2019). Signs of enhanced formation of gist memory in children with autism spectrum disorder–a study of memory functions of sleep. *Journal of Child Psychology and Psychiatry, 60*(8), 907–916.

Lavenex, P., & Lavenex, P. B. (2013). Building hippocampal circuits to learn and remember: insights into the development of human memory. *Behavioural Brain Research, 254*, 8–21.

Luongo, A., Lukowski, A., Protho, T., Van-Vorce, H., Pisani, L., & Edgin., J. O. (in press). Sleep's role in memory consolidation: What can we learn from atypical development? *Advances in Child Development and Behavior, 60*.

Marcus, C. L., Keens, T. G., Bautista, D. B., von Pechmann, W. S., & Ward, S. L. D. (1991). Obstructive sleep apnea in children with Down syndrome. *Pediatrics, 88*(1), 132–139.

McClelland, J. L., McNaughton, B. L., & O'reilly, R. C. (1995). Why there are complementary learning systems in the hippocampus and neocortex: insights from the successes and failures of connectionist models of learning and memory. *Psychological Review, 102*(3), 419.

Mednick, S., Nakayama, K., & Stickgold, R. (2003). Sleep-dependent learning: a nap is as good as a night. *Nature Neuroscience, 6*(7), 697.

Miano, S., Bruni, O., Elia, M., Trovato, A., Smerieri, A., Verrillo, E.,… & Ferri, R. (2007). Sleep in children with autistic spectrum disorder: a questionnaire and polysomnographic study. *Sleep Medicine, 9*(1), 64–70.

Mindell, J. A., & Moore, M. (2018). Does Sleep Matter? Impact on Development and Functioning in Infants. *Pediatrics, 142*(6), e20182589.

Mölle, M., Marshall, L., Gais, S., & Born, J. (2004). Learning increases human electroencephalographic coherence during subsequent slow sleep oscillations. *Proceedings of the National Academy of Sciences, 101*(38), 13963–13968.

Nadel, L. (2003). Down's syndrome: A genetic disorder in biobehavioral perspective. *Genes, Brain and Behavior, 2*(3), 156–166.

Noack, H., Schick, W., Mallot, H., & Born, J. (2017). Sleep enhances knowledge of routes and regions in spatial environments. *Learning & Memory, 24*(3), 140–144.

Ohayon, M. M., Carskadon, M. A., Guilleminault, C., & Vitiello, M. V. (2004). Metaanalysis of quantitative sleep parameters from childhood to old age in healthy individuals: Developing normative sleep values across the human lifespan. *Sleep, 27*(7), 1255–1273.

Orban, P., Rauchs, G., Balteau, E., Degueldre, C., Luxen, A., Maquet, P., & Peigneux, P. (2006). Sleep after spatial learning promotes covert reorganization of brain activity. *Proceedings of the National Academy of Sciences, 103*(18), 7124–7129.

Pisch, M., Wiesemann, F., & Karmiloff-Smith, A. (2019). Infant wake after sleep onset serves as a marker for different trajectories in cognitive development. *Journal of Child Psychology and Psychiatry, 60*(2), 189–198.

Poe, G. R. (2017). Sleep is for forgetting. *Journal of Neuroscience, 37*(3), 464–473.

Prehn-Kristensen, A., Göder, R., Fischer, J., Wilhelm, I., Seeck-Hirschner, M., Aldenhoff, J., & Baving, L. (2011). Reduced sleep-associated consolidation of declarative memory in attention-deficit/hyperactivity disorder. *Sleep Medicine, 12*(7), 672–679.

Rosanova, M., & Ulrich, D. (2005). Pattern-specific associative long-term potentiation induced by a sleep spindle-related spike train. *Journal of Neuroscience, 25*(41), 9398–9405.

Rothschild, G., Eban, E., & Frank, L. M. (2017). A cortical–hippocampal–cortical loop of information processing during memory consolidation. *Nature Neuroscience, 20*(2), 251.

Sandoval, M., Leclerc, J. A., & Gómez, R. L. (2017). Words to sleep on: Naps facilitate verb generalization in habitually and nonhabitually napping preschoolers. *Child Development, 88*(5), 1615–1628.

Scholle S., Beyer U., Bernhard M., Eichholz S., Erler T., Graness P., et al. (2011) Normative values of polysomnographic parameters in childhood and adolescence: quantitative sleep parameters. *Sleep Medicine, 12*(6):542–549.

Simon, K. N., Werchan, D., Goldstein, M. R., Sweeney, L., Bootzin, R. R., Nadel, L., & Gómez, R. L. (2017). Sleep confers a benefit for retention of statistical language learning in 6.5 month old infants. *Brain and Language*, *167*, 3–12.

Spanò, G., Gómez, R. L., Demara, B. I., Alt, M., Cowen, S. L., & Edgin, J. O. (2018). REM sleep in naps differentially relates to memory consolidation in typical preschoolers and children with Down syndrome. *Proceedings of the National Academy of Sciences*, *115*(46), 11844–11849.

Takashima, A., Nieuwenhuis, I. L., Jensen, O., Talamini, L. M., Rijpkema, M., & Fernández, G. (2009). Shift from hippocampal to neocortical centered retrieval network with consolidation. *Journal of Neuroscience*, *29*(32), 10087–10093.

Takashima, A., Petersson, K. M., Rutters, F., Tendolkar, I., Jensen, O., Zwarts, M. J., ... & Fernandez, G. (2006). Declarative memory consolidation in humans: a prospective functional magnetic resonance imaging study. *Proceedings of the National Academy of Sciences*, *103*(3), 756–761.

Tamminen, J., Payne, J. D., Stickgold, R., Wamsley, E. J., & Gaskell, M. G. (2010). Sleep spindle activity is associated with the integration of new memories and existing knowledge. *Journal of Neuroscience*, *30*(43), 14356–14360.

Tononi, G., & Cirelli, C. (2014). Sleep and the price of plasticity: From synaptic and cellular homeostasis to memory consolidation and integration. *Neuron*, *81*(1), 12–34.

Urbain, C., De Tiège, X., De Beeck, M. O., Bourguignon, M., Wens, V., Verheulpen, D., ... & Peigneux, P. (2016). Sleep in children triggers rapid reorganization of memory-related brain processes. *NeuroImage*, *134*, 213–222.

Werchan, D. M., & Gómez, R. L. (2014). Wakefulness (not sleep) promotes generalization of word learning in 2.5-year-old children. *Child Development*, *85*(2), 429–436.

Werk, C. M., Harbour, V. L., & Chapman, C. A. (2005). Induction of long-term potentiation leads to increased reliability of evoked neocortical spindles in vivo. *Neuroscience*, *131*(4), 793–800.

Wilhelm, I., Rose, M., Imhof, K. I., Rasch, B., Büchel, C., & Born, J. (2013). The sleeping child outplays the adult's capacity to convert implicit into explicit knowledge. *Nature Neuroscience*, *16*(4), 391.

Yonelinas, A. P., Ranganath, C., Ekstrom, A. D., & Wiltgen, B. J. (2019). A contextual binding theory of episodic memory: Systems consolidation reconsidered. *Nature Reviews Neuroscience*, *1*.

Zeamer, A., Heuer, E., & Bachevalier, J. (2010). Developmental trajectory of object re-cognition memory in infant rhesus macaques with and without neonatal hippocampal lesions. *Journal of Neuroscience*, *30*(27), 9157–9165.

Zimmerman, M. E., & Aloia, M. S. (2012). Sleep-disordered breathing and cognition in older adults. *Current Neurology and Neuroscience Reports*, *12*(5), 537–546.

18

THE DEBATE ON SCREEN TIME: AN EMPIRICAL CASE STUDY IN INFANT-DIRECTED VIDEO

Tim J. Smith, Parag K. Mital, and Tessa M. Dekker

Annette Karmiloff-Smith cared deeply about engaging with the debate around screen time. As expressed in the title of her book chapter *'TV is bad for children' – less emotion, more science please!'*, she was outraged by the blinkered and emotive proclamations made in the press and by some academics, on the apparent damaging effect of screen time on child development (Christakis, & Zimmerman, 2006; Sigman, 2012, Greenfield, 2015), and by the similarly unfounded claims from infant-directed media producers about the beneficial effects of their products on infant development (Karmiloff-Smith, 2012). With screen exposure occurring at increasingly younger ages (Anderson, & Hanson, 2010; Courage, & Howe, 2010; Cristia, & Seidl, 2015; Bedford, Saez de Urabain, Cheung, Karmiloff-Smith, & Smith, 2016), parents are further confused by highly contradictory re-commendations on what constitutes "healthy" screen time for infants and toddlers from national and international authorities (e.g., American Academy of Pediatrics, Radesky et al., 2016; WHO, 2019; The Royal College of Pediatricians and Child Health, Viner, Davie, & Firth, 2019).

Annette was a great pragmatist who understood the importance of working with parents, media producers, and policy-makers, to find the best ways for infants to benefit from scientific insights into development. Recalling that some regions of human cortex are initially very immature (Hill et al., 2010) and that infants are born into a world filled with screen-based media (Anderson, & Hanson, 2010; Courage, & Howe, 2010; Wartella, Richert, & Robb, 2010), she argued it is crucial to ensure that infant screen experiences are as developmentally and edu-cationally appropriate and as scientifically informed as possible. She believed this required empirically investigating the bidirectional relationship between how the developing visual system responds to screen media and how designers intuitively tailor their media to developmental constraints. Her argument was that the discussion of whether screen time is good or bad for infants is unhelpful and simplistic; rather than debating whether screen exposure itself is intrinsically 'good' or 'bad' we should investigate empirically what differentiates positive from negative screen experiences (see Courage, 2017 for a recent review). Annette therefore decided to collaborate with Abbey Media to develop infant-directed DVD content designed specifically with infant cognitive research in mind. Echoing her approach in other areas of research, Annette argued that understanding the effects of video on the developing infant requires careful investigation of how video content affects infant behaviour. For decades, infancy researchers have been using stimuli shown on screens, to carefully measure infant knowledge and learning at different post-natal ages. She believed that infant-directed videos and apps would do well to draw upon this extensive body of knowledge about the infant visual system and its changing limitations and demands. While never a central focus of her research, Annette's work on screen time spanned many years and encompassed various approaches including public education (e.g., public lectures, TV including *Baby It's You!*, her work with Proctor & Gamble and Nursery World), collaboration with industry to develop scientifically inspired infant-directed video (with Abbey Home Media; more details below), understanding how infants watch screen media, and latterly studying the long-term associations between screen media use and infant development (see Discussion). This symbiotic approach to science, policy work and public engagement illustrates Annette's dedication to "impactful" research before the buzzword even appeared on the radar of most UK academics.

When and what can infants learn from video?

Infancy studies have used video to investigate, among others, which visual features (fine detail, motion, colour, and depth) infants of different ages can detect, how infants learn to remember and retrieve hidden objects, recognise impossible or unexpected events, learn language and imitate actions, with varying degrees of success depending on age and manipulations (Bremner, & Wachs, 2010; Braddick, & Atkinson, 2011). For example, after showing 11-month-old infants several hours of video that incorporated stimuli classically used to measure infant's cognitive control and flexibility, one study showed significant improvements in the ability to sustain and flexibly deploy attention 3 days later (Wass, Porayska-Pomsta, & Johnson, 2011). Unsurprisingly, several studies have reported that infants learn better from real life than identical content presented on a screen (DeLoache, 1991). However, this screen-learning deficit may be smaller in young infants (<6 months) who may possess fewer cognitive capacities needed to distinguish between real-life and screen presentations (Barr, Muentener, & Garcia, 2007), and in older infants, learning from screens improves with repetition, spacing of learning trials, applying standard techniques of video production and, at least by 2 years of age, experience with television as a medium in general (Troseth, 2003; Anderson, & Pempek, 2005; Barr et al., 2007).

This large body of scientific research clearly shows that, while television content cannot and should not replace the richness of real-life, even young infants can learn from this medium. Moreover, the dynamic content of video and the interactive features of apps offer clear benefits over static books when similar levels of parental engagement and interaction are involved. Annette therefore frequently suggested that video and apps might be used like a dynamic book (although this view did get her in hot water when misquoted and then retracted by the Sunday Times; June 14th, 2015).

To set an example for how other developmental scientists may directly shape the future direction of infant-directed video Annette extracted 7 key recommendations from research on infant development, which guided her collaboration with Abbey Home Media on Baby Bright (these recommendations are collated from Annette's various unpublished and published texts including Karmiloff-Smith, 2012):

1. **Allow clearer differentiation between those aspects that are required for the processing message (the 'signal') and those that are not (the 'noise')**. Infant brains are noisier, less sensitive and more naive information processing systems. This need for simplification of the sensory environment is evident in the exaggeration and accentuation of speech by caregivers (i.e., Motherese; Ferguson, 1964) and the same simplification is required for visual communication. Most adult screen media involves social interactions and action sequences that are incomprehensible to the infant brain as they assume knowledge and perceptual skills yet to be developed (Valkenburg, & Vroone, 2004).

2. **Foreground shapes on screen need to "stand-out" from their background**. Our adult visual system has sufficient top-down knowledge to separate the two grounds easily even when borders are hard to decode, but the infant system finds this more difficult (Baker, Tse, Gerhardstein, & Adler, 2008). Infant gaze is more driven by visual saliency than adult gaze with a rapidly increasing bias towards semantically relevant features such as faces over the first year of life (Frank et al., 2009; Saez De Urabain, 2015; Franchak, Heeger, Hasson, & Adolph, 2016), increasing throughout childhood (Rider et al., 2018). Infants and young children do not look differently at video clips that have their semantics violated via shuffling compared to unshuffled versions (Pempek et al., 2010; Kirkorian, & Anderson, 2018) suggesting that designers cannot rely on semantics to guide infant attention and must instead use visual saliency.

3. **Avoid passive central fixation**. In adult TV, salient characters and action in most programmes feature at the screen centre, and movement of these characters is conveyed by movement of the background as a camera tracks their movement (Smith, 2013). Adult viewers re-interpret this as movement of the foregrounded central character, despite the fact that the viewers' eyes are fixated on the centre keeping eye movements to a minimum. By contrast, real movement rather than inferred movement is likely to attract the infant visual

system; yet, to focus on the central character of most adult programmes, infants hardly need to move their eyes. This makes them very passive observers.

4. **Sound can be cognitively distracting**. What would seem like mere auditory decoration in the background of screen exposure may add significant cognitive load to the young children, particularly when it does not meaningfully connect visual and auditory content. A nice example of this is in the number domain. Background music while showing pairs or trios of objects to demonstrate the numbers '2' or '3', may be more disruptive than playing two drum beats or three drum beats, which match the number of visual objects on the screen, the latter procedure perhaps enhancing learning by providing multi-modality audio-visual representations of the same content.

5. **Use frequent repetitions**. Adult programming involves relatively few repetitions of event sequences; adults can learn from a single presentation, whereas infants' cognitive systems need repetition (Fiser, & Aslin, 2002; Pelucchi, Hay, & Saffran, 2009).

6. **Use exaggerated visual action**. A large proportion of new information in adult TV is conveyed through dialogue or nuanced facial expressions rather than clear visual action, which both require perceptual abilities that develop slowly over the first few years of life.

7. **Slow Down!** When new information is provided visually it is generally at a pace faster than can be processed by the sluggish attention/perceptual systems of infants, leading to missed comprehension or fatigue. The average time between overt attention shifts in infants whilst watching videos is almost twice as long (M = 644 ms, SD = 142) as in adults (M = 327 ms, SD = 45; Saez de Urabain, Nuthmann, Johnson, & Smith, 2017).

Annette was hopeful that the incorporation of these recommendations in infant-directed TV (I-DTV) might result in a screen experience more suitable to the developing mind. However, she was keenly aware that the translation of these recommendations into a commercial product comes with challenges, because other factors beyond the experience of the infant, such as attractiveness to the parent who buys and also watches the DVD, may also play a role in the design process. She also noted that failures of adult-directed TV may also be addressed intuitively by I-DTV creators with no scientific consultant on board. Indeed, I-DTV content analyses by Wass and Smith (2015) reveal that much I-DTV optimizes the audiovisual stimulus to simplify their infant viewers' task of deciding what to attend to (i.e., the signal) over the irrelevant background features (e.g., noise). This is done, for example, by aligning peaks in low-level visual features (e.g., luminance, colours, edges, flicker, and motion) with the location of a speaking face (the *signal*), increasing shot length, and reducing edits to give young children more time to locate the focal object within a frame. Interestingly, these formal differences between I-DTV and Adult TV are not entirely due to the I-DTV being animated, as they persisted when comparing animations directed at infants versus those directed at adults, suggesting that their creators understood the need to tailor the flow of audiovisual information to the age of their respective audiences. Being an empiricist, Annette therefore viewed the incorporation of these scientifically informed considerations as hypotheses about how infants may respond to I-DTV and, as such she endeavoured to test these hypotheses in an experimental study with author TD and later, TS.

The following section presents some detailed empirical work that illustrates the kinds of methods Annette believed were needed to ground an evidence-based approach to understanding the impact of screen time on infant development. For readers who do not want to descend into this level of detail, feel free to skip ahead to the end of this section for the take-home points.

In a novel empirical case study, Annette aimed to use eye tracking to test whether a clip derived from an infant-directed DVD designed according to earlier versions of the recommendations above (*Baby Bright*, Abbey Home Media) would hold infant attention, simplify the viewing process and guide infant gaze to the intended concepts, in this case the numbers "2" and "3" better than a matched control clip derived from a different infant-directed DVD covering the same concepts. Drafts of the study account included below were written by Annette and the present authors (TD, PKM and TS) from 2011 to 2015. For this festschrift we welcome the opportunity to help Annette's empirical contribution on the screen time debate finally reach its audience.

Empirical case study in I-DTV: Methods

Participants

Sixteen 6-month-olds (mean age = 27.4 weeks, SD = 2 weeks; 9 girls), 16 12-month-olds (mean age = 50 weeks, SD = 6 weeks; 6 girls), and 16 adults were tested. The age of the infants studied here (6 and 12-months) is considerably younger than the officially recommended age at which children should be first exposed to screens (AAP suggest 18 months; Radesky et al., 2016) but at which the majority of UK infants are actually receiving daily exposure (Bedford et al., 2016).

Stimuli

Two 130-second clips were extracted from 30-minute, commercially available DVDs, specifically targeted at infants. A scientifically informed clip (video-SI) was extracted from Baby Bright (Abbey Home Media), developed with the deliberate aim of incorporating earlier-listed recommendations. A control clip matched on topic (video-C) was taken from Baby Einstein (Disney). Both clips were on the numbers two and three, and used a combination of brightly coloured infant animation, simple photographed sequences, and naturalistic scenes. In each clip, number was represented as the quantity of objects presented on the screen e.g., three cups (Figure 18.1, shot 15) or three lambs (Figure 18.2, shot 2). Video-SI used movement across the screen and an accompanying narration to count objects.

Video-C showed pairs or triplets of objects, accompanied by a classical music soundtrack. The two clips also differed slightly on other stylistic dimensions such as the amount of naturalistic photographed sequences used (video-C > video-SI),

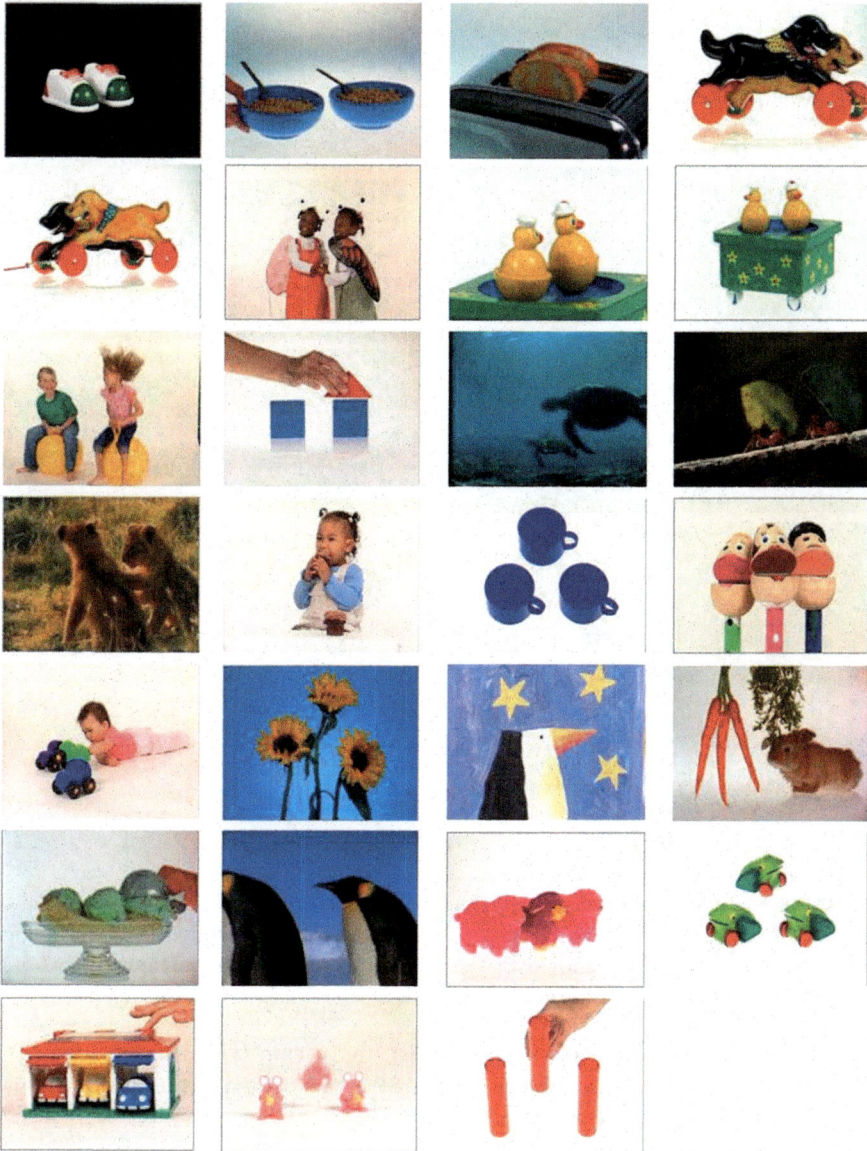

FIGURE 18.1   Individual shots presented in video C (control) taken from Baby Einstein (Disney). Shots were presented in order with an accompanying baby-friendly music. Shots are intended to represent quantities of two or three as represented by the number of objects on the screen.

FIGURE 18.2 Individual shots presented in video SI (science inspired) taken from Baby Bright (Abbey Home Media). Shots were presented in order with an accompanying narration. Individual lambs and cows (shots against green background) spun across the screen as the narrator counted them.

use of classical music (video C > video SI), the pacing (video C > video SI), repetition of visually similar objects (video-SI > video-C) and the use of multimodal counting (video-SI > video-C).

Procedure

The infants sat on their caregiver's lap in front of a 17-inch LCD screen at a set height and at 60-cm distance from the screen. The infants' looking behaviour was registered with a Tobii 1750 Infrared Eye Tracker. Before the experiment was run, an infant-friendly five-point calibration was run. Each clip was presented separated by a short break of 1–3 minutes. Order of clip presentation was counter-balanced across infants. Volume was kept constant. All parents gave informed consent before the study commenced and received detailed debriefing afterwards.

The procedure and setup was identical for the adults except for the fact that they sat on a seat positioned with their heads in a similar position to that of the infants. Adults were instructed to freely view the video clips.

Results

To test how actively engaged infants were with the videos, we first examined how each video held infants interest, by quantifying overall looking time to the screen. We then tested how successful the videos were in directing infants' attention to the intended number of objects, a prerequisite of subsequent comprehension. Finally, we examined how these differences may have been influence by compositional differences between the two videos by looking how well gaze of all infants was attracted successfully to key focal points in the scenes. We did this by comparing infant gaze distributions to that of an adult control group, whose gaze patterns we took as ground-truth for active visual processing of meaningful scene-content.

Interest in the videos

We can ascertain whether the infants are interested in the videos by using an eye-tracking proxy of looking time: whether the eye tracker can detect they are looking at the screen. To examine the loss of interest over time the two 130-second clips were divided into 13 10-second time blocks. Overall, infants looked significantly more at Video SI than at Video C, $(F(1,30) = 7.314, p = 0.011$; see Figure 18.3). This pattern held at both 6 and 12 months of age (not shown). Three interesting patterns emerged: (1) infants *started* by looking for an equally long time at both Video SI and Video C (during first quarter of clips: $F(1,30) = 0.085, p = 0.773$); (2) looking decreased more over the course of Video C than over the course of Video SI and (3) by the final quarter infants looked significantly longer at Video SI than Video C $(F(1,30) = 12.773, p = 0.001)$.

Looking at number

To explore what drives differences in infant's looking at the two videos we next explore how successfully infant's gaze was directed to the intended number of objects in each scene.

Objects in the videos were coded in each frame using dynamic Regions of Interest (dROI). An object counter was incremented every time a new object was looked at within a scene. At the end of each number scene, the total count was compared to the intended number of objects. The intended number was clear to

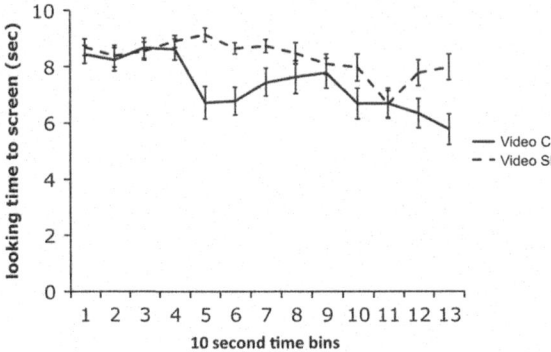

FIGURE 18.3 **Time spent** looking **at the** screen **over time** (split into 10 second bins) for each video (Video SI vs. Video C). Error bars + /-1 SE. Decreasing looking time can be interpreted as an increase in blinking/looking away from the screen.

adult observers because the first half of each clip was devoted to two-object scenes and the second half to three-object scenes. However, a significant number of shots actually contained four or more objects, potentially distracting from intended content (see Figure 18.1). For example, in the scene of the toddler eating two cupcakes, a match would be recorded if the infant had fixated both of the cupcakes – the intended two objects for the scene – but not if they had fixated one cupcake and the toddler.

Figure 18.4 displays the average probability that the intended number of items was attended within each clip for the 2 age groups. Video SI resulted in a greater probability of fixating the correct number than Video C, $F(1,29) = 21,465$, $p < .001$, confirmed at all ages and number with bonferroni-corrected t-tests. There was also a main effect of the number of objects ($F(2,29) = 76.87$, $p < .001$) with the intended number of objects more likely to be attended in the 2-object than 3-object scenes, although this effect was mostly driven by Video C. The greater accuracy for 2-object scenes observed overall may be due to the fact that the two potential targets for attention can be easily identified and shifted back-and-forth between, without competition from other salient items.

Gaze similarity

The analyses above suggest that Video SI held infant attention more, and was more successful in directing gaze to the intended number of objects. These results begin to suggest that Annette's developmentally informed design considerations may have been successful. Next, we explored how two composition elements, namely visual salience from rapid luminance changes (*flicker*), and the presence of a face in a scene, helped infants actively engage with key plot points of the videos. We tested this by measuring the effects of these two factors on infant and adult *gaze similarity*, an index of how well gaze positions correspond across viewers and

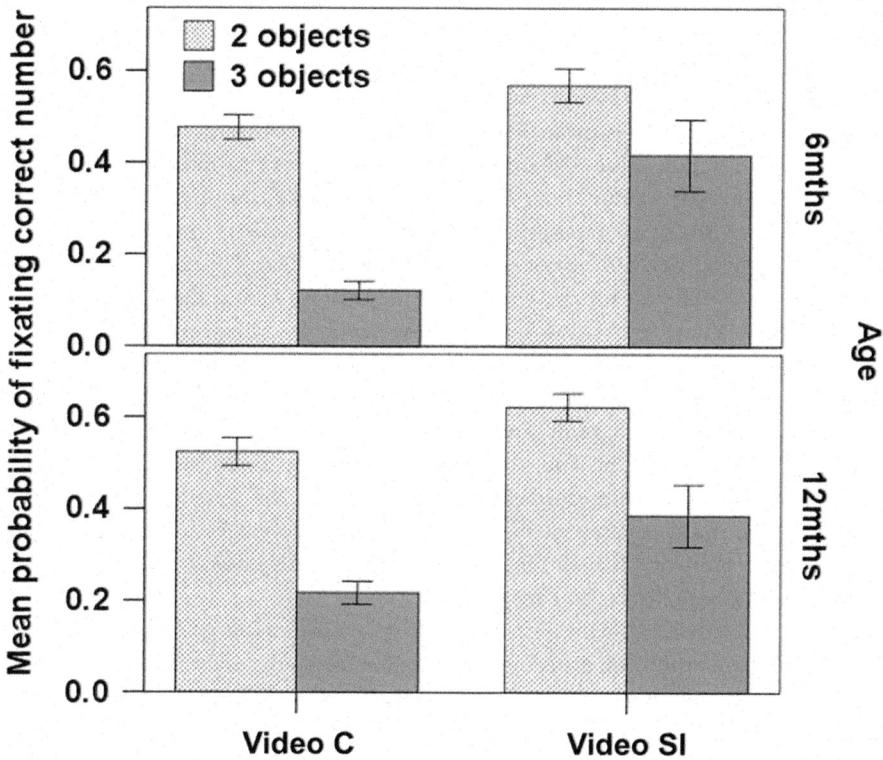

FIGURE 18.4 Probability of fixating the intended number of objects during a scene as a function of age and video. Error bars + /-1 SE.

scenes. We used this measure to quantify how much infant gaze patterns deviated from adult gaze patterns across each video, for scenes with faces and varying flicker. This approach follows the reasoning that while we cannot be sure which gaze pattern for infants reflects active processing of scene content, the extent to which they look at the same focal points as adults provides a reasonable bench-mark. Gaze of younger infants is more influenced by visual salience than older children and adults (Frank et al., 2009; Franchak et al., 2016), which may both help and hinder infants in fixating key plot points.

Measuring gaze similarity

To compute gaze similarity, we first removed rapid eye movements (saccades) to identify gaze positions (fixations). We then used a modification of a common technique used to measure attentional synchrony, by expressing the observers' gaze on each frame as a probability distribution (for review see Le Meur, & Baccino, 2013). How our metric, *gaze similarity* is computed, is described in detail in Loschky, Larson, Magliano, and Smith (2015). In brief: for adults, we used a leave-one-out approach: fixations of all but one adult within a 225ms window were replaced with 2D circular fovea-sized Gaussian distributions (μ = xy$_{\text{fixation}}$, σ = 1.2° visual angle). For pixels covered by multiple Gaussians, the multiple intensity values were summed. The result is a probabilistic gaze map for each timeframe, with high values reflecting high likelihoods of fixation, and low values reflecting low likelihoods. These distributions were then normalized relative to the mean and SD of all intensities across the entire video, resulting in spatio-temporal map of gaze similarity z-scores across the video. This map was used to compute gaze similarity for the left-out participant, by taking the z-score corresponding to their gaze position in each time window. This leave-one-out procedure was repeated for all adults in the group, until each participant had gaze similarity z-scores for each frame. These z-scores reflect (1) how well the gaze position of an individual adult fits within the group for a given scene (below-average values = poorer fit) and (2) how the mean gaze similarity across all adults varies across the video: A z-score close to zero indicates the scene has average gaze similarity compared to the rest of the movie, negative values indicate less than average gaze similarity (i.e., more varied looking across observers), and positive values indicate more gaze similarity.

We next quantified infant gaze similarity relative to adult gaze similarity. For each infant gaze point, the probability that it belongs to the adult gaze distribution for the corresponding video time window is identified by sampling the value at that location from the adult gaze probability distribution (this time leave-one-out is not used as the gaze does not belong to the same distribution so cannot be sampled twice). The resulting raw probabilities are then normalized to the reference adult distribution. Of critical importance, if gaze distributions are identical across age, the average z-scored similarity for infants should overlap with the adult gaze similarity index, expressing more (positive z-score) or less (negative z-score) attentional synchrony at the same moments. However, a lower similarity score in infants does not necessarily mean lower synchrony of gaze – it only means that the infant distribution differs more from the adult distribution than itself. This could reflect weaker clustering of infant gaze around the same screen location (lower attentional synchrony), that infant gaze is tightly clustered but focused on a different screen location, or a combination of the two. Of interest are (1) whether infant gaze similarity is identical to that of adults, suggesting adult-like viewing behaviour and (2) whether infant gaze synchrony fluctuates in the same way as adult gaze even if the overall score deviates, suggesting that the same aspects of the video drive gaze patterns across age.

Quantifying flicker entropy

Flicker is the luminance change from one frame to another that correlates with perceptual features such as motion and optic flow (Mital, Smith, Hill, & Henderson, 2011). Mital et al. (2011) demonstrated that flicker was highly predictive of adult gaze when gaze was highly clustered (i.e., had large synchrony) and hypothesised that this was due to *low entropy* in this feature at those moments (e.g., all points of high flicker are clustered together rather than spread randomly throughout the image) rather than *high entropy*. In other words, higher gaze similarity is predicted for scenes high in flicker but low *Flicker Entropy*, when flicker is spatially clustered. To compute Flicker Entropy, two-dimensional luminance flicker maps (i.e., ignoring the colour channels) were computed for both videos, and entropy across these maps was calculated (see Mital *et al.* 2011, for details). High entropy values (measured in *bits*) indicate a fairly uniform distribution of flicker across the frame (like an old grainy TV signal), and low values indicate clear peaks/clusters of flicker, e.g., a small light switching on/off.

Identifying faces

Previous studies have shown that the presence of semantically rich information (i.e., a face) helps guide infants attention to important semantic content in scenes of varied complexity (Frank *et al.* 2012). To investigate if faces were helping infants direct their gaze to important plot elements in the video, we identified scenes with faces and investigated effects of flicker entropy separately for these scenes.

Comparing gaze similarity as function of scene composition across both videos

First, we provide a qualitative analysis of gaze similarity, flicker entropy fluctuations, and use of scenes with faces across the two videos, to understand how and why specific video content causes adult and infant looking patterns to deviate.

Both videos are designed for young infants and share some common visual features including the use of simple bold colours, objects photographed against block colour or white backgrounds and sequentially presented objects or actions. However, the videos differ considerably in the complexity of their images and the pace at which new images are presented. Video C has a higher mean flicker entropy (4.20 bits), smaller variance of flicker entropy (1.934 bits) and faster shot rate (mean shot length is 5.26 s) compared to Video SI (1.70 bits, 2.60 bits and 10.54 s, respectively). These differences in complexity can be seen in the screenshots included in Figures 18.5 and 18.6 and the associated plots of Flicker entropy over time.

Video C uses occasional nature videos shot against a normal background that create a lot of flicker, low contrast between the objects and the background and peaks in flicker entropy (see shots 4 and 7 in Figure 18.5). These shots do not pose a problem to the adults who are used to parsing such scenes and can easily locate the centres of interest, typically the heads of the animals depicted. For the infants, such low contrast / high entropy images may be difficult to parse (i.e., to locate foreground objects of interest). This results in low gaze similarity as infants attend to the images in different ways. By comparison, the shots composed against a white background, usually depicting a relatively static child or animal interacting

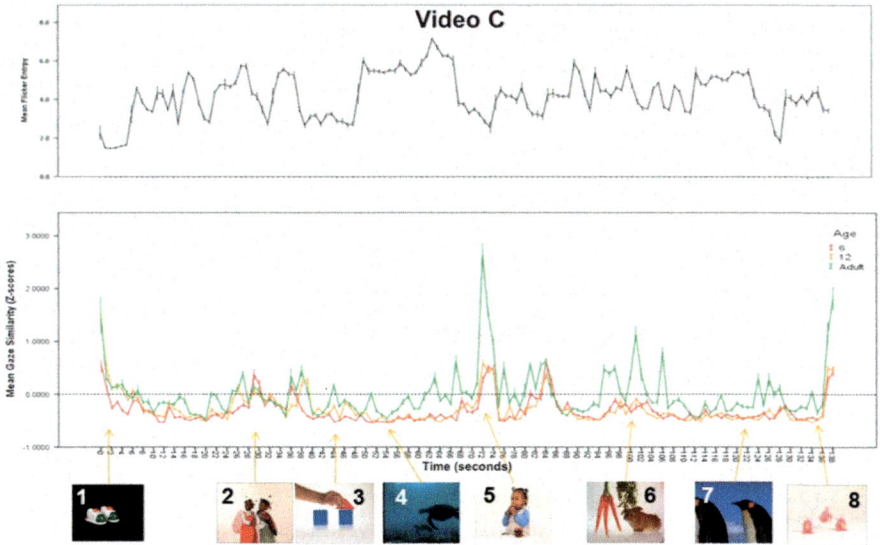

FIGURE 18.5 Top graph = Flicker entropy during every second of Video C (measured in bits). Bottom graph = Gaze Similarity (Z-scored relative to Adult gaze distribution) across the three age groups (6 mth = red line, 12 mth = orange line, Adult = green line) over time. Error bars represent + /-1 SE.

with objects (e.g., shot 5 = the child eating a muffin or shot 6 = the rabbit eating the carrot), therefore produce clear peaks in gaze similarity (bottom chart; Figure 18.5). The subtle movement of the child's hands or the face of the child and animal create points of low flicker entropy relative to the static background that attracts attention across all age groups. This interest in faces and hands has previously been confirmed using similar videos in infants (Frank *et al.* 2012). Critically, the motion is restricted to a small area of the screen allowing even the younger infants to identify it as a saccade target, move their eyes to it and fixate it before the shot has changed.

Video SI utilises some similar nature scenes to video C (shots 1,5 and 6 in Figure 18.6), and these produce similar peaks in Flicker Entropy and low gaze similarity. However, the majority of Video SI's shots involve either a single object or a series of objects presented against a plain background. Instead of using editing to change the centre of interest as in Video C, Video SI uses animations to present objects sequentially by having them either fly on to the screen or suddenly grow from nothing. These presentations match the audio narration: "Look, a baby sheep! A baby sheep is called a lamb [shot 1]. Lamb. Lamb. Lamb [shot 2]. Baa. Baa. Baa. [first repetition of shot 2] One. Two. Three. [second repetition of shot 2]". Each repetition of "Lamb", "Baa" or the number corresponds with the appearance of a lamb flying on to the screen (shot 2, Figure 18.6). These audiovisual correspondences are intended to attract the infant's eyes to the object being named and reinforce the naming and subsequent counting through repetition. The sequential

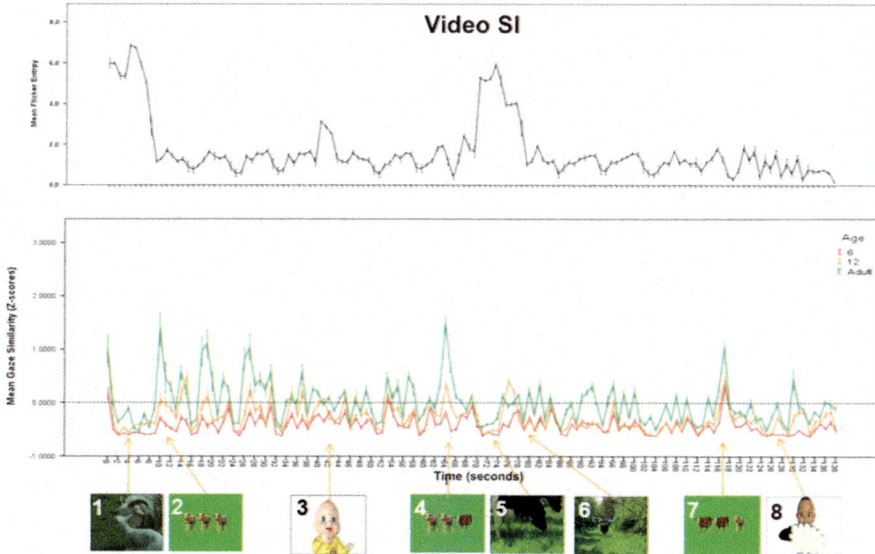

FIGURE 18.6 Top graph = Flicker entropy during every second of Video SI (measured in bits). Bottom graph = Gaze Similarity (Z-scored relative to Adult gaze distribution) across the three age groups (6 mth = red line, 12 mth = orange line, Adult = green line) over time. Error bars represent + /-1 SE.

presentation results in peaks and troughs of gaze clustering at all ages (e.g., Figure 18.6, 10 to 32 seconds) as the sudden appearance of an object on screen attracts all attention then viewers return to exploring the frame once the object has stopped. However, the reduced gaze similarity trace indicates that the 6-month and 12-month olds do not respond as quickly to the sudden appearance of the object or pursue it as accurately as the adults as it moves across the screen.

The descriptive age differences visible in gaze similarity in Figures 18.5 and 18.6 are confirmed by a repeated-measures ANOVA (Age: $F(1,45) = 43.158$, $p = .001$, $\eta_p^2 = .657$). There is no main effect of Video ($F(1,45) = 1.674$, $p = .202$, $\eta_p^2 = .036$) or interaction ($F(2,45) = 1.443$, $p = .247$, $\eta_p^2 = .06$), suggesting, on average both videos create the same variance in attentional synchrony. Collapsing across Videos, the main effect of Age can be attributed to significantly lower gaze similarity at 6 months (mean = -0.363, sd = .06) than for Adults (mean = .00, sd = .155; $p = .001$) and a marginal difference relative to 12 months (mean = -0.273, sd = .110). The 12-month-old gaze also showed significantly less similarity to the adults ($p < .001$).

Gaze similarity as function of flicker entropy and faces

To test for age differences is in how flicker entropy and the presence of the face in the scene affects gaze, video frames with and without face were binned by Flicker

Entropy (0 bits-low to 6 bits-high entropy) for subsequent analyses. The two videos were analysed separately as the simpler composition of Video SI meant that we could not identify shots containing a face and high flicker entropy like in Video C.

Video C

A repeated-measures ANOVA on gaze similarity within each age group (6mth, 12mth and Adults) with factors Flicker Entropy (0, 2, 4, 6 bit bins) and whether the shot contains at least one human face (face vs. no face) within Video C revealed a main effect of Flicker Entropy ($F(3,135) = 37.782$, $p = .001$, $\eta_p^2 = .456$), a main effect of Face ($F(1,45) = 41.645$, $p = .001$, $\eta_p^2 = .481$), an effect of Age ($F(2,45) = 9.362$, $p = .001$, $\eta_p^2 = .294$) and Face x Entropy interaction ($F(3,135)=6.245$, $p = .001$, $\eta_p^2 = .122$). The effect of Flicker Entropy can clearly be seen in Figure 18.7. As Flicker Entropy decreases (i.e., flicker becomes more concentrated) gaze similarity increases from $-.246$ (SE $= .024$) when entropy is large (6 bits) to $+ .287$ (0 bits; SE $= .081$; $p = .001$). This pattern is similar

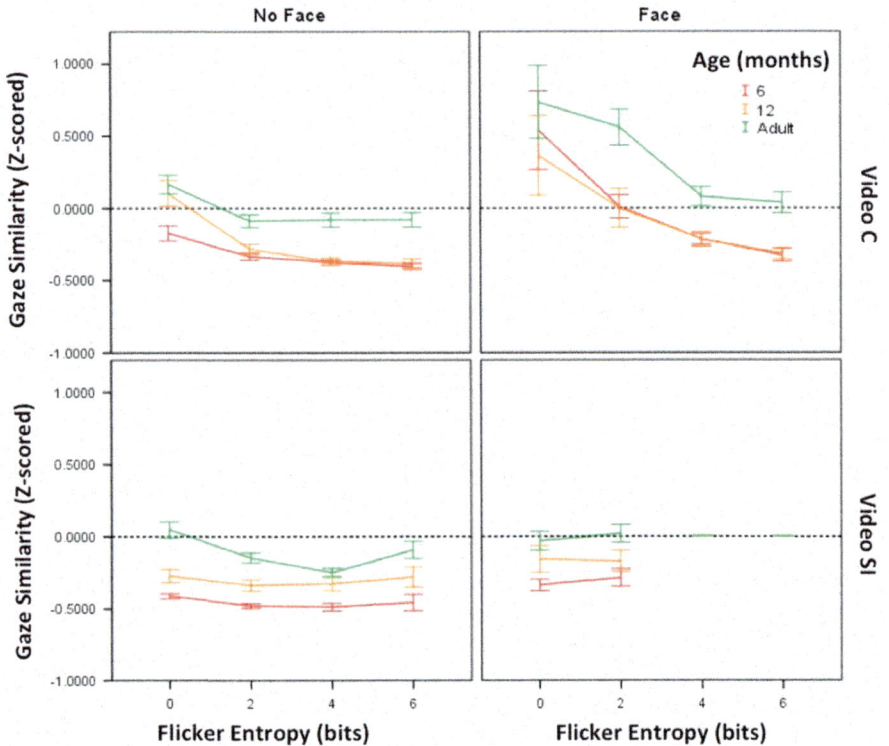

FIGURE 18.7 Gaze similarity (expressed as a Z-score relative to the within-video adult distribution) across the three age groups (6 mth=red line, 12 mth= orange line, Adult=green line) and mean flicker entropy for a frame (binned into 2 bit bins; x-axis) split into shots containing at least one human face ('Face') and shots not containing faces ('no face'). Error bars represent $+ /-2$ SE.

across all three age groups suggesting a universal influence of image features on gaze behaviour.

Splitting the data into shots with and without a human face, the main effects of Age (F(1,45) = 19.089, p = .001, η_p^2 = .459) and Flicker Entropy (F(1.44,64.975) = 58.096, p = .001, η_p^2 = .564; Greenhouse Geisser corrected) are very large within the No Face shots suggesting that both age and flicker entropy strongly influence gaze. There is also an interaction between Age and Flicker Entropy (F(2.88,64.975) = 4.262, p = .009, $\eta_p^2\eta_p^2$ = .159) due to the difference in gaze similarity between 12mths and Adults disappearing in the lowest Flicker Entropy bin (0 bits) and 12 months displaying greater gaze similarity than the 6 months (t(30)=−2.694, p = .011). This indicates that by 12 months, infants can display adult-like gaze behaviour if non-social video sequences are designed to guide their eyes to the centre of interest using low-level visual features (e.g., low flicker entropy; Wass & Smith, 2015).

When human faces are present in Video C, the main effects of Age (F(2,45) = 5.182, p = .009, $\eta_p^2\eta_p^2$ = .187) and Flicker Entropy (F(1.26, 56.73) = 18.914, p = .001, $\eta_p^2\eta_p^2$ = .296; Greenhouse Geisser corrected) remain but the interaction disappears (F < 1) as all ages show the same increase in gaze similarity for the lowest Flicker Entropy bin (see Figure 18.7, top right, 0 bits). In the presence of a face, the difference between 6- and 12-month Gaze Similarity disappears for all entropy bins and in the lowest bin (0 bits) even the difference between 6 months, 12 months and Adults disappears (all ts < 1). This suggests that when social scenes are composed with only one area of visual change at a time even the youngest infants are able to exhibit adult-like viewing behaviour.

Video SI

For confirmation of the influence of Flicker Entropy and Faces on gaze we can now turn to Video SI. Given the simpler composition of Video SI there were no shots containing faces with a high degree of Flicker Entropy so the analyses focused on only the lowest two entropy bins (0 and 2 bits). A repeated-measures ANOVA reveals main effects of Age (F(2,45) = 15.764, p = .001, $\eta_p^2\eta_p^2$ = .412), Faces (F(1,45) = 17.276, p = .001, $\eta_p^2\eta_p^2$ = .277), and Flicker Entropy (F(1,45) = 4.184, p = .047, $\eta_p^2\eta_p^2$ = .085). The effect of Face is similar to Video C in that shots containing faces produce greater gaze similarity than without faces.

The effect of Flicker Entropy is considerably weaker than in Video C probably due to the effect being measured over fewer levels of entropy. There is also an interaction between Face and Flicker Entropy (F(1,45) = 12.143, p = .001, $\eta_p^2\eta_p^2$ = .212) due to the Flicker Entropy effect being absent in the Face shots (F < 1) but present in the No Face shots (the other two levels of entropy can be added to this analysis; F(1.72,77.27) = 10.566, p = .001, $\eta_p^2\eta_p^2$ = .190). Within the No Face shots there is also an interaction between Flicker Entropy and Age (F(3.43,77.269)=3.021, p = .029, $\eta_p^2\eta_p^2$ = .118) but unlike in Video C this is not because the Age difference at 12 months old disappears at the lowest flicker bin. Instead, the interaction is caused by the Flicker Entropy effect, which is only present in Adults, F(1.95,29.23)=13.905, p = .001, $\eta_p^2\eta_p^2$ = .481) and not 6 mths or 12 mths (both Fs < =1). This may be because, unlike in Video C which has much more varied shot content there are only two types of non-face scene, all the high entropy shots are natural scenes which the infants are unable to parse (e.g., shots 1, 5 and 6, Figure 18.6) and all of the low entropy scenes depict a series of animal images flying into shot sequentially (e.g., shots 2, 4 and 7, Figure 18.6). The infant gaze is unable to exactly match the location of the adult gaze during these shots as their anticipation and pursuit of moving targets is sluggish. However, as was evident in the number ROI analysis (Figure 18.4), this sequential presentation helps the infants arrive at the correct number of objects even if they may get there at a later time than the adults (leading to low gaze similarity scores).

In sum, our gaze similarity analysis reveals that infants visually engage with infant-directed video in different ways than adults do, and highlights the importance of using salience in the scenes in a manner that helps infants process meaningful content. Specifically, in visually complex scenes, a large amount of image change (i.e., Flicker Entropy) caused by object or camera motion may prevent infant gaze from locating key plot elements in the scene (e.g., animals moving through naturalistic scenes). The inclusion of faces, which are highly salient to infants and adults alike, can help mitigate this. Differences in looking patterns can also occur in low-entropy scenes with clear points of saliency, when the more sluggish developing visual system is engaged in tracking dynamic content.

Discussion

The empirical case study presented here suggests that by carefully designing I-DTV to simplify the process of parsing the visual *signal* – whether it be a particular number of objects or a human face – from the background visual *noise*, the gaze behaviour of even 6-month-old infants can resemble that of adults. By comparing a commercially available I-DTV clip, Baby Einstein (Video C) to a clip depicting the same concept but using scientifically informed design, Baby Bright (Video SI), Annette aimed to demonstrate the importance of empirical tests in understanding how a scientifically informed video can lead to better engagement, more accurate attending to the number of objects and a simpler active viewing process . The results above show that whilst Baby Bright indeed engaged infant gaze more, and attracted more looks at the intended object numbers, results for active viewing measures were mixed across the two videos. On the one hand, the greater variation in shot content and compositional style used in Baby Einstein led to scenes with high flicker entropy that overwhelmed infant gaze, reducing their ability to identify key content. On the other hand, it created moments during which low flicker entropy and the presence of faces within complex scenes guided infant gaze to the same point as adult gaze. These findings confirm our earlier predictions derived from feature-analysis of I-DTV (Wass & Smith, 2015) and prior eye-tracking studies (Kirkorian & Anderson, 2018; Franchak *et al.* 2016, Frank *et al.* 2012). By comparison, the much simpler sequential presentations used in Baby Bright were successful in representing the concept of number, encouraging infants to actively seek out objects and maintain their interest in the screen. The infants looked at the correct number of objects and matched the gaze pattern of the adults (see the matching peaks and troughs of gaze similarity in Figure 18.6). However, the high speed of the moving objects caused the sluggish infant gaze to lag behind the objects and the adult gaze, giving rise to lower gaze synchrony overall. This highlights the importance of striking a balance between Annette's recommendation 3 (avoid passive fixation) and 7 (slow down!) where movement across the screen might be stimulating for the developing infant's visual system, but moving too fast may prevent infants from processing key plot elements.

The most notable dissociation between infant and adult viewing overall, was for nature shots of animals included in both videos. Anecdotally, during testing parents often commented on how much they liked the natural scenes and believed they would be educational for their infants. However, the reduced looking time overall and the lower gaze similarity results indicate that the high flicker entropy in these scenes meant that infants found them very hard to watch (Figures 18.5 and 18.6). This is a clear example of how adult designers must inhibit their own preferences when designing for a developmentally immature visual system[1].

These results confirm the importance of *content* and stimulus design for understanding how the developing mind responds to the stimulus and how it may encourage learning. These results also demonstrate the importance of appreciating *how* infants engage with screen media when trying to understand the long-term consequences of regular screen time on infant neurocognitive and behavioural development. In order to combat propaganda against screen time and non-evidence-based guidelines, detailed longitudinal studies of associations between different types of screen use and infant development are required. This work fits in with the emerging perspective in the developmental cognitive neuroscience of screen use, that the *content* and *context* of screen media use is important for understanding how we can maximise the benefits of screen time for supporting learning, education, and social connection in childhood and adolescence and avoiding negative impacts around displacement of other activities such as sleep and physical exercise (Courage, 2017).

Conclusions and legacy

Annette believed that that "the dynamics of screen exposure may be an important supplementary stimulation to the young infant's visuo-cognitive system, alongside images in books and mobiles rotating in central vision, provided that the content and design of videos are based on infant scientific research and encourage active screen exploration, rather than passive central fixation" (Annette Karmiloff-Smith unpublished manuscript, 23rd January, 2012). As such, she cautioned against using screen time as baby-sitter to replace social interaction, or displace other developmentally critical activities such as physical exploration, sleep, and feeding. Our findings endorse Annette Karmiloff-Smith's view that rather than the "mindless" passive viewing assumed by some critics of I-DTV (e.g., Greenfield, 2015), screen exposure can be designed to encourage saccadic exploration, rapid visuospatial orienting, anticipation, figure/ground separation, and attention to number of intended targets making the infant a highly active viewer.

Annette's pioneering perspective on how screens may best be used to support infant and child development was invaluable when she co-founded the Toddler Attentional Behaviours and Learning with Touchscreens (TABLET) Project with author TS and Dr. Rachael Bedford (King's College London). TABLET is ongoing project aiming to test whether infant exposure to touchscreen devices at 12 months-of-age is associated with long-term developmental differences at 18 and 42 months-of-age in domains including attention, temperament, developmental milestones, language, executive function and sleep. The intention is to triangulate effects using parent-report questionnaires in a large online sample (N > 700), and behavioural, questionnaire, EEG and eye-tracking measures in an intensive lab sample (N = 60). Annette's contribution to the TABLET project was critical for guiding its future direction (see her APS interview on the project[2]) and she contributed to the first findings establishing associations between infant touchscreen use and sleep problems (Cheung, Bedford, Saez De Urabain, Karmiloff-Smith, & Smith, 2017) and earlier fine-motor development (Bedford et al., 2016).

The TABLET project is testament to how Annette's rigorous experimental approach to studying developmental cognitive neuroscience has broad implications for many areas of development and their societal implications, such as the pressing issue of childhood screen time, and the need to understand the mechanisms of visual processing to understand visual impairments and how to best treat them.

Annette may not have had the opportunity to see her empirical contributions on this topic published during her lifetime but her influence will be long-lasting for those of us who had the opportunity to collaborate with and be inspired by her.

In closing, we can think of no better way to end our exploration of Annette's work onscreen time than with her own words:

> "In sum, screen exposure for young children should be far more than a display of coloured patterns and music to mesmerize babies; they should be a scientifically designed effort to stimulate babies. We live in a media-saturated world. It is far better that parents know how to choose the right television or DVD programmes for their children than to make them ashamed at even thinking of ever using screen exposure."
>
> *(Karmiloff-Smith, 2012)*

Notes

1 Annette would wish us to document that she actually advised against using such natural video scenes in Baby Bright precisely because she predicted these difficulties!
2 https://www.psychologicalscience.org/publications/observer/obsonline/toddlers-and-touchscreens-a-science-in-development.html

References

Anderson, Daniel R., & Katherine G. Hanson. (2010) From blooming, buzzing confusion to media literacy: The early development of television viewing. *Developmental Review, 30*(2), 2010: 239–255.

Anderson, D. R., & Pempek, T. A. (2005). Television and very young children. *American Behavioral Scientist, 48*(5), 505–522.

Baker, T. J., Tse, J., Gerhardstein, P., & Adler, S. A. (2008). Contour integration by 6-month-old infants: Discrimination of distinct contour shapes. *Vision Research, 48*(1), 136–148.

Barr, R., Muentener, P., & Garcia, A. (2007). Age-related changes in deferred imitation from television by 6-to 18-month-olds. *Developmental Science, 10*(6), 910–921.

Bedford, R., Saez de Urabain, I. R., Cheung, C. H., Karmiloff-Smith, A., & Smith, T. J. (2016) Toddlers' fine motor milestone achievement is associated with early touchscreen Scrolling. *Frontiers in Psychology 7*, 1108, doi:10.3389/fpsyg.2016.01108.

Braddick, O., & Atkinson, J. (2011). Development of human visual function. *Vision Research, 51*(13), 1588–1609.

Bremner, J. G., & Wachs, T. D. (Eds.). (2010). *The Wiley-Blackwell Handbook of Infant Development, Basic Research* (Vol. 1). Chichester: John Wiley & Sons.

Cheung, C., Bedford, R., Saez De Urabain, I. R., Karmiloff-Smith, A., & Smith, T. J. (2017) Daily touchscreen use in infants and toddlers is associated with reduced sleep and delayed sleep onset. *Scientific Reports 7*, p. 46104. ISSN 2045-2322.

Christakis, D. A., & Zimmerman, F. J. (2006). Media as a public health issue. *Archives of Pediatrics & Adolescent MedicinePAM, 160*(4), 445–446.

Brown, A. (2011) Media use by children younger than 2 years: Policy statement of the AAP, *Pediatrics 128*(5), 1–6.

Courage, M. L. (2017). Screen media and the youngest viewers: Implications for attention and learning. In *Cognitive Development in Digital Contexts* (pp. 3–28). London: Academic Press.

Courage, M. L., & Howe, M. L. (2010). To watch or not to watch: Infants and toddlers in a brave new electronic world. *Developmental Review, 30*(2), 101–115.

Cristia, A. & Seidl, A. (2015) Parental reports on touch screen use in early childhood. *PLoS One 10*, e0128338, doi:10.1371/journal.pone.0128338.

Davies S. C., Atherton F., Calderwood C.,McBride M. (2019) United Kingdom Chief Medical Officers' commentary on 'Screen-based activities and children and young people's mental health and psychosocial wellbeing: a systematic map of reviews'. Department of Health and Social Care.

DeLoache, J. S. (1991). Symbolic functioning in very young children: Understanding of pictures and models. *Child Development, 62*(4), 736–752.

Dickson K., Richardson M., Kwan I., MacDowall W., Burchett H., Stansfield C., Brunton G., Sutcliffe K.1., Thomas J. (2018) *Screen-based activities and children and young people's mental health: A Systematic Map of Reviews*, London: EPPI-Centre, Social Science Research Unit, UCL Institute of Education, University College London.

Ferguson, Charles A. (1964) Baby talk in six languages. *American Anthropologist, 66*(2/6), 103–114.

Fiser, J., & Aslin, R. N. (2002). Statistical learning of new visual feature combinations by infants. *Proceedings of the National Academy of Sciences, 99*, 15822–15826.

Franchak, John M., David J. Heeger, Uri Hasson, & Karen E. Adolph. (2016) Free viewing gaze behavior in infants and adults. *Infancy: the official journal of the International Society on Infant Studies, 21*(3), 262–287.

Frank, M. C., Edward, V., & Scott, P. J. (2009). Development of infants' attention to faces during the first year. *Cognition*, *110*(2), 160–170.

Frank, Michael C., Edward Vul, & Rebecca Saxe. (2012) Measuring the development of social attention using free-viewing. *Infancy: the official journal of the International Society on Infant Studies*, *17*(4), 355–375.

Greenfield, S. (2015). *Mind Change: How digital technologies are leaving their mark on our brains*. New York: Random House.

Hill, J., Inder T., Neil J., Dierker D., Harwell J., & Van Essen D. (2010). Similar patterns of cortical expansion during human development and evolution. *Proceedings of the National Academy of Sciences of the United States of America 107*(29), 13135–13140.

Karmiloff-Smith, Annette, & House, R. (2015). ICT - for & against: Toddlers, TV and touchscreens. *Nursery World*, ISSN 0029-6422.

Karmiloff, K., & Karmiloff-Smith, Annette (2011) A unique child: cognitive development: TV times. *Nursery World*, ISSN 0029-6422.

Karmiloff, K., & Karmiloff-Smith, A. (2011-12). Series of 6 articles on Infant TV and DVDs; Growing Up Multilingual; Gender Differences; Special Talents in Early Development; Handedness in Humans, Apes & Prehistoric Man; Infant Predictors of Reading Abilities. *Nursery World*.

Karmiloff-Smith, A. (1994). *Baby It's You: A unique insight into the first three years of the developing baby*. London: Ebury Press, Random House.

Karmiloff-Smith, A. (2012). 'TV is bad for children' – less emotion, more science please!, In Philip Adey & Justin Dillon, *Bad Education* (pp. 231–244), Open University Press.

Kirkorian, H. L., & Anderson, D. R. (2018). Effect of sequential video shot comprehensibility on attentional synchrony: A comparison of children and adults. *Proceedings of the National Academy of Sciences*, *115*(40), 9867–9874.

Le Meur, O., & Baccino, T. (2013). Methods for comparing scanpaths and saliency maps: strengths and weaknesses. *Behavior Research Methods*, *45*(1), 251–266.

Loschky, L. C., Larson, A. M., Magliano, J. P. Smith, Tim J. (2015) What would Jaws do? The tyranny of film and the relationship between gaze and higher-level narrative film comprehension. *PLoS One 10*(11), e0142474. ISSN 1932-6203.

Mital, P. K., Smith, Tim J., Hill, R. L., & Henderson, J. M. (2011) Clustering of gaze during dynamic scene viewing is predicted by motion. *Cognitive Computation 3*(1), 5–24. ISSN 1866-9956.

Ofcom (2018, February 21) *Children and Parents: Media Use and Attitudes Report*. Retrieved from https://www.ofcom.org.uk/research-and-data/media-literacy-research/childrens/children-and-parents-media-use-and-attitudes-report-2018

Pelucchi, B., Hay, J. F., & Saffran, J. R. (2009). Statistical learning in a natural language by 8-month-old infants. *Child Development*, *80*(3), 674–685.

Pempek, Tiffany A., Heather L. Kirkorian, John E. Richards, Daniel R. Anderson, Anne F. Lund, and Michael Stevens. (2010) Video comprehensibility and attention in very young children. *Developmental Psychology*, *46*(5), 1283.

Radesky, J., Christakis, D., Hill, D., Ameenuddin, N., Chassiakos, Y., Cross, C.,... & Moreno, M. A. (2016). Media and young minds. *Pediatrics*, *138*(5).

Rider, A. T., Coutrot, A., Pellicano, E., Dakin, S. C., & Mareschal, I. (2018). Semantic content outweighs low-level saliency in determining children's and adults' fixation of movies. *Journal of Experimental Child Psychology*, *166*, 293–309.

Saez De Urabain, I. R. (2015) *Investigating the mechanisms underlying fixation durations during the first year of life: a computational account*, Unpublished Doctoral Thesis, Birkbeck: University of London. http://bbktheses.da.ulcc.ac.uk/148/1/cp_thesis_irati.pdf

Saez de Urabain, Irati R., Nuthmann, A., Johnson, Mark H., & Smith, Tim J. (2017). Disentangling the mechanisms underlying infant fixation durations in scene perception: a computational account. *Vision Research 134*, 43–59. ISSN 0042-6989.

Sigman, A. (2012). Time for a view on screen time. *Archives of disease in childhood* 97(11), 935–942.

Smith, Tim J. (2013). Watching you watch movies: Using eye tracking to inform film theory. In: Shimamura, A. (ed.) *Psychocinematics: Exploring Cognition at the Movies* (pp. 165–191). New York: Oxford University Press. ISBN 9780199862139.

Stiglic N., & Viner, R. M. (2019). The effects of screentime on the health and wellbeing of children and adolescents: A systematic review of reviews. *BMJ Open 9(1)*. doi:10.1136/bmjopen-2018-023191.

Strasburger, V. C., & Hogan, M. J. (2013). Children, adolescents, and the media: Policy statement of the AAP. *Pediatrics 132*, 958–961.

Troseth, G. L. (2003). TV guide: Two-year-old children learn to use video as a source of information. *Developmental Psychology, 39*(1), 140.

Valkenburg, P. M., & Vroone, M. (2004). Developmental changes in infants' and toddlers' attention to television entertainment. *Communication Research, 31*, 288-31.

Viner, R., Davie, M., & Firth, A. (2019). The health impacts of screen time: a guide for clinicians and parent. *Royal College of Pediatrics and Child Health*

Wartella, E., Richert, R. A., & Robb, M. B. (2010). Babies, television and videos: How did we get here? *Developmental Review, 30*, 116–127.

Wass, S., Porayska-Pomsta, K., & Johnson, M. H. (2011). Training attentional control in infancy. *Current Biology, 21*(18), 1543–1547.

Wass, S. V., & Smith, Tim J. (2015). Visual motherese? Signal-to-noise ratios in toddler-directed television. *Developmental Science, 18*(1), 24–37.

WHO (2019). *Guidelines on physical activity, sedentary behaviour and sleep for children under 5 years of age.* Geneva: World Health Organization. Licence: CC BY-NC-SA 3.0 IGO.

Zimmerman, Frederick J., Dimitri A. Christakis, & Andrew N. Meltzoff. (2007). "Television and DVD/Video Viewing in Children Younger than 2 Years." *Archives of Pediatrics & Adolescent Medicine, 161*(5), 473–479.

19

NEVER MISSING THE WHOLE PICTURE: INTELLECTUAL DEVELOPMENT FROM A NEUROCONSTRUCTIVIST PERSPECTIVE

Luca Rinaldi

Introduction

Historically, there has always been a vibrant debate surrounding the construct of intelligence. Identifying a widely shared definition, as well as its neurocognitive counterparts, has proven inherently difficult, despite the fact that huge scientific efforts on this topic can be traced back to the beginning of the past century or even before (Deary, 2012; Neisser et al., 1996). This is especially true when attempting to describe the apparently subtle changes that characterize the individual's intellectual developmental trajectory. Indeed, until recently, most research was dedicated to describing the key factors responsible for human intelligence across the whole life-span, losing focus on developmental time (Baron & Leonberger, 2012; Kamphaus, 2005). The reasons for this depend on a host of factors that make it to capture the dynamic nature of intelligence at early stages of development difficult. By definition, intelligence encompasses the mental abilities essential for survival and advancement in any environmental context[1] (Deary, Penke, & Johnson, 2010). This concept is also widely used to refer to individual differences as measured by means of psychometric tests, which cover a range of various cognitive domains such as problem-solving, executive function, memory, processing speed, verbal and spatial abilities. Intelligence tests should thus show marked individual differences, which are considered to be quite stable in rank order throughout development and even over long time spans. Yet, how the individual *develops* such abilities represents one of the most puzzling but fascinating questions that is currently captivating both psychologists and neuroscientists.

In this chapter, I briefly summarize recent advancements in the field and describe how this evidence can be framed under the theoretical scaffold of neuroconstructivism (Karmiloff-Smith, 1995, 1998; Mareschal et al., 2007), thus reflecting the general influence of Karmiloff-Smith's inspirational view. The neuroconstructivist model, indeed, conceives development as a trajectory that is dynamically modulated by multiple interacting biological and environmental constraints (Sirois et al., 2008; Westermann et al., 2007; Westermann, Thomas, & Karmiloff-Smith, 2010). Intelligence is thus possibly the ideal construct that can be framed within the neuroconstructivist model.

I will first review evidence pointing to intelligence as a primarily dynamic – or, provocatively speaking, *developmental* – trait, rather than merely stable. The concept of dynamicity is, in fact, a milestone of Karmiloff-Smith's concept of cognitive development, which harks back to the Piagetian notions of stage, interaction, and transitions. Next, I will bring together recent empirical findings and describe how they perfectly fit within the neuroconstructivist model, providing support to the idea that intelligence profiles are the result of developmental mechanisms and principles that operate and interact at different levels of description (i.e., genetic, cellular, neural, behavioural and environmental). This will illustrate how Karmiloff-Smith's theory still echoes in the state-of-the-art models of intellectual development.

Intelligence as a dynamic construct

The view that intelligence is a stable trait of the individual is a general assumption in many cultures, which is, in turn, reflected in many cognitive theories. This view is substantiated by the relatively high correlation between IQ measurements made at different life periods (Deary, 2014; Deary et al., 2010; Moffitt, Caspi, Harkness, & Silva, 1993). However, high values of correlations in longitudinal studies do not imply that two IQ scores measured at different developmental times are almost identical (or that the individual has the same mental abilities over time; Valsiner, 1984). For instance, even when two intelligence scores show a strong correlation coefficient of 0.7, over 50% of the variation remains unexplained. In this sense, intelligence may be less stable than has been previously believed, as supported by the evidence reviewed in the next paragraphs.

First, the degree of correlation seems to be strictly dependent on the life period in which measurements are taken (e.g., Watkins, & Smith, 2013). Indeed, the consistency between scores is not optimal if one measurement is obtained during the preschool age (Bishop et al., 2003; McCall, Hogarty, & Hurlburt, 1972), with cognitive abilities becoming relatively stable only from childhood through adulthood (Chen, & Siegler, 2000; Deary, 2014; W. Johnson, Gow, Corley, Starr, & Deary, 2010). By reviewing a number of studies employing standard IQ tests, Schuerger and Witt (1989) observed that about 13% of the 6-year-olds changed scores at least by one standard deviation (i.e., a huge change, considering that standard intelligence tests typically have a mean of 100 and a standard deviation of 15). The same was true for 7% of the 30 years old (see also Gottfried, Gottfried, & Guerin, 2006; Schneider, Niklas, & Schmiedeler, 2014). This testifies to drastic increases in reliability with age, with less stable scores in childhood and adolescence compared to adulthood.

Similarly, a recent study revealed that despite a good test–retest correlation across a 2-year interval, 25% of typically developing children and adolescents showed changes of 9 IQ points or more across this interval (Waber, Forbes, Almli, & Blood, 2012). This may be accounted by non-linear developmental trajectories of IQ profiles.

Second, higher correlations are typically found for shorter intervals, thus indicating that the stability of intelligence varies as a function of the interval of measurement (Schuerger & Witt, 1989). Accordingly, stability coefficients of IQ scores have been shown to drastically decline from the .80 to .90 range in short-term test-retest investigations to the .50 to .90 range in longer retest intervals (e.g., 3 years or more) (Canivez, & Watkins, 1998; Oakman, & Wilson, 1988; Watkins, & Smith, 2013). The relatively low stability of intelligence profiles may also be attributed to the use of different standardized tests for measuring IQ at different ages. Indeed, despite intelligence testing has enormously evolved during the past century, there is no single IQ test suited for all ages. Thus, different kinds of IQ tests (including a wide variety of item content) are used at different ages, and this may contribute to the disparity in intelligence profiles observed particularly at greater age intervals.

Third, many studies that have classified individuals into subgroups initially showing high, average and low levels of intellectual ability, reported that the stability of the measurement differs depending on the level of IQ. In particular, it has been shown that low scores for young children are more stable than high scores. For instance, IQ scores obtained for those children with an initially low IQ turned out to be more stable over time than those found for the two other subgroups (i.e., average and high) from preschool age onwards (Mortensen, Andresen, Kruuse, Sanders, & Reinisch, 2003; Schneider et al., 2014; Schneider, Wolke, Schlagmüller, & Meyer, 2004). Remarkably, and despite this evidence, it is important to pinpoint that some of these children with low IQ (i.e., < 85) subsequently reached scores above 120 at the age of 17, thus again suggesting the occurrence of drastic changes, at least at the individual level, during development (Schneider et al., 2014).

Finally, a large number of cognitive interventions have been shown to exert profound effects on IQ and, more generally, on academic achievement, thus speaking against a predetermined trajectory of intelligence. In particular, there is clear evidence that formal education has a great impact on IQ (Ceci, 1991; Neisser et al., 1996). When children are deprived of school for a protracted period of time they can exhibit deficits in IQ of as much as 2 SD. Such effects of schooling on intelligence measures can even be twice the size of effect of age (Cahan & Cohen, 1989) and may depend on the age in which schooling begins (Bedard & Dhuey, 2006; Ritchie & Tucker-Drob, 2018).

Interestingly, these behavioural fluctuations in IQ are paralleled by neural changes, suggesting that cognitive abilities may be more associated to the magnitude and timing of changes in brain structure during development than to brain structure per se (Schnack et al., 2015). In a seminal longitudinal investigation, Shaw and colleagues (Shaw et al., 2006) showed that patterns of correlation between intelligence and brain structure (e.g., cortical thickness) vary as a function of the age of the participants. This implies a brain–behaviour link that is highly sensitive to development. Further, developmental trajectories of increase and subsequent thinning of the cortex during developmental time differ between high and low intelligent children and adolescents, with more intelligent children demonstrating a particularly plastic cortex (Shaw et al., 2006). Another neuroimaging study corroborates this possibility more directly. In the study by Ramsden and colleagues (Ramsden et al., 2011), a sample of neurologically normal adolescents had structural and functional brain scans, along with an IQ measurement, at two different times. Results showed that changes in Verbal IQ, observed between the two assessments, were positively correlated with changes in grey matter density (and volume) in a region of the left motor cortex, which is activated by the articulation of speech. In striking contrast, changes in Non-Verbal IQ, observed between the two assessments, were positively correlated with grey matter density in the anterior cerebellum, which is associated with motor movements of the hand (Ramsden et al., 2011). The idea that changes in intelligence measures across development can reflect meaningful changes in general cognitive abilities and have a neuroanatomical substrate is further supported by a recent study based on a sizeable sample of children and adolescents (Burgaleta, Johnson, Waber, Colom, & Karama, 2014).

Putting neuroconstructivism at the disposal of intellectual development

Perhaps the biggest challenge for developmental psychologists trying to understand intelligence is to integrate individual observations into a developmental trajectory and to consider the multiple sources that influence this process in order to explain the possible causes of developmental change. A broad theoretical framework was therefore needed to better understand intelligence over developmental time. In this section, I will argue that neuroconstructivism (Karmiloff-Smith, 1998, 2015; Westermann et al., 2007, 2010) provided such a parsimonious developmental framework that can account for the dynamic essence of intelligence, as briefly reviewed earlier.

The neuroconstructivist approach proposed by Karmiloff-Smith characterizes development as a trajectory that is shaped by multiple interacting biological and environmental constraints. As such, neuroconstructivism integrates different views of brain and cognitive development, including: (a) probabilistic epigenesis, which underscores the interactions between experience and gene expression (Gottlieb, 2007); (b) neural constructivism, which emphasizes the role of experience on neural development (Quartz & Sejnowski, 1997); (c) the "interactive specialization" view of brain development, according to which the shift from distributed to more localized processing would be due to activity-dependent interactions between brain regions (M. H. Johnson, 2011); (d) embodiment approaches, which maintain that bodily states are necessary for cognition, especially in the developmental time (Clark, 1999; Smith, 2005); (e) views recognizing a crucial role of the social environment for the developing child (Bradley & Corwyn, 2002).

These five views encompass different levels of constraints (i.e., genetic, cellular, neural, behavioural and environmental) that deeply influence development. Crucially, and contrary to a static view of development, neuroconstructivism implies a strict interdependency between these levels (Westermann et al., 2010). Thus, a change in one level has the potential to affect the others. For instance, the constraints that shape the developing neural system necessarily alter the context in which the individual develops, affecting consequently the developmental trajectory itself and the specific outcome that is measured (Annaz, Karmiloff-Smith, & Thomas, 2008; Karmiloff-Smith, 2015).

There is now converging evidence pointing to intelligence as a flexible, plastic concept, sensitive to environmental influences at the level of gene expression, brain, cognition and behaviour. The evidence collected in the past years, thus, seems to conform well to the multilevel contextual dependency implied by the neuroconstructivist approach put forward by Karmiloff-Smith. In the following paragraphs, I will describe how the different levels of constraints implied by neuroconstructivism have shaped whole lines of research and the underlying theoretical models that tried to explain intellectual development.

Probabilistic epigenesis

The notion that development can be conceptualized as a predetermined process has slowly given way over the past decades. The role of environmental and behavioural influences on the developing organism has been increasingly recognized. Accordingly, the unidirectional relationship between genetic activity and functions has been completely revisited, by postulating bidirectional influences within and between levels of analysis (Gottlieb, 2007). This probabilistic epigenetic view of development emphasizes the reciprocity of influences between genetic activity, neural activity, behaviour and the physical and social environments of an organism (Karmiloff-Smith et al., 2012; Lickliter, & Honeycutt, 2003).

Such a probabilistic framework nicely adapts to explain how the genes may interact with the environment in shaping intellectual development (Caspi et al., 2007). The heritability of intelligence, i.e., the proportion of observed variation that can be ascribed to genetic factors, in any naturally occurring population is in fact neither zero nor one. The acceptance of the importance of both genetic and environmental influences has led to the empirical exploration of the interplay between genes and environment, such as their interaction and correlation in the development of complex traits (Plomin, & Deary, 2015). However, the problem of identifying the specific genetic variants that contribute to the high heritability of intelligence is far from being resolved (Butcher, Davis, Craig, & Plomin, 2008; for a review see Savage et al., 2018). As complex behavioural traits, individual differences in intellectual development are likely to be determined by many different genes (Davies et al., 2011). Yet, the causes and consequences of variance at different levels of cognitive specificity-generality are hard to understand (Davies et al., 2018). Perhaps more critically, such an integrated approach may suffer from the risk of conflating statistical predictors of IQ variations (i.e., genes, environment and their relative interaction) with the very same developmental causal mechanisms.

Neural constructivism

Contrary to popular selectionist models that rely on regressive mechanisms, according to "neural constructivism," the representational features of cortex are dynamically built from the interaction between neural growth mechanisms and environmentally derived neural activity (Elman et al., 1996). By adopting a constructive process of growth, the protracted period of postnatal growth assumes, therefore, a primary role in affecting the domain-specificity of the developing neocortex (Karmiloff-Smith, 1998). A clear example of such a complex relationship comes from research on environmental enrichment, investigating the influence of the environment on brain structure and function (Sale, Berardi, & Maffei, 2009). Effects of environmental enrichment have been observed at the anatomical level, with robust increases in cortical thickness and weight, but also in synaptic transmission and plasticity, especially in the first years of life (Sale et al., 2009).

The relation between childhood socioeconomic status (SES) and individual differences in intelligence nicely captures such interactive process. Research indicates that SES has reasonably specific neurocognitive effects, with children who grow up in poverty tending to have lower IQ scores (Brooks-Gunn, & Duncan, 1997; Farah et al., 2006; Gottfried, Gottfried, Bathurst, Guerin, & Parramore, 2003). Interestingly, a similar influence of the socioeconomic environment on brain structure has been recently reported. A number of studies have found a relationship between environmental factors and neurobehavioral functioning in children (Noble, Norman, & Farah, 2005; Rao et al., 2010; Tomalski et al., 2013), with socioeconomic status appearing to moderate patterns of age-related cortical thinning (Piccolo, Merz, He, Sowell, & Noble, 2016). In one of the largest studies to date to characterize associations between socioeconomic factors and children's brain structure, parental education and family income accounted for individual variation in independent characteristics of brain structural development (Noble et al., 2015). In particular, parental education was linearly associated with children's total brain surface area over the course of childhood and adolescence. Conversely, surface area mediated the link between family income and children's performance on certain executive function tasks. These findings add to the emerging literature suggesting that SES relates to structural brain variation in the hippocampus (Noble, Houston, Kan, & Sowell, 2012), amygdala (Noble et al., 2012) and prefrontal cortex (Lawson, Duda, Avants, Wu, & Farah, 2013; for a full review see Hackman, Farah, & Meaney, 2010, or Ursache & Noble, 2016).

Interactive specialization

The interactive specialization view maintains that, during postnatal development, the refinement of specific computational abilities by brain regions is the result of a process of neuronal competition and gradual specialization that takes place in relationship with other cortical regions (Johnson, 2001, 2011; Karmiloff-Smith, 1995). This narrowing process also involves the re-organization of inter-regional interactions (Johnson, & De Haan, 2015). As a consequence, cortical areas are much more highly interconnected in the infant's brain than in the adult's brain, with the ratio between white and grey matter changing over developmental time (Johnson, 2011).

The development of intelligence seems to be closely related to and comprehensively described by this process of neural specialization. First, a strong relationship between specific intellectual abilities of children and the level of integration of their structural network organization has recently been reported (Kim et al., 2016). These findings indicate that children with high scores on measures of intelligence also exhibit greater global efficiency of structural brain networks. This is in line with the hypothesis that superior performance of children on tests of intelligence is associated with the efficient transfer of local information among all nodes in the network (see also Khundrakpam et al., 2016). Hence, an increase in intellectual abilities seems to be related with a higher interactive specialization (see also Koenis et al., 2015).

Second, individual differences in intelligence are strictly related to the pattern of cortical growth during childhood and adolescence (Narr et al., 2007; Van Leeuwen et al., 2009). Detecting changes in grey and white matter volume, as well as in cortical thickness, in the first years of life seems, therefore, to be essential for better-grasping fluctuations in intelligence scores over developmental time.

Embodiment (perceiving and acting)

Cognitive development is clearly triggered by a proactive exploration of the surrounding environment through our body (Barsalou, 2008). In general, extensive amounts of learning is made possible through embodied action-perception loops, with the developing body that acts as a real filter from the outer world to the brain. For instance, the infant's ability to gather information from the surrounding world is mainly dependent on visual orienting (Johnson, & De Haan, 2015). Similarly, motor development in infancy relies on general and basic psychological functions necessary for survival, such as response selection, behavioural adaptation, and categorization. The development of attentional and sensorimotor systems has thus been linked to individual differences in general cognitive abilities.

The possibility that visual habituation and dishabituation may be promising candidates for predicting intelligence at later stages of life is certainly not new (Fagan, & McGrath, 1981; Lewis, & Brooks-Gunn, 1981; McCall, 1981). Visual habituation is defined as a decrement of attention to a repeatedly or continuously presented stimulus, whereas dishabituation refers to the reactivation of attention towards a novel stimulus following habituation. Both these two basic information-processing skills rely on the gradual construction of a memory trace, the recall of information from the memory trace itself, and the comparison of the representation in memory with the online visual input. There is evidence in the literature that both habituation and dishabituation predict cognitive abilities later in child development, such as later intelligence scores (see for a review Kavšek, 2004) or academic achievement (Fagan, Holland, & Wheeler, 2007). That is, core abilities from infancy and toddlerhood can predict IQ performance at 2 and 3 years, with this predictive power extending into pre-adolescence. Critically, and according to the neuroconstructivist approach, the way information is gathered from the surrounding environment can be mediated by environmental variables. Accordingly, infants whose mothers encourage them more frequently to attend to objects and events in the home environment in the first 6 months of life outperform on a conventional psychometric assessment of intelligence at 4 years those infants with less encouraging caretakers (Bornstein, 1985). This is further complemented by recent evidence showing that inherited family environments can affect children's educational attainment (Kong et al., 2018).

Another piece of evidence in favour of the important role of embodiment comes from the fact that individual differences in cognitive abilities are tightly related to infant motor development (Murray, Jones, Kuh, & Richards, 2007). The rate of progression of infant motor development, especially age of walking onset, is associated with IQ at 64 months (Hamadani, Tofail, Cole, & Grantham-McGregor, 2013), 3 years (Capute, Shapiro, Palmer, Ross, & Wachtel, 1985) and at 6–11 years (Piek, Dawson, Smith, & Gasson, 2008), and even with educational attainment (Taanila, Murray, Jokelainen, Isohanni, & Rantakallio, 2005) and adult brain structure (Ridler et al., 2006). Further, during adolescence, motor coordination is associated with academic outcomes (Rigoli, Piek, Kane, & Oosterlaan, 2012b) and working memory (Rigoli, Piek, Kane, & Oosterlaan, 2012a). In this case, these relations seem to interact dynamically with environmental factors. Indeed, the relationship between early motor development and intelligence is stronger in infants of low parental social status than in those of high parental social status (Flensborg-Madsen, & Mortensen, 2015). Similarly, the association between body mass index in preschool children and IQ appears to be mediated by SES (Tabriz et al., 2015).

Ensocialment

The social environment, referring to the individual's physical surrounding, social relationship and community resources, can also have a profound impact on development. There is a real relationship between socio-economic background and intelligence scores, reflected, for instance, by the fact that higher levels of parents' education (Lemos, Almeida, & Colom, 2011) and family income (Ganzach, 2014) are associated with higher IQ. Their possible interactive effects with the genetic, neural and cognitive counterparts have been widely described before (for a review see Plomin & von Stumm, 2018). Indeed, socioeconomic status and parental attitudes play a determining role in engaging the child in those tasks that, in turn, promote those abilities that are ultimately measured in intelligence tests. For instance, child neglect (i.e., one of the most prevalent forms of child maltreatment) has been associated with significantly lower intellectual development in young children (De Bellis, Hooper, Spratt, & Woolley, 2009).

Toward an integrated framework

To account for the continuous fluctuations of intelligence over development, in a recent article, Karmiloff-Smith and I proposed a theoretical framework in which intellectual development is conceived of as a progressive increase in the efficiency of mental representations, defined as the neural activation patterns that sustain adaptive behaviour in the environment (Rinaldi & Karmiloff-Smith, 2017). There is huge individual variability, in fact, in the way brains of different people become adapted to the surrounding environment, with more intelligent individuals displaying more plastic brains and more efficient neural activations on a range of cognitive tasks (Deary et al., 2010). The specific interaction between the different levels previously described (i.e., genetic, cellular, neural, behavioural and environmental) would determine an individual's initial intellectual level, with the plasticity of neural structures allowing the further development of intelligence depending on the context in which the same structures develop (i.e., principle of *context dependence*) (see Figure 19.1). In this sense, all constraints imposed interactively by these different levels contribute to the dynamic equilibrium-characterized by alternating periods of stability and instability-supporting intellectual functioning over development (Rinaldi, & Karmiloff-Smith, 2017).

Three key domain-general mechanisms would sustain intellectual development (i.e., the realization of more efficient representations): *competition, cooperation* and *chronotopy* (Westermann et al., 2007, 2010) In particular, *competition* would contribute to the increase of specialization and, consequently, in efficiency of the mental representations. *Cooperation,* would lead to the integration of components and sub-functions, a process that would also be adopted for carrying knowledge at higher levels. This process of neuronal competition and gradual specialization over development would induce the brain to produce a more efficient activation for certain kinds of inputs over others. The timing of these mechanisms is well captured by *chronotopy*, which refers to the temporal dynamic in which the *context-dependence* principle constrains the emergence of representations. These constraints are supposed to operate mainly during the early phases of life, since intelligence profiles that are expected to be more stable in adulthood compared to infancy and early childhood, as suggested by the evidence reviewed above (see Figure 19.1). We further hypothesized that a positive *context-dependence* between different levels (i.e., genes, brain, cognition, and environment) may promote the functioning and the progress of the three mechanisms and affect consequently intellectual abilities. In particular, the fact that high intelligence is related to more plastic cortical development may indicate that the brains of these children adapt more efficiently to the requests from the environment.

Concluding remarks

In this chapter, I have reviewed recent advances in the fields of intelligence research and highlighted how such progress was actually triggered by the pioneering

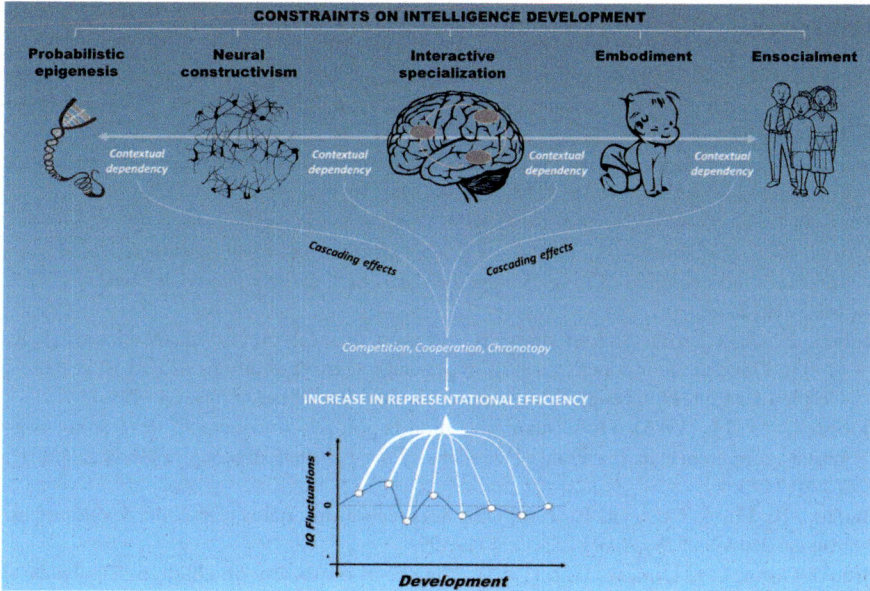

FIGURE 19.1 The multiple interacting constraints that influence the development of intelligence, conceived as the construction of efficient mental representation (i.e., neural activation patterns that sustain adaptive behaviour). The principle of *context-dependence* constrains intellectual development by means of three general mechanisms: *cooperation*, *competition* and *chronotopy*. This last mechanism reflects the developmental essence of intelligence, as constraints are supposed to operate mainly during early phases of life, with intelligence profiles expected to be more stable in adulthood compared to infancy and early childhood. Together, all constraints imposed interactively by genes, brain, cognition and environment are viewed as responsible for the fluctuations of intelligence over developmental time

work of Annette Karmiloff-Smith, dating back to 20 years ago. The interdisciplinary approach pursued by many recent studies, at the basis of the neuroconstructivist model, speaks in favour of a strict interplay between genes, brain, cognition and environment over developmental time to influence intelligence profiles. Under this view, intellectual functioning is characterized by a fluctuating interaction between the developing system itself and the environmental factors involved at different times across ontogenesis. Together, this attests that the revolutionary idea offered by Karmiloff-Smith is being operationalized and realized in current research trends.

Note

1 Despite on some occasions being differentiated, in the present chapter I will use terms like intelligence quotient (IQ), Spearman's *g* [6] and mental ability in an interchangeable way to generally describe the concept of intelligence.

References

Annaz, D., Karmiloff-Smith, A., & Thomas, M. S. C. (2008). The importance of tracing developmental trajectories for clinical child neuropsychology. *Child Neuropsychology: Concepts, Theory and Practice*, 7.

Baron, I. S., & Leonberger, K. A. (2012). Assessment of intelligence in the preschool period. *Neuropsychology Review*, 22(4), 334–344.

Barsalou, L. W. (2008). Grounded cognition. *Annual Review of Psychology*, 59, 617–645.

Bedard, K., & Dhuey, E. (2006). The persistence of early childhood maturity: International evidence of long-run age effects. *The Quarterly Journal of Economics*, 121, 1437–1472.

Bishop, E. G., Cherny, S. S., Corley, R., Plomin, R., DeFries, J. C., & Hewitt, J. K. (2003). Development genetic analysis of general cognitive ability from 1 to 12 years in a sample of adoptees, biological siblings, and twins. *Intelligence*, 31(1), 31–49.

Bornstein, M. H. (1985). How infant and mother jointly contribute to developing cognitive competence in the child. *Proceedings of the National Academy of Sciences*, 82(21), 7470–7473.

Bradley, R. H., & Corwyn, R. F. (2002). Socioeconomic status and child development. *Annual Review of Psychology*, 53(1), 371–399.

Brooks-Gunn, J., & Duncan, G. J. (1997). The effects of poverty on children. *The Future of Children*, 7, 55–71.

Burgaleta, M., Johnson, W., Waber, D. P., Colom, R., & Karama, S. (2014). Cognitive ability changes and dynamics of cortical thickness development in healthy children and adolescents. *NeuroImage*, 84, 810–819.

Butcher, L. M., Davis, O. S. P., Craig, I. W., & Plomin, R. (2008). Genome-wide quantitative trait locus association scan of general cognitive ability using pooled DNA and 500K single nucleotide polymorphism microarrays. *Genes, Brain and Behavior*, 7(4), 435–446.

Cahan, S., & Cohen, N. (1989). Age versus schooling effects on intelligence development. *Child Development*, 60, 1239–1249.

Canivez, G. L., & Watkins, M. W. (1998). Long-term stability of the Wechsler Intelligence Scale for Children—Third Edition. *Psychological Assessment*, 10(3), 285.

Capute, A. J., Shapiro, B. K., Palmer, F. B., Ross, A., & Wachtel, R. C. (1985). Cognitive-motor interactions the relationship of infant gross motor attainment to IQ at 3 Years. *Clinical Pediatrics*, 24(12), 671–675.

Caspi, A., Williams, B., Kim-Cohen, J., Craig, I. W., Milne, B. J., Poulton, R., ... Moffitt, T. E. (2007). Moderation of breastfeeding effects on the IQ by genetic variation in fatty acid metabolism. *Proceedings of the National Academy of Sciences*, 104(47), 18860–18865.

Ceci, S. J. (1991). How much does schooling influence general intelligence and its cognitive components? A reassessment of the evidence. *Developmental Psychology*, 27(5), 703.

Chen, Z., & Siegler, R. S. (2000). Across the great divide: Bridging the gap between understanding of toddlers' and older children's thinking. *Monographs of the Society for Research in Child Development*.

Clark, A. (1999). An embodied cognitive science? *Trends in Cognitive Sciences*, 3(9), 345–351.

Crone, E. A., & Elzinga, B. M. (2015). Changing brains: How longitudinal functional magnetic resonance imaging studies can inform us about cognitive and social-affective growth trajectories. *Wiley Interdisciplinary Reviews: Cognitive Science*, 6(1), 53–63.

Davies, G., Lam, M., Harris, S. E., Trampush, J. W., Luciano, M., Hill, W. D., ... & Liewald, D. C. (2018). Study of 300,486 individuals identifies 148 independent genetic loci influencing general cognitive function. *Nature Communications*, 9(1), 1–16.

Davies, G., Tenesa, A., Payton, A., Yang, J., Harris, S. E., Liewald, D., … Luciano, M. (2011). Genome-wide association studies establish that human intelligence is highly heritable and polygenic. *Molecular Psychiatry, 16*(10), 996–1005.

De Bellis, M. D., Hooper, S. R., Spratt, E. G., & Woolley, D. P. (2009). Neuropsychological findings in childhood neglect and their relationships to pediatric PTSD. *Journal of the International Neuropsychological Society, 15*(06), 868–878.

Deary, I. J. (2012). Intelligence. *Annual Review of Psychology, 63*(1), 453–482. https://doi.org/doi:10.1146/annurev-psych-120710-100353

Deary, I. J. (2014). The stability of intelligence from childhood to old age. *Current Directions in Psychological Science, 23*(4), 239–245.

Deary, I. J., Penke, L., & Johnson, W. (2010). The neuroscience of human intelligence differences. *Nature Reviews Neuroscience, 11*(3), 201–211.

Elman, J. L., Bates, E. A., Johnson, M. H., Karmiloff-Smith, A., Parisi, D. & Plunkett, K. (1996). *Rethinking innateness: A connectionist perspective on development.* MIT Press.

Fagan, J. F., Holland, C. R., & Wheeler, K. (2007). The prediction, from infancy, of adult IQ and achievement. *Intelligence, 35*(3), 225–231.

Fagan, J. F., & McGrath, S. K. (1981). Infant recognition memory and later intelligence. *Intelligence, 5*(2), 121–130.

Farah, M. J., Shera, D. M., Savage, J. H., Betancourt, L., Giannetta, J. M., Brodsky, N. L., … Hurt, H. (2006). Childhood poverty: Specific associations with neurocognitive development. *Brain Research, 1110*(1), 166–174.

Flensborg-Madsen, T., & Mortensen, E. L. (2015). Infant developmental milestones and adult intelligence: A 34-year follow-up. *Early Human Development, 91*(7), 393–400.

Ganzach, Y. (2014). Adolescents' intelligence is related to family income. *Personality and Individual Differences, 59*, 112–115.

Gottfried, A. W., Gottfried, A. E., Bathurst, K., Guerin, D. W., & Parramore, M. M. (2003). Socioeconomic status in children's development and family environment: Infancy through adolescence. In: M. H. Bornstein, & R. H.Bradley (Eds.), *Monographs in parenting series. Socioeconomic status, parenting, and child development* (pp. 189–207). Lawrence Erlbaum Associates Publishers.

Gottfried, A. W., Gottfried, A. E., & Guerin, D. W. (2006). The Fullerton longitudinal study: A long-term investigation of intellectual and motivational giftedness. *Journal for the Education of the Gifted, 29*(4), 430–450.

Gottlieb, G. (2007). Probabilistic epigenesis. *Developmental Science, 10*(1), 1–11.

Hackman, D. A., Farah, M. J., & Meaney, M. J. (2010). Socioeconomic status and the brain: mechanistic insights from human and animal research. *Nature Reviews Neuroscience, 11*(9), 651–659.

Hamadani, J. D., Tofail, F., Cole, T., & Grantham-McGregor, S. (2013). The relation between age of attainment of motor milestones and future cognitive and motor development in Bangladeshi children. *Maternal & Child Nutrition, 9*(S1), 89–104.

Johnson, M. H. (2001). Functional brain development in humans. *Nature Reviews Neuroscience, 2*(7), 475–483.

Johnson, M. H. (2011). Interactive specialization: a domain-general framework for human functional brain development? *Developmental Cognitive Neuroscience, 1*(1), 7–21.

Johnson, M. H., & De Haan, M. (2015). *Developmental cognitive neuroscience: An introduction.* John Wiley & Sons.

Johnson, W., Gow, A. J., Corley, J., Starr, J. M., & Deary, I. J. (2010). Location in cognitive and residential space at age 70 reflects a lifelong trait over parental and environmental circumstances: the Lothian Birth Cohort 1936. *Intelligence, 38*(4), 402–411.

Kamphaus, R. W. (2005). *Clinical assessment of child and adolescent intelligence*. Springer Science+ Business Media.

Karmiloff-Smith, A. (1995). *Beyond modularity: A developmental perspective on cognitive science*. MIT press.

Karmiloff-Smith, A. (1998). Development itself is the key to understanding developmental disorders. *Trends in Cognitive Sciences*, 2(10), 389–398.

Karmiloff-Smith, A. (2015). An alternative to domain-general or domain-specific frameworks for theorizing about human evolution and ontogenesis. *AIMS Neuroscience*, 2(2), 91.

Karmiloff-Smith, A., D'Souza, D., Dekker, T. M., Van Herwegen, J., Xu, F., Rodic, M., & Ansari, D. (2012). Genetic and environmental vulnerabilities in children with neurodevelopmental disorders. *Proceedings of the National Academy of Sciences*, 109 (Supplement 2), 17261–17265.

Kavšek, M. (2004). Predicting later IQ from infant visual habituation and dishabituation: A meta-analysis. *Journal of Applied Developmental Psychology*, 25(3), 369–393.

Khundrakpam, B. S., Lewis, J. D., Reid, A., Karama, S., Zhao, L., Chouinard-Decorte, F., … Group, B. D. C. (2016). Imaging structural covariance in the development of intelligence. *Neuroimage*.

Kim, D.-J., Davis, E. P., Sandman, C. A., Sporns, O., O'Donnell, B. F., Buss, C., & Hetrick, W. P. (2016). Children's intellectual ability is associated with structural network integrity. *NeuroImage*, *124*, 550–556.

Kingsbury, M. A., & Finlay, B. L. (2001). The cortex in multidimensional space: where do cortical areas come from? *Developmental Science*, 4(2), 125–142.

Koenis, M. M., Brouwer, R. M., van den Heuvel, M. P., Mandl, R. C., van Soelen, I. L., Kahn, R. S., … Hulshoff Pol, H. E. (2015). Development of the brain's structural network efficiency in early adolescence: a longitudinal DTI twin study. *Human Brain Mapping*, *36*(12), 4938–4953.

Kong, A., Thorleifsson, G., Frigge, M. L., Vilhjalmsson, B. J., Young, A. I., Thorgeirsson, T. E., … Gudbjartsson, D. F. (2018). The nature of nurture: Effects of parental genotypes. *Science*, *359*(6374), 424–428.

Lawson, G. M., Duda, J. T., Avants, B. B., Wu, J., & Farah, M. J. (2013). Associations between children's socioeconomic status and prefrontal cortical thickness. *Developmental Science*, 16(5), 641–652.

Lemos, G. C., Almeida, L. S., & Colom, R. (2011). Intelligence of adolescents is related to their parents' educational level but not to family income. *Personality and Individual Differences*, 50(7), 1062–1067.

Lewis, M., & Brooks-Gunn, J. (1981). Visual attention at three months as a predictor of cognitive functioning at two years of age. *Intelligence*, 5(2), 131–140.

Lickliter, R., & Honeycutt, H. (2003). Developmental dynamics: toward a biologically plausible evolutionary psychology. *Psychological Bulletin*, 129(6), 819.

Mareschal, D., Johnson, M. H., Sirois, S., Spratling, M. W., Thomas, M., & Westermann, G. (2007). *Neuroconstructivism: How the brain constructs cognition*. Oxford: Oxford University Press.

McCall, R. B. (1981). Early predictors of later IQ: The search continues. *Intelligence*, 5(2), 141–147.

McCall, R. B., Hogarty, P. S., & Hurlburt, N. (1972). Transitions in infant sensorimotor development and the prediction of childhood IQ. *American Psychologist*, 27(8), 728.

Moffitt, T. E., Caspi, A., Harkness, A. R., & Silva, P. A. (1993). The Natural History of Change to Intellectual Performance: Who Changes? How Much? Is it Meaningful? *Journal of Child Psychology and Psychiatry*, *34*(4), 455–506.

Mortensen, E. L., Andresen, J., Kruuse, E., Sanders, S. A., & Reinisch, J. M. (2003). IQ stability: The relation between child and young adult intelligence test scores in low-birthweight samples. *Scandinavian Journal of Psychology, 44*(4), 395–398.

Murray, G. K., Jones, P. B., Kuh, D., & Richards, M. (2007). Infant developmental milestones and subsequent cognitive function. *Annals of Neurology, 62*(2), 128–136.

Narr, K. L., Woods, R. P., Thompson, P. M., Szeszko, P., Robinson, D., Dimtcheva, T., … Bilder, R. M. (2007). Relationships between IQ and regional cortical gray matter thickness in healthy adults. *Cerebral Cortex, 17*(9), 2163–2171.

Neisser, U., Boodoo, G., Bouchard Jr., T. J., Boykin, A. W., Brody, N., Ceci, S. J., … Sternberg, R. J. (1996). Intelligence: knowns and unknowns. *American Psychologist, 51*(2), 77.

Noble, K. G., Houston, S. M., Brito, N. H., Bartsch, H., Kan, E., Kuperman, J. M., … Libiger, O. (2015). Family income, parental education and brain structure in children and adolescents. *Nature Neuroscience, 18*(5), 773–778.

Noble, K. G., Houston, S. M., Kan, E., & Sowell, E. R. (2012). Neural correlates of socioeconomic status in the developing human brain. *Developmental Science, 15*(4), 516–527.

Noble, K. G., Norman, M. F., & Farah, M. J. (2005). Neurocognitive correlates of socioeconomic status in kindergarten children. *Developmental Science, 8*(1), 74–87.

Oakman, S., & Wilson, B. (1988). Stability of WISC-R intelligence scores: Implications for 3-year reevaluations of learning disabled students. *Psychology in the Schools, 25*(2), 118–120.

Piccolo, L. R., Merz, E. C., He, X., Sowell, E. R., & Noble, K. G. (2016). Age-Related Differences in Cortical Thickness Vary by Socioeconomic Status. *PLoS One, 11*(9), e0162511.

Piek, J. P., Dawson, L., Smith, L. M., & Gasson, N. (2008). The role of early fine and gross motor development on later motor and cognitive ability. *Human Movement Science, 27*(5), 668–681.

Plomin, R., & Deary, I. J. (2015). Genetics and intelligence differences: Five special findings. *Molecular Psychiatry, 20*(1), 98–108.

Plomin, R., & von Stumm, S. (2018). The new genetics of intelligence. *Nature Reviews Genetics, 19*(3), 148–159.

Quartz, S. R., & Sejnowski, T. J. (1997). The neural basis of cognitive development: A constructivist manifesto. *Behavioral and Brain Sciences, 20*(04), 537–556.

Ramsden, S., Richardson, F. M., Josse, G., Thomas, M. S. C., Ellis, C., Shakeshaft, C., … Price, C. J. (2011). Verbal and non-verbal intelligence changes in the teenage brain. *Nature, 479*(7371), 113–116.

Rao, H., Betancourt, L., Giannetta, J. M., Brodsky, N. L., Korczykowski, M., Avants, B. B., … Detre, J. A. (2010). Early parental care is important for hippocampal maturation: Evidence from brain morphology in humans. *NeuroImage, 49*(1), 1144–1150.

Ridler, K., Veijola, J. M., Tanskanen, P., Miettunen, J., Chitnis, X., Suckling, J., … Isohanni, M. K. (2006). Fronto-cerebellar systems are associated with infant motor and adult executive functions in healthy adults but not in schizophrenia. *Proceedings of the National Academy of Sciences, 103*(42), 15651–15656.

Rigoli, D., Piek, J. P., Kane, R., & Oosterlaan, J. (2012a). An examination of the relationship between motor coordination and executive functions in adolescents. *Developmental Medicine and Child Neurology, 54*(11), 1025–1031.

Rigoli, D., Piek, J. P., Kane, R., & Oosterlaan, J. (2012b). Motor coordination, working memory, and academic achievement in a normative adolescent sample: Testing a mediation model. *Archives of Clinical Neuropsychology, 27*(7), 766–780.

Rinaldi, L., & Karmiloff-Smith, A. (2017). Intelligence as a developing function: A neuro-constructivist approach. *Journal of Intelligence.* https://doi.org/10.3390/jintelligence5020018

Ritchie, S. J., & Tucker-Drob, E. M. (2018). How much does education improve intelligence? A meta-analysis. *Psychological Science, 29*(8), 1358–1369.

Sale, A., Berardi, N., & Maffei, L. (2009). Enrich the environment to empower the brain. *Trends in Neurosciences, 32*(4), 233–239.

Savage, J. E., Jansen, P. R., Stringer, S., Watanabe, K., Bryois, J., De Leeuw, C. A., ... & Grasby, K. L. (2018). Genome-wide association meta-analysis in 269,867 individuals identifies new genetic and functional links to intelligence. *Nature Genetics, 50*(7), 912–919.

Schnack, H. G., Van Haren, N. E. M., Brouwer, R. M., Evans, A., Durston, S., Boomsma, D. I., ... Pol, H. E. H. (2015). Changes in thickness and surface area of the human cortex and their relationship with intelligence. *Cerebral Cortex, 25*(6), 1608–1617.

Schneider, W., Niklas, F., & Schmiedeler, S. (2014). Intellectual development from early childhood to early adulthood: The impact of early IQ differences on stability and change over time. *Learning and Individual Differences, 32,* 156–162.

Schneider, W., Wolke, D., Schlagmüller, M., & Meyer, R. (2004). Pathways to school achievement in very preterm and full term children. *European Journal of Psychology of Education, 19*(4), 385–406.

Schuerger, J. M., & Witt, A. C. (1989). The temporal stability of individually tested intelligence. *Journal of Clinical Psychology, 45*(2), 294–302.

Shaw, P., Greenstein, D., Lerch, J., Clasen, L., Lenroot, R., Gogtay, N., ... Giedd, J. (2006). Intellectual ability and cortical development in children and adolescents. *Nature, 440*(7084), 676–679.

Sirois, S., Spratling, M., Thomas, M. S. C., Westermann, G., Mareschal, D., & Johnson, M. H. (2008). Précis of neuroconstructivism: how the brain constructs cognition. *The Behavioral and Brain Sciences, 31*(3), 321–356. https://doi.org/10.1017/S0140525X0800407X

Smith, L. B. (2005). Cognition as a dynamic system: Principles from embodiment. *Developmental Review, 25*(3), 278–298.

Taanila, A., Murray, G. K., Jokelainen, J., Isohanni, M., & Rantakallio, P. (2005). Infant developmental milestones: A 31-year follow-up. *Developmental Medicine and Child Neurology, 47*(9), 581–586.

Tabriz, A. A., Sohrabi, M. R., Parsay, S., Abadi, A., Kiapour, N., Aliyari, M., ... Roodaki, A. (2015). Relation of intelligence quotient and body mass index in preschool children: a community-based cross-sectional study. *Nutrition & Diabetes, 5*(8), e176.

Tomalski, P., Moore, D. G., Ribeiro, H., Axelsson, E. L., Murphy, E., Karmiloff-Smith, A., ... Kushnerenko, E. (2013). Socioeconomic status and functional brain development–associations in early infancy. *Developmental Science, 16*(5), 676–687.

Ursache, A., & Noble, K. G. (2016). Neurocognitive development in socioeconomic context: Multiple mechanisms and implications for measuring socioeconomic status. *Psychophysiology, 53*(1), 71–82.

Valsiner, J. (1984). Conceptualizing intelligence: From an internal static attribution to the study of the process structure of organism-environment relationships. *International Journal of Psychology, 19*(1–4), 363–389.

Van Leeuwen, M., Peper, J. S., van den Berg, S. M., Brouwer, R. M., Pol, H. E. H., Kahn, R. S., & Boomsma, D. I. (2009). A genetic analysis of brain volumes and IQ in children. *Intelligence, 37*(2), 181–191.

Waber, D. P., Forbes, P. W., Almli, C. R., & Blood, E. A. (2012). Four-year longitudinal performance of a population-based sample of healthy children on a neuropsychological battery: the NIH MRI study of normal brain development. *Journal of the International Neuropsychological Society, 18*(02), 179–190.

Watkins, M. W., & Smith, L. G. (2013). Long-term stability of the Wechsler Intelligence Scale for Children—Fourth Edition. *Psychological Assessment, 25*(2), 477.

Westermann, G., Mareschal, D., Johnson, M. H., Sirois, S., Spratling, M. W., & Thomas, M. S. C. (2007). Neuroconstructivism. *Developmental Science, 10*(1), 75–83.

Westermann, G., Thomas, M. S. C., & Karmiloff-Smith, A. (2010). Neuroconstructivism. In U. Goswami (Ed.), *The Wiley-Blackwell handbook of cognitive development* (2nd ed., Vol. 10, pp. 723–748). Chichester: Wiley-Blackwell.

20

TRANSLATION OF SCIENTIFIC INSIGHTS INTO BETTER NAPPIES AND CONSUMER EDUCATION FOR HEALTHY INFANT DEVELOPMENT AND BETTER SLEEP

Frank Wiesemann

> *We can study every aspect of infant development and then tell parents about our discoveries. And these discoveries lead to the invention of new products, that actually. meet babies' needs at every stage of their development.*
> *Annette Karmiloff-Smith, Pampers "Ask the expert" video, 2003*

Annette Karmiloff-Smith was an exceptional scientist widely recognized in the academic world. Furthermore, she was also passionate about bringing new scientific insights about infants' development to parents and baby-care professionals and so enable them to better care for their babies. Annette believed that the discoveries she made about infants' development should be applied to improve products that are used every day by parents for their babies, so these products would meet the changing needs of infants during their development. The collaboration between Annette and Procter & Gamble's (P&G's) Baby Care division worked towards both objectives: educating parents through Pampers' advertising and brand communication and influencing the design of nappies and wipes to better answer to babies' needs. But Annette did more than that for Pampers: She taught P&G scientists how to learn *directly from babies*, leveraging methods that were only evolving at the beginning of the millennium, such as eye-tracking or actigraphy. These methods have enabled insights about how we could not only prevent leaks and keep babies' skin dry, but also how to support their development and their sleep quality. In this chapter I would like to show how this collaboration developed and give some examples of the innovations inspired by this work.

Understanding babies' development

Around the year 2000 Annette became a member of the "Pampers institute", a group of experts from different scientific areas related to babies' health and development. This group was setup by the Pampers team in order to keep updated on the latest scientific research in multiple fields of interest including neonatal care, pediatric dermatology or developmental psychology. This team was also consulted to review the scientific support for statements made in advertising and in information provided to parents, and to help answer questions from parents or caregivers. Annette visited P&G several time and gave lectures to the Pampers team, but also recommended reading of some of her books as well as other publications in the field. She also reviewed and commented on communication materials developed by the Pampers team. Despite the fact that at that time the Pampers team had been working with parents to develop better nappies for 40 years, Annette's insights resulted in a new way for Pampers to think about infant development. The simple chart below (Figure 20.1) was created by P&G designers based on Annette's insights. It shows an overview of the development of an infant during the first two years of life. It was added to P&G trainings provided to new employees as well as to business partners, referred to as "Baby Stages of Development". It is still in use to provide trainees with a quick overview of the changing needs of infants during the time they use nappies. It has inspired many innovations in the past two decades, such as the examples shared in this chapter.

The chart nicely summarizes what Annette has taught us about how a baby develops, physically and cognitively. Annette inspired us to consider not only how babies grow during the first 2 years of life, but also the many skills and capabilities they develop: how they start moving, shaping their sense of self and owning their autonomy, for example developing the ability to actively indicate or even request the need for a nappy change. All these aspects have inspired the Pampers team and other manufacturers to approach the development of products for babies differently.

"See the World through Babies' Eyes" – An advertising campaign which celebrated babies' development"

Learning about the amazing progress a baby makes in the first years of life inspired us to tell parents about these interesting facts. For more than 10 years "Baby Stages

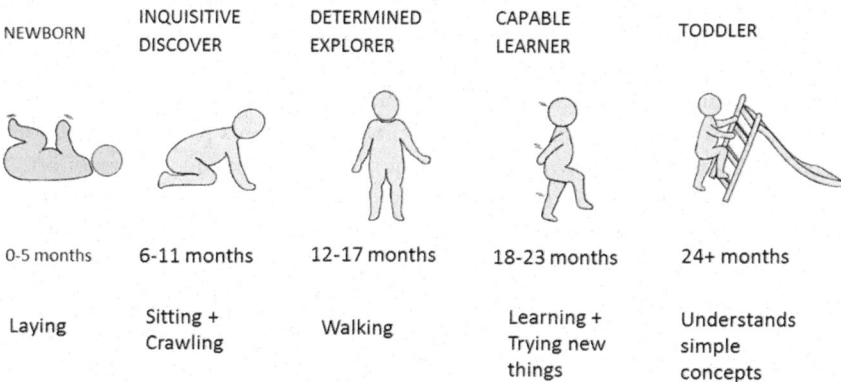

NEWBORN	INQUISITIVE DISCOVER	DETERMINED EXPLORER	CAPABLE LEARNER	TODDLER
0-5 months	6-11 months	12-17 months	18-23 months	24+ months
Laying	Sitting + Crawling	Walking	Learning + Trying new things	Understands simple concepts

FIGURE 20.1 Baby stages of development

of Development" was the main theme of Pampers advertising and communication materials. When watching such an advert or reading information from Pampers parents might learn that a newborn's sight is still blurred but he would recognize his mother's voice, how fast a 9-month-old child would crawl, or by when babies develop the ability to understand cause and effect. Many interesting facts discovered by Annette and others became available to the public through Pampers communications, and helped parents to better understand their babies' behaviours and reactions. One key theme she taught us was the importance of learning and exploration for the development of babies. Importantly, Annette has not only focused on her own research when she was pointing out such insights, but it was also important to her to make the broad knowledge of developmental psychology available to the public.

Here are some examples of how Annette's input translated into Pampers' advertising: One TV copy for Pampers New Baby, a diaper for newly born infants, started with a blurred view of a hand of a baby exploring a women's face – watchers realize this is taken from the view of a baby. A speaker says: "You are five weeks old …" Next the baby is nursed and then the smiling face of the baby, with droplets of milk, is shown. – "… Your digestive system can only handle liquids, which has its advantages …" The mother is cuddling the baby, and a "pooping" sound can be heard. – "… As for the disadvantages, there is Pampers New Baby, specially designed to handle pee and also soft poo …" In this way mums were not only informed about the performance of the Pampers New Baby nappy, they could learn that the view of very young babies is blurred, they explore their world with their sense of touch and why their poo is soft.

Another copy shows a mum playing with her baby. The baby is crawling on the floor, trying to catch a balloon. A voice says: "You are nine months old. You can cover up to a quarter of a mile in under 20 minutes, on your hands and knees" It is shown how the new nappy is particularly stretchy to support the active baby. The copy closes with "When you see the world through a baby's eyes you see how to make it better". The development of a child's sense of autonomy was shown in an advertisement for Pampers pants. It shows a toddler running around. "You are fourteen months old. Now that you can stand on your own two feet you like to keep it that way". The mother approaches the baby with open arms and says: "Let me change your nappy". However, the toddler shakes her head, runs away and shouts "No, no!" Next, it's shown how the mother pulls up a pant-nappy without the need to lay the child down, and after a quick change the toddler moves on.

In order to further teach the public how babies develop, Pampers built an exhibition called "See the world through a baby's eye". All parts of this were inspired by Annette and allowed visitors to experience how babies see and explore the world. Annette not only provided insights for this, but she consulted during the design of this exhibition to ensure that the information given was scientifically correct. The exhibition was setup in the P&G European headquarter in Geneva and later in the P&G Innovation Center in Frankfurt. Journalists, midwives, doctors, but also Procter & Gamble's employees were invited to a tour through the exhibition to help them better understand the changes a baby goes through. Pampers product designers also learned how these insights could be used to make better nappies and wipes. It was a tour through the first 2 years of a baby's life and through the world that expands with the growing capabilities of a baby.

The tour started with a video explaining why it was so important to see the world as a baby does, and how scientific insights into infant perception and cognition are developed. In the video, Annette was shown during an infant eye-tracking study and she explained, how eye-tracking enables infant researchers to learn about babies' development. She says in the video: "We can even study babies' language before they learn to speak", and she emphasizes that such.

The first booth in the exhibition was a newborn's bedroom. Visitors came into an oversized bed with a large stuffed teddy bear so they felt they were only a few weeks old again. Over the bed were a mobile and the face of a woman, both enlarged. When laying down on the mattress visitors could feel as if they were a baby again, perhaps feeling a bit scared by all the huge things around them.

In the next booth visitors learned about important developmental steps over the first few months of life, in particular about the development of the senses. There was a wall with soft and hard, smooth and rough materials and facts relating to the development of touch. Next to a blurred mirror the development of sight was explained, and on another poster, it was explained how babies use their sense of hearing in the womb and therefore recognize their mother's voice already when they are born. The journey continued into a living room showcasing the growing world toddlers explore when they become active. A floor made from wobbly foam material simulated the instability of a 1-year-old just learning to walk. An oversized table with gigantic fruits on it (Figure 20.2) let people understand the temptation to try and reach onto the tabletop when you are just big enough to reach up.

Next to the living room, visitors learned how increased mobility allows 1-year-old children to experiment and understand the world. Visitors were surprised when they tried to pick up a 1-liter pack of milk that was filled with lead and therefore actually weighed 11 kilograms. They learned how adults connect what they see and what they expect to feel when they touch it because they did this many time as a child.

The final part of the exhibition was the garden as an environment where toddlers train their developing mobility skills, such as kicking a ball. In the booth next door, a mirror represented the growing self-awareness of toddlers.

Finally, for each of these stages of baby development some examples were shown of how nappies can support this development, which is explored in the next paragraph.

Better nappies through insights about babies' development combined with innovative technology

What we learned from Annette has inspired our communication to consumers. The insights into infant development have also opened a new way for us to improve our products through meaningful product innovation. Figure 20.3 shows

FIGURE 20.2 The author of this article in the exhibition "See the world through a baby's eyes", here cradling a giant table leg and grabbing giant fruit from the tabletop!

the typical innovation strategy we follow. There are two ways in which a new innovation can be initiated. One way is through the development of a new technology or new material that offers a better way to address a known consumer need. For example, the development of a new "superabsorbent" material allowed us to reduce the amount of absorbent material needed in a nappy and so make nappies much thinner and more comfortable, while still having sufficient absorption. The second way is a newly discovered consumer need leading to an improved product. For example, the innovation of hook-and-loop-fasteners for nappies was triggered by comments from mothers who complained that the adhesive tape fasteners were often not sticking well because they had come in touch with baby cream during the nappy change.

In both cases we match the technology and the consumers' needs; only in the sweet spot, where technology addresses a need better than did previous products, does the innovation becomes relevant.

Annette showed us new ways to understand what our consumers needed. Historically, we typically learned from parents or pediatricians. We asked those groups what they needed to take care of babies, and what they thought a baby would need. We learned from Annette that we could learn directly from babies, by observing them or by applying methods like eye-tracking. Not only has this led to new research techniques within P&G, such as a "baby lab" in our innovation centre, it also motivated research projects in collaboration with developmental researchers,

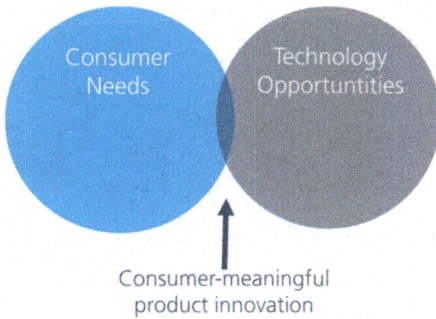

Consumer-meaningful
product innovation

FIGURE 20.3 Matching new technology with an unmet consumer need leads to meaningful innovations

as described later in this chapter. We also learned that the needs of babies change over time. Historically, we had thought about baby development mainly in terms of physical growth. Therefore, nappies for a newborn or a toddler had different sizes, but they were made with the same materials, properties, and features. Now we have learned how much development happens in the first years of life and how babies' needs change during this time. We found that different aspects in this development are relevant for the design of nappies. For example, the increasing ability and motivation to move around leads to changes in the body shape, requiring different nappy designs for walking babies than for less mobile babies, but it also leads to the need for faster and simpler nappy changes as babies are less willing to stop their play and lay down for a nappy change. Importantly we learned that there are large individual differences in patterns of development and that physical growth and cognitive development are not always happening in parallel, so that you could have smaller children already walking or larger children still crawling.

Following our innovation approach, we combined this new knowledge with our expertise in nappy technology. This led to a new design approach for nappy design. In 2001 we developed a line called "Pampers Baby Stages of Development" (Pampers BSOD). While previous product lines had different sizes of the same product design, the new nappy line consisted of nappies which were differently designed for different developmental stages of early childhood. Those nappies did not only differ by size but incorporated features and technologies specially designed for the needs of babies at the respective developmental stage. To reflect the individual differences in developmental patterns, some sizes were available for different stages. Figure 20.4 shows the different products in this line.

The Pampers BSOD product line started with "Pampers New Baby", a nappy for newborns, featuring especially soft materials to provide the comfort and security newborns need, and a special liner to absorb soft poo. These features were chosen as we learned about the particular importance of the sense of touch for small infants, and the fact that at this age infants usually have soft poo due to being primarily fed with liquid. The next stage, "Pampers Active Baby", was a nappy

FIGURE 20.4 The "Pampers Baby Stages of Development" line

designed for crawlers and walkers, with very stretchy tabs and a particularly thin and flexible core to prevent restrictions while crawling and walking, allowing babies to explore their environment. This was inspired by insights about how babies and toddlers develop the motor skills to become mobile, which then allows them to explore and learn about their environment. "Pampers Easy-Up pants" were the next product type in the range. They were developed for toddlers. We had learned that many toddlers disliked laying down for a nappy change. They would be more cooperative if they could keep standing and help with the change or even pull-up nappies on their own. To enable this, Easy-Up pants were designed as a pant rather than an "open nappy". On top they featured graphics appealing to toddlers, as we had also learned that toddlers communicate their preferred graphics and parents used that as a way of interacting with them. The last product type in this range were "Pampers Feel'n Learn". These were a special toilet training product with an underwear-like look and designs that were "gender-typical", so toddlers could feel like a "big girl" or a "big boy". These products also featured a liner that would stay slightly wet and cool if babies peed in the nappy instead of their potty, and it had graphics that disappeared when wet. Both of these innovations helped with toilet training. The product line has evolved over the nearly 20 years since Pampers BSOD was first launched, but many technologies and features are still present, and insights into baby development are still guiding nappy innovation.

All of this was inspired by Annette's lessons about the development babies go through in their first years. Annette visited P&G several times and provided lectures about the latest insights from developmental research or consulted on specific projects to help with these developments. Even more important is that Annette's knowledge has been converted into training materials for Pampers team members, such as the chart (Figure 20.1) or the exhibition (Figure 20.2) shown above, which enable us to teach these insights to new colleagues even today.

Educating the public about the importance of sleep for babies' development

Another area where Annette and Pampers worked together was understanding of the role of sleep for infant development. We both learned that infants' sleep is highly relevant for parents and infants. In the context of our nappy research, parents used to tell Pampers researchers how much the sleep quality of their baby impacts the sleep of the whole family, and so the whole of family life. They also reported frequently that nappy performance and quality can make a difference to their babies' sleep, and that therefore "overnight dryness" is an important criterion for their nappy choice. Annette had pointed us to an increasing number of studies on infant sleep. We started to investigate how sleep is related to the development of infants. We learned, for example, how sleep supports the consolidation of learning and memory, and how it is crucial for cognitive development in the early years of life. Not only did we learn the critical impact sleep can have on baby development, but we also realized that parents rarely thought about baby sleep in these terms. Most parents wanted undisturbed sleep for their baby because it allowed them and their families to sleep through the night. They did not understand the importance of sleep quality for the development of their babies.

We decided to educate parents about these scientific insights. With this purpose in mind we developed a series of adverts that showed how good quality sleep can support infant development. For example, in 2003 we aired an advert showing a baby playing with a toy that was a hollow wooden cube with holes in different shapes. The baby was trying to fit a ball into a square hole, which of course didn't work. The next scene showed the baby sleeping while wearing a Pampers nappy. Then the baby was shown waking up and playing with the ball again. After a night of good quality sleep, the baby found a creative solution for the problem: it threw the ball into a bucket instead of fitting it into the cube. Annette not only inspired these advertising copies, her advice also helped us to ensure we were only sharing scientifically supported information. When our creative team develops an idea, it goes through an approval process. If this idea was related to baby development we would consult with Annette to get her opinion and ask for scientific references. She would typically cite relevant publications and often propose some changes or clarifications, that were then considered in revisions. Over time she also taught some of P&G's scientists how to assess those ideas, so today this can be done by them. Annette commented several times that she appreciated the rigour of Pampers in supporting their advertising.

Understanding the connection between sleep, babies' development and nappies

The subject of sleep and baby development was very important for Annette as well as for the Pampers mission to support the healthy and happy development of babies. In 2011 we therefore decided to extend our collaboration. Following an invitation from Annette, Procter & Gamble joined an application from Birkbeck University for a EU Marie Curie funded PhD program called TRACKDEV. Anette's idea was to supervise a PhD project together to learn more about the impact of sleep on infants' development.

Initially there was some hesitation on both sides. For a company like P&G, a long-term project like a PhD-thesis is a big commitment, as the outcome is uncertain and research interests could change over several years. From Annette's perspective, having to select a student not only based on academic skills but also with a skillset enabling the student to work in a business environment was different from what she and her colleagues were used to. Understanding and trust built through our longstanding relationship, and Annette's conviction that science had to be useful for the people, helped to overcome our skepticism, and we agreed to move on with the collaboration.

We found a student, Manuela Pisch, with the right academic qualifications and a skillset meeting both of our needs. As part of her PhD project Manuela spent six months at P&G. She designed a longitudinal study following infants over six months. She collected data about sleep and development from 40 babies who were assessed at the age of 4, 6, 8 and 10 months. This work was conducted in the Consumer Discovery Center of Pampers in Schwalbach, Germany. Here we have contact with about 1,000 families with infants at any point in time to test new nappy prototypes. Conducting the study in Germany allowed recruitment from this cohort, which would have been difficult at Birkbeck. Manuela measured sleep with actigraphy and diaries, she also applied eye-tracking to measure progress in the cognitive development of the children; in particular memory, attention and number processing (Pisch, Wiesemann, Karmiloff-Smith 2018)[1]. One interesting finding was that sleep efficiency, which is the ratio of time asleep compared to total time in bed, was more strongly correlated to memory development than sleep duration (Pisch, Wiesemann, & Karmiloff-Smith, 2018). This is an insight we are following up on, with the goal to support good sleep quality by minimizing disturbance due to nappies. Manuela's work has also taught P&G researchers how to measure good sleep quality. This has enabled a clinical study amongst 100 infants (age 8–12 m) over 3 weeks comparing the impact of different nappies on sleep quality. This was a randomized study, all babies used all nappies, each for one week, in randomized order. We have developed a new way to analyse Actigraphy data to assess the sleep quality of infants by translating the activity over a full night into a continuous curve and calculating the area under this curve. This measure was then compared with the traditional sleep measures *sleep efficiency* and *sleep quality rating*. We concluded that babies sleeping in Pampers nappies experience good sleep quality. This work was recently presented on a conference of the International Pediatric Sleep Association[2], a publication is in preparation.

The final project Annette initiated with Pampers is still ongoing. After our collaboration on the TRACKDEV program was finished successfully, we decided to collaborate in another Marie Curie program, INTERLEARN. Two PhD students, Alicja Brzozowska and Louisa Gossé, are conducting studies in our Schwalbach Innovation Center, spending 18 months (Sept. 2017-Feb. 2019), half of their PhD time, with Procter & Gamble and will finish their PhD project by Feb. 2021 at Birkbeck University. One student is exploring the role of sleep for development, building on the work from the previous project. She is exploring the impact of sleep on development between 4 and 14 months using Actigraphy, eye-tracking questionnaires and electroencephalograms (EEG). The second student is looking into the impact of the amount of social touch infants receive on their development – another important aspect in babies' lives. In her first study, being conducted in P&G's German Innovation centre, she is looking at the impact of parents' touch on infants' arousal using observation, questionnaires, eye-tracking and saliva sampling for hormone assessments. We want to understand the impact of touch better because it is important in the relationship between parents and infants. As nappies are touching the baby all day long we want to learn if this touch is also relevant for infant development.

Both projects were inspired by Annette, who planned to be the supervisor and helped to develop the projects. Unfortunately, she passed away before the students had started their work, but the new findings that undoubtedly will stem from this work and the new scientists who are trained will be a continuing legacy of our collaboration.

A collaboration over so many years and in so different projects and ways like the one with Annette is an exception for industry-academia collaborations. It is often difficult to bring the interests of a company together with the interests of academia. A company like P&G needs to show that investment into research and scientific communication leads to an increased profit. The scientific communication is directed to the users of the products, which are usually laymen, and therefore science needs to be communicated clear, and in a re-levant way to gain interest from these consumers. The traditional role of academia is to create new knowledge and exchange this with other scientists to develop insights to serve society. Their communication needs to be accurate and detailed to allow their peers to build on it. A collaboration requires un-derstanding and respect for each other's interests, the different recipients of the scientific communications, and the ability to find common interest. Annette was able to understand these different needs and she had an interest to move beyond the academic community.

The recipe for the long-standing, successful collaboration was an open and transparent communication from both parties about commonalities and differ-ences. We recognized our common goal to help parents to support their babies' development. But we also were clear on different priorities. For example the time by when P&G would communicate publicly about a new learning is determined by the business need, for example, the deadline for a new advertising campaign, while in academia it might rather be driven by the time a PhD-thesis has to be finished. Clarifying those needs and addressing them is essential for a successful collaboration. This can best be done by starting "small", for example, with some consulting on a single project, and then build to larger collaborations.

Annette Karmiloff-Smith not only improved our understanding of the devel-opment of infants, she truly helped to improve the everyday lives of babies and by making her own insights, and those of other developmental scientists, available to the public. She has opened the eyes of millions of parents to the amazing development of their babies, and she has inspired better nappies. Together we have helped babies to develop better.

Acknowledgements

The PhD project as part of TRACKDEV is supported by the EC Marie Curie Initial Training Network FP7-PEOPLE-2010-ITN PART B.

The PhD projects as part of INTERLEARN are supported from the EU Horizon 2020 Research and Innovation programme under the Marie Sklodowska-Curie Grant agreement no 721895.

Notes

1 Pisch, M., Wiesemann, F., Karmiloff-Smith, A., J Child Psychol Psychiatry. 2018 Jul 10. doi: 10.1111/jcpp.12948.
2 F. Wiesemann, Sleep quality of 8-12 months old infants using Actigraphy and sleep diaries, Presentation at IPSA2018, Paris, April 27-29, 2018. Presentation available from the author (wiesemann.f@pg.com).

INDEX

Note: Page numbers in italics refer to figures. Page numbers in bold refer to tables. Page numbers followed by 'n' refer to notes.

312 Index